AF883033

# Subcellular Biochemistry

Volume 115


**Series Editor**

Tapas K. Kundu, Transcription and Disease Laboratory, JNCASR, Bangalore, Karnataka, India

**Advisory Editor**

J. Robin Harris, Institute of Molecular Physiology, University of Mainz, Mainz, Germany

The book series SUBCELLULAR BIOCHEMISTRY is a renowned and well recognized forum for disseminating advances of emerging topics in Cell Biology and related subjects. All volumes are edited by established scientists and the individual chapters are written by experts on the relevant topic. The individual chapters of each volume are fully citable and indexed in Medline/Pubmed to ensure maximum visibility of the work.

Kaushik Sengupta • Antoine Muchir
Editors

# Lamins in Genome Organization and Diseases

Springer

*Editors*
Kaushik Sengupta
Biophysics and Structural Genomics Division
Saha Institute of Nuclear Physics
Kolkata, West Bengal, India

Antoine Muchir
Sorbonne Université, INSERM U974
Institute of Myology
Center of Research in Myology
Paris, France

ISSN 0306-0225 ISSN 2542-8810 (electronic)
Subcellular Biochemistry
ISBN 978-3-032-00536-6 ISBN 978-3-032-00537-3 (eBook)
https://doi.org/10.1007/978-3-032-00537-3

© The Editor(s) (if applicable) and The Author(s), under exclusive license to Springer Nature Switzerland AG 2025

This work is subject to copyright. All rights are solely and exclusively licensed by the Publisher, whether the whole or part of the material is concerned, specifically the rights of translation, reprinting, reuse of illustrations, recitation, broadcasting, reproduction on microfilms or in any other physical way, and transmission or information storage and retrieval, electronic adaptation, computer software, or by similar or dissimilar methodology now known or hereafter developed.
The use of general descriptive names, registered names, trademarks, service marks, etc. in this publication does not imply, even in the absence of a specific statement, that such names are exempt from the relevant protective laws and regulations and therefore free for general use.
The publisher, the authors and the editors are safe to assume that the advice and information in this book are believed to be true and accurate at the date of publication. Neither the publisher nor the authors or the editors give a warranty, expressed or implied, with respect to the material contained herein or for any errors or omissions that may have been made. The publisher remains neutral with regard to jurisdictional claims in published maps and institutional affiliations.

This Springer imprint is published by the registered company Springer Nature Switzerland AG
The registered company address is: Gewerbestrasse 11, 6330 Cham, Switzerland

If disposing of this product, please recycle the paper.

# Foreword

One of the first papers I read when I started research on the nuclear envelope as a postdoctoral fellow was Günter Blobel's "Gene Gating: A Hypothesis" (*Proc Natl Acad Sci U S A* 1985;82:8527–8529). As part of the gene-gating hypothesis, Günter proposed "that the lamina participates in the organization of the compacted portion of the genome." While the overall gene-gating hypothesis remains unproven and perhaps even unlikely, the nuclear lamina—a meshwork on noncanonical intermediate filaments composed of proteins called lamins—likely plays a role in chromatin organization and the regulation of gene expression. This role must be better defined.

Since starting my independent research on the nuclear envelope in 1990, the biggest scientific surprise of my career occurred when I received a message from colleagues in France along the lines of "the lamin A/C gene (*LMNA*) that you characterized is mutated in autosomal dominant Emery-Dreifuss muscular dystrophy." The publication of that finding by Gisèle Bonne, Ketty Schwartz, and colleagues (*Nat Genet* 1999;21:285–288) was the first in an onslaught of subsequent discoveries that mutations in *LMNA* cause a range of diseases now often referred to as laminopathies. Depending upon the mutation, affected organs are striated muscle, peripheral fat, peripheral nerve, or multiple systems generating phenotypic features of accelerated aging. Broadly speaking, the respective diseases are dilated cardiomyopathy with variable skeletal muscle involvement (including autosomal dominant Emery-Dreifuss muscular dystrophy), partial lipodystrophy, Charcot-Marie-Tooth disease type 2, or Hutchinson-Gilford progeria syndrome and related progeroid disorders. Altered expression of nuclear lamins also variably occurs in some types of cancer; however, the functional significance is mostly unclear. Hence, while once of interest to only a small specialist group of cell biologists, physicians, geneticists, and many other scientists have jumped on the lamin research bandwagon.

This book entitled *Lamins in Genome Organization and Diseases* brings together these two exciting aspects of nuclear lamin research. The chapters cover a range of topics. Some are focused on lamins and chromatin organization in diseases, including cardiomyopathy caused by *LMNA* mutation and cancer. Others cover more fundamental aspects of how lamins may organize the nucleus and genome. The emerging role of the nuclear envelope and lamins in transmission of forces from outside the cell to the nucleus and how such forces may impact genomic organization are also addressed.

Four decades have passed since the gene-gating hypothesis was proposed. However, how the nuclear envelope and lamins' impact on normal and possibly pathological chromatin organization remains mostly a mystery. A complete understanding of the precise function of lamins in chromatin organization and disease is likely more than 40 more years into the future. The chapters in this book provide thought-provoking ideas and a framework for further investigation in this exciting and important field.

Howard J. Worman
Columbia University
New York, NY, USA

# Preface

Lamins are intermediate filament proteins of nuclear origin. These proteins play diverse roles in nuclear homeostasis, impart structural rigidity to the nucleus, and tether chromosomes. Numerous mutations in the A-type lamin protein give rise to a plethora of diseases termed as laminopathies which affect primarily the muscle, cardiac, adipose, nervous tissues, and a combination of all in the form of progeroid syndromes. On the other hand, differential expression of lamins has been implicated in a wide variety of cancers. Over the past two decades, there has been an upsurge of interesting findings relating to the perturbation of the signaling cascades and transcriptome analysis in the pathogenesis of laminopathies. Few reports have also shed light on the epigenetic changes associated with these diseases. Rapid advancement of chromosome conformation capture methods has provided a new impetus in delineating the finer details of chromosome association with lamins at the topological level. We feel it is an opportune time to bring together most of these exciting findings in the form of a book that will attract a wide range of readers from the lamin as well as chromatin biology. At this level, we envisage a marriage of the two important fields of biology that can better explain and revolutionize the pathogenesis of the diseases, which in turn can accentuate the research and development in molecular medicine significantly.

Kolkata, West Bengal, India — Kaushik Sengupta
Paris, France — Antoine Muchir

# Contents

# Chapter 1
# Role of Nuclear Lamins in the Regulation of the Genome: Focus on CardioLaminopathy

**Marie Kervella and Antoine Muchir**

**Abstract** Nuclear lamins, the unique type 5 intermediate filaments of the nucleus, form a structural network beneath the nuclear envelope. Through their strategic localization, they interact with numerous partners and regulate a wide range of biochemical and biophysical processes. In load-bearing tissues such as the heart, nuclear lamins are crucial for maintaining mechanical integrity and genome stability during cardiac contraction. By associating with chromatin through lamina-associated domains, they play a fundamental role in genome organization and gene expression. Disruption of the nuclear lamin meshwork in CardioLaminopathy leads to chromatin remodeling and dysregulated gene expression. Although the precise cellular and molecular mechanisms driving CardioLaminopathy remain incompletely understood, alterations in genome organization are emerging as key contributors to disease pathogenesis and progression.

**Keywords** Nuclear lamins · Genome organization · CardioLaminopathy · Lamina-associated domains · Cardiomyocytes

M. Kervella
PhyMedExp, University of Montpellier, INSERM, CNRS, Montpellier, France

Sorbonne Université, INSERM U974, Institute of Myology, Center of Research in Myology, Paris, France

A. Muchir (✉)
Sorbonne Université, INSERM U974, Institute of Myology, Center of Research in Myology, Paris, France
e-mail: a.muchir@institut-myologie.org

© The Author(s), under exclusive license to Springer Nature Switzerland AG 2025 
K. Sengupta, A. Muchir (eds.), *Lamins in Genome Organization and Diseases*, Subcellular Biochemistry 115, https://doi.org/10.1007/978-3-032-00537-3_1

## CardioLaminopathy

Laminopathies are rare diseases described for the first time in the late 1990s. Laminopathies are tissue-specific disorders affecting striated muscles, peripheral nerves, and adipose tissue or leading to premature aging [1]. Diagnosis becomes particularly challenging due to the overlapping phenotypes among different laminopathies, and phenotypes often differ among the same disorders. Hence, classifying these diseases and establishing a link between genotype and phenotype seem difficult. Laminopathies can still be classified into two categories. Primary laminopathies bring together all the laminopathies associated with mutations in *LMNA* gene. In contrast, secondary laminopathies are related to mutations in B-type lamin-encoding genes (*LMNB1* and *LMNB2*) or lamin-binding protein-encoding genes. A non-exhaustive list of primary and secondary laminopathies is given in Table 1.1.

Among the various phenotypes affecting striated muscles, cardiac involvement is one of the most common and severe manifestations, serving as a defining feature of several laminopathies. The genetic cause of CardioLaminopathy was solved in 1999 when the groups of Dr. Ketty Schwartz and Pr. Catherine Seidman identified a point mutation in *LMNA* gene, encoding nuclear lamins, in large families with patients suffering from dilated cardiomyopathy (DCM) [2, 39]. CardioLaminopathy is associated with more than 150 different *LMNA* mutations, scattered along the entire gene [40]. DCM is characterized by an increase in both myocardial mass and volume. The ventricular walls become thin and stretched, compromising cardiac contractility and resulting in poor left ventricular function. CardioLaminopathy is associated with complications including arrhythmia and heart failure, surpassing the risks observed in most other cardiac diseases, with the probability of sudden cardiac death [41]. Given the increased awareness among physicians, cardiologists are now facing difficult queries regarding patient management. These queries concern the use of defibrillators to avoid sudden death from aggressive ventricular arrhythmias and pharmacological interventions to improve heart failure symptoms. Upon clinical detection of DCM, the management of CardioLaminopathy adheres to the standard protocols for heart failure care [42]. It remains uncertain whether the early initiation of these therapeutic agents, before detectable cardiac dysfunction, can alter the aggressive nature of CardioLaminopathy. Currently, there is no definitive treatment for the progressive cardiac enlargement and decline in contractile function observed in CardioLaminopathy, apart from considering heart transplantation [43]. The precise cellular and molecular mechanisms underlying the pathogenesis and progression of CardioLaminopathy are not yet fully understood and are still under investigation.

**Table 1.1** Non-exhaustive list of primary and secondary laminopathies

| Class | Type | Proteins | Disorders | Type of disorders | Biological characteristics | Clinical characteristics | Refs. |
|---|---|---|---|---|---|---|---|
| Striated muscle dystrophy | Primary | Emerin and A-type lamins | Emery-Dreifuss muscular dystrophy (types 2 and 3) | Autosomal dominant (type 2) and recessive (type 3) | Impaired expression of mechanosensitive genes in response to strain and abnormal control of cell differentiation or proliferation | Gradual and progressive atropatrophy and wasting of skeletal muscles, particularly in the shoulder girdle and distal leg muscles, coupled to cardiac muscle dystrophy | [2, 3] |
| | Primary | A-type lamins | Dilated cardiomyopathy (type 1A) | Autosomal dominant | Chromatin reorganization, abnormal p38a mitogen-activated protein kinase signaling, senescence, abnormal nuclear shape, and cytoskeletal signaling | Left ventricular or biventricular dilatation, cardiac conduction defects, resulting in arrhythmia and heart failure, associated with the probability of sudden cardiac death | [4–9] |
| | Primary | A-type lamins | Limb-girdle muscular dystrophy type 1B | Autosomal dominant | Nuclear envelope alteration | Progressive limb-girdle muscular dystrophy resulting in disturbances of atrioventricular cardiac conduction and the lack of early contracture | [10, 11] |
| | Primary | A-type lamins | Congenital muscular dystrophy | Autosomal dominant | Synthesis of full-length or quasi-full-length mutant lamin A/C proteins or small in-frame deletion leading to both decreased expression and mislocalization of the mutated lamin A/C | Muscle weakness, delayed or rapid loss of motor development, cardiac onset, orthopedic complication | [12] |
| | Primary | A-type lamins | Heart-hand syndrome | Autosomal dominant | Formation of a new cryptic splice site, leading to the retention of 11 intronic nucleotides in the LMNA mRNA | Combination of a congenital cardiac disease and limb malformations | [13] |
| | Primary | A-type lamins | Malouf syndrome | Autosomal recessive | | Congestive or dilated cardiomyopathy, ovarian dysgenesis in females or primary testicular failure in males, mental retardation, broad nasal base, blepharoptosis, skin lesions, bone abnormalities, and, sporadically, marfanoid habitus | [14, 15] |
| | Secondary | Lamin B receptor | Greenberg dysplasia | Autosomal recessive | Defect in sterol metabolism – loss of the sterol reductase function of the lamin B receptor | Skeletal dysplasia, fetal hydrops, short limbs, and abnormal chondro-osseous calcification | [16] |

(continued)

**Table 1.1** (continued)

| Class | Type | Proteins | Disorders | Type of disorders | Biological characteristics | Clinical characteristics | Refs. |
|---|---|---|---|---|---|---|---|
| Neuropathy | Primary | A-type lamins | Charcot-Marie-Tooth disease type 2B1 | Autosomal recessive | Strong reduction of axon density, axonal enlargement, and the presence of nonmyelinated axons | Sensory and motor involvement associated with a gradual disability, displaying diverse patterns of inheritance and distinct electrophysiological classes (axonal, demyelinating, intermediate) | [17] |
| | Secondary | Lamin B1 | Autosomal dominant leukodystrophy | Autosomal dominant | Demyelination of the central nervous system | Onset of autonomic dysfunction, followed by pyramidal and cerebellar abnormalities affecting the white matter or myelin tracts of the central nervous system | [18–20] |
| | Secondary | Lamin B2 | Epilepsy and ataxia | Autosomal recessive | Impaired neuronal migration | Severe action myoclonus, progressive neurological decline, and ataxia | [21] |
| | Secondary | Lamin B1 | Ataxia-telangiectasia | Autosomal recessive | Demyelination of the central nervous system, nuclear shape alterations, and senescence | Voluntary movements' lack of coordination, permanent widening of blood vessels | [22] |
| | Secondary | Lamins B1 and B2 | Microcephaly | Autosomal dominant | Interference with normal assembly of nuclear lamina, alteration of lamin B1 conformation, weakness of nuclear lamina, impaired nucleokinesis, and decreased neuronal survival | Brain developmental defect caused by significant smaller occipital-frontal head circumference | [23, 24] |
| | Secondary | Lamin B1 | Neural tube closure defects | | Nuclear dysmorphology, altered cell cycle progression, and premature senescence | Severe malformations termed neural tube defects, spina bifida, and anencephaly | [25] |
| | Secondary | Lamin B1 | Alzheimer's disease | Autosomal dominant | Relaxation of constitutive heterochromatin and activation of cell cycle in post-mitotic neurons resulting in neurons' death | Neurodegeneration | [26, 27] |
| | Secondary | Lamin B receptor | Pelger-Huet anomaly | Autosomal dominant | Altered shape of nuclei in granulocytes neutrophils and eosinophils | Epilepsy, developmental delay, and skeletal abnormalities | [28] |

| | | | | | | | |
|---|---|---|---|---|---|---|---|
| Lipodystrophy | Primary | A-type lamins | Familial partial lipodystrophy type 2 | Autosomal dominant | | Absence or reduction of subcutaneous adipose tissue from childhood or early adulthood, coupled with metabolic complications, more or less pronounced depending on the extent of fat loss | [29] |
| | Secondary | Lamin B2 | Barraquer-Simons syndrome | | Presence of auto-immune-mediated loss of adipose tissue | Imbalance between the upper and lower parts of the patient's body resulting from a gradual onset of bilaterally symmetrical loss of subcutaneous fat tissue accompanied with metabolic complications, membranoproliferative glomerulonephritis, and immune disorders | [30, 31] |
| Premature aging | Primary | A-type lamins | Hutchinson-Gilford progeria syndrome | Autosomal dominant | Generation of a cryptic splice site in *LMNA* exon 11 leading to an in-frame deletion in pre-lamin A, resulting in abnormal protein, called progerin | Alopecia, abnormalities in bone and joint, a loss of subcutaneous fat, and severe atherosclerosis | [32] |
| | Secondary | BANF1 | Nestor-Guillermo progeria syndrome | Autosomal recessive | Misshapen nuclei and mislocalization of nuclear proteins—BANF1-LEMD2 complex—may fail to recruit lamin A to the nuclear rupture site, preventing the stabilization of the rupture | Less severe form of HGPS | [33] |
| | Secondary | LEMD2 | LEMD2-associated progeria syndrome | Autosomal dominant | | | [33] |
| | Primary | A-type lamins | Atypical Werner syndrome | Autosomal recessive | Might perturb intermolecular interactions between lamin A/C dimers—weak activation of the same cryptic splicing site as in HGPS | Late onset of HGPS symptoms | [34] |
| | Primary | A-type lamins | Mandibuloacral dysplasia (type A) | Autosomal recessive | Nuclear dysmorphism, loss of peripheral heterochromatin, and nuclear lamina thickening | Mandibular hypoplasia, dental crowding, clavicular resorption, acral osteolysis, skin abnormalities, partial lipodystrophy, and accelerated aging | [35] |
| | Secondary | ZMPSTE24 | Mandibuloacral dysplasia (type B) | Autosomal recessive | Effect on pre-lamin A processing in mature lamin A | Mandibular hypoplasia, dental crowding, premature tooth eruption, clavicular resorption, acral osteolysis, skin abnormalities including absent or sparse hair (total or subtotal alopecia), generalized lipodystrophy, short stature, and accelerated aging | [36] |
| | Secondary | ZMPSTE24 | Restrictive dermopathy | Autosomal recessive | mRNA frameshifts—specific defect in pre-lamin A processing caused by the absence of functional ZMPSTE24 | Intrauterine growth retardation, tight and rigid skin with prominent superficial vessels, bone mineralization defects, dysplastic clavicles, and arthrogryposis, which result in stillbirth or neonatal death | [37, 38] |

## Nuclear Lamins

Lamins are proteins found exclusively in multicellular metazoans, exhibiting a high degree of conservation. Lamins are the main component of the nuclear lamina, a meshwork that lines the inner membrane of the nuclear envelope [44]. Lamins belong to the category of type 5 intermediate filaments, which is the distinctive class of intermediate filaments located within the cell nucleus. Two types of lamins are distinguished based on their structural and protein features and their expression patterns. A- and B-type lamins share identical N-terminal head and central rod domains. However, A-type lamins distinguish themselves by having an additional 90 amino acids within their C-terminal tail [45]. A- and B-type lamins possess different isoelectric points (neutral and acidic, respectively). With its lower isoelectric point, B-type lamins stay associated with the nuclear envelope during mitosis, whereas A-type lamins are soluble [46].

A-type lamins are primarily limited to differentiated cells [47]. A-type lamins yield a minimum of four isoforms, which include two primary forms (A and C) and two less prevalent forms (C2 and AΔ10) [48]. Lamin A and C isoforms are obtained through an alternative splicing of the *LMNA* gene, located on chromosome 1q11-q23. Both A and C lamins share an identical sequence for the initial 566 amino acids, but lamin C lacks the final 98 amino acids found in pre-lamin A (before post-translational processing) (Fig. 1.1a).

The precursor pre-lamin A undergoes post-translational processing, facilitating binding to the nuclear envelope and ultimately forming the mature lamin A protein of 646 residues. Pre-lamin A contains a CaaX box that undergoes farnesylation, followed by cleavage of the last three residues, and carboxymethylation of the terminal cysteine. Subsequently, the endoprotease ZMPSTE24 (FACE-1) cleaves the last 15 C-terminal residues, resulting in mature lamin A. Instead, lamin C possesses a distinct six-amino-acid sequence at its carboxyl terminus (Fig. 1.1a).

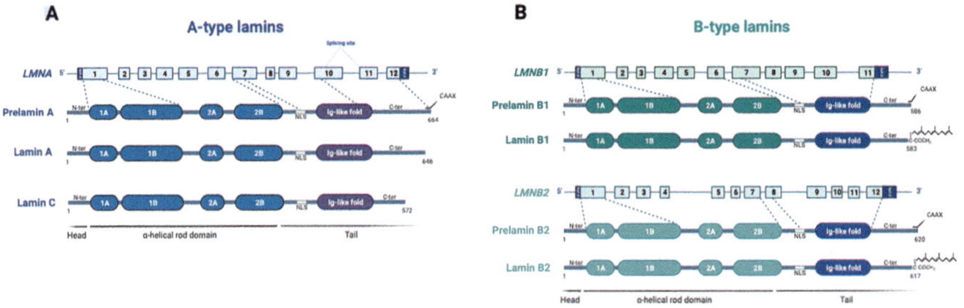


**Fig. 1.1** Lamin-encoding genes and protein structure. (**a**) *LMNA* gene encodes the two isoforms lamin A and lamin C, obtained through an alternative splicing. Pre-lamin A undergoes post-translational processing to give mature protein. (**b**) *LMNB1* gene encodes the pre-lamin B1, and LMNB2 encodes the pre-lamin B2. Both B-type pre-lamins undergo post-translational processing to give mature proteins and remain farnesylated. Created with BioRender

B-type lamins are present in a wide range of cell types and are essential during tissue development [47]. In mammals, B-type lamins are composed of two major isoforms, namely, lamin B1 and lamin B2. Lamins B1 and B2 are encoded by *LMNB1* and *LMNB2* genes located on chromosomes 5q23.2-q31.3 and 19p13.3, respectively. The *LMNB2* gene also gives rise to a less common isoform known as lamin B3, which is specific to germ cells and arises through alternative splicing [49]. All B-type lamin isoforms are initially synthesized as pre-lamins and require post-translational modifications (PTMs) on their C-terminal CaaX box. These PTMs include farnesylation, carboxymethylation, and partial cleavage to generate the mature forms of lamin B1 and lamin B2. In contrast to lamin A, mature B1 and B2 lamins remain farnesylated [50] (Fig. 1.1b).

The abundance of each lamin isoform in vivo is unknown, but the relative proportions of lamins are non-stoichiometric [51]. Lamins form an independent but interactive network with the establishment of extensive high-order polymers by the parallel association of two or more lamin dimers in a head-to-tail fashion [45]. The interaction between A-type and B-type lamins is possible, and pre-lamin A, lamin A, lamin C, and lamin B1 can form homo- or heterodimers [52] (Fig. 1.2).

Lamins underlie the inner nuclear membrane, providing essential support to the nucleus (Fig. 1.3). Each lamin type has its own structural and biochemical properties. Lamin B1, positioned peripherally in the nucleus, plays a role in stabilizing the nuclear shape. It actively engages in significant deformation of the nuclear envelope, exploiting its elastic properties to restrict the outward protrusions of the lamin A/C network [53, 54]. In contrast, lamin A/C confers deformation-resistant viscous stiffness to the nuclear envelope. While both lamin B1 and lamin A/C contribute to nuclear stiffness, only lamin A/C participates in the viscosity of the nucleus [54, 55].

Lamins interact with the cytoskeleton in the cytoplasm through to the linker of the nucleoskeleton and cytoskeleton (LINC) complex while also establishing interactions with the chromatin in the nucleoplasm (Fig. 1.3). In light of these interactions, lamins play a role in a wide range of biophysical and biochemical processes in cells.

Several processes involving lamins depend on their proper functionalization and structural organization, which are regulated by PTMs. Lamins are first farnesylated and carboxymethylated (except for lamin C—see above) to produce a mature protein. The most common PTM of lamins is phosphorylation, which mostly takes place in the rod domain, and is involved in a plethora of structural and biological processes, such as lamin polymer formation and solubility [56, 57], cell cycle regulation [58, 59], viral infection [60–62], as well as contributing to the development [63, 64]. Lamins can also be acetylated, mostly in the rod domain, to play a role in cell cycle, DNA repair [65], and nuclear architecture and genome integrity [66]. Lamins can be post-translational modified by small ubiquitin-like modifiers (SUMOs), referred to as SUMOylation, to play a key role in cell cycle progression [67] and nucleophagy [68]. SUMOylation is also involved in diseases such as DCM [69] and familial partial lipodystrophy [70]. The modification of lamin A by SUMOylation on specific sites may interfere with homotypic interactions necessary for the assembly of the nuclear lamina. SUMO sites appear to be adjacent to the

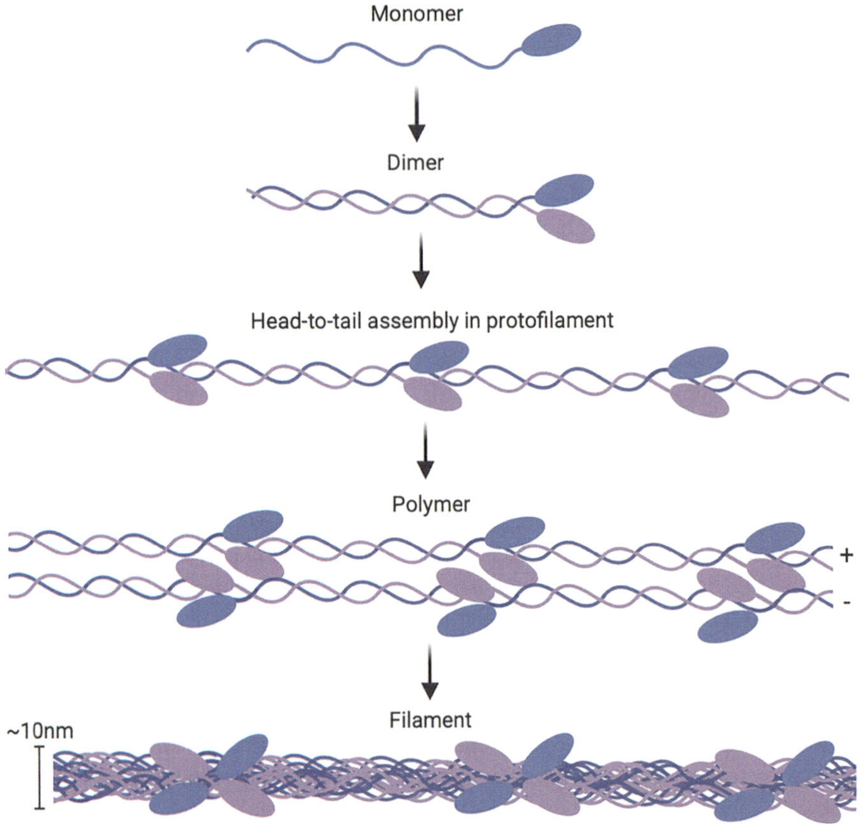


**Fig. 1.2** Structural organization of lamin meshwork. Lamins are structured in monomer that can associate with another monomer to form a homo- or heterodimer. Dimers can be associated together from head to tail, to form a protofilament. Polymers contain several protofilaments associating in a parallel manner. Polymers assemble to form a filament structure. Created with BioRender

*LMNA* mutation sites responsible for laminopathies, suggesting their potential involvement in the nuclear lamina assembly and thus the pathogenesis of the disease [71]. Few more lamin PTMs exist, including ubiquitination for disease progression [72, 73], O-linked β-N-acetylglucosamine (O-GlcNAc) modification (only mature lamin A) [74], as well as methylation involved in the lamins' maturation (carboxymethylation—see above), and myoblast fusion [75].

Lamins, particularly A-type lamins, are actively involved in mechanotransduction signaling, controlling the stiffness and shape of the nucleus [76, 77], and maintaining the genome integrity [78]. Mechanotransduction is the cellular capacity to sense, integrate, and convert external mechanical stimuli into biochemical signals, leading to specific cellular responses. This adaptive process enables cells to continuously respond to dynamic changes in their surrounding environment [79].

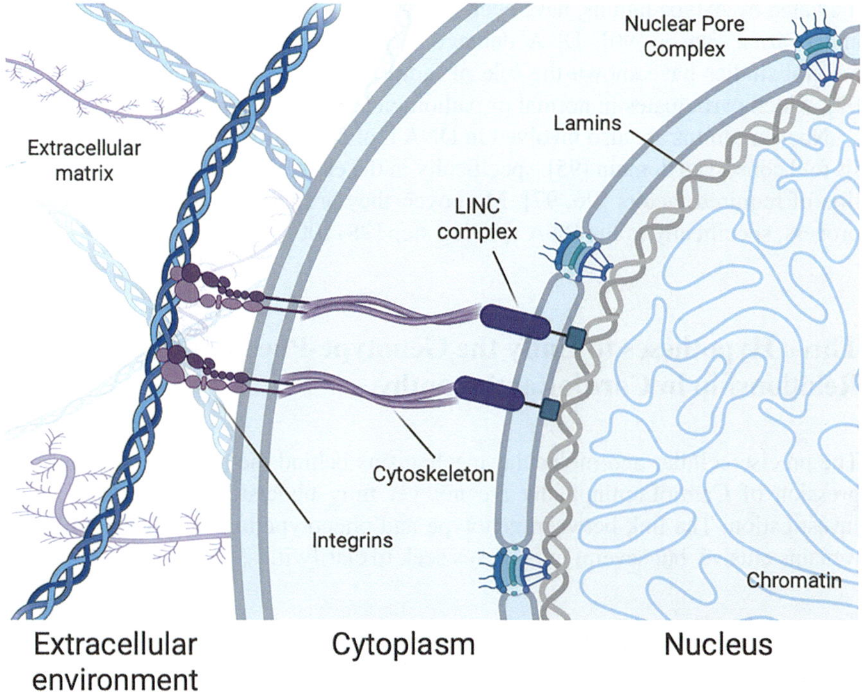


**Fig. 1.3** Physical continuum existing between the extracellular environment and the nuclear interior. The extracellular environment connects to the cytoskeleton via integrins, which, in turn, associate with the nuclear interior—particularly lamins and chromatin—through the LINC complex. Integrins sense changes in the cellular microenvironment and transmit these signals through this physical continuum, ultimately modulating gene expression in response

Biophysical investigations have provided new perspectives on the distinct mechanical characteristics of lamins. On the one hand, A-type lamins play the key role of structural support of the nucleus and appear to be viscoelastic, contributing to nuclear stiffness. On the other hand, the elastic properties of B-type lamins facilitate nuclear deformability and integrity [55, 80]. Moreover, the expression levels of lamin A correlate with the stiffness of the cellular environment. Altering this environment affects lamin A levels, and conversely, changes in lamin A levels can influence the cellular environment [81]. These findings highlight the role of lamin A in the mechanotransduction of environmental cues from the surrounding environment to the nucleus, through the physical continuum between the extracellular matrix, the cytoskeleton, the LINC complex, and the chromatin.

Several studies have suggested a link between the nuclear envelope and DNA damage response [82, 83], making lamins one of the safeguards of genetic integrity. Lamins are involved in the DNA damage response process by sequestering essential factors [84, 85], activation of the base-excision repair [86], or genome integrity maintenance during replication fork stalling [87], for instance. Defects in genome integrity maintenance, particularly through DNA damage response pathways mediated by A-type lamins, have been implicated in diseases such as DCM [88, 89] and ovarian cancer [90]. DNA damage response is also related to senescence. Several studies have shown the role of lamins in telomere biology alteration [91, 92], which participates in normal or pathological aging [93, 94].

Nuclear lamins are also involved in DNA replication, thanks particularly to their Ig-fold conserved domain [95], specifically in the elongation phase and the distribution of required factors [96, 97]. Moreover, they are involved in the transcription process, specifically in the RNA splicing step [98–100].

## Three Hypotheses to Unify the Genotype-Phenotype Relationship in CardioLaminopathy

The precise cellular and molecular mechanisms behind the pathogenesis and progression of CardioLaminopathy are not yet fully understood and remain under investigation. The link between genotype and phenotype in these diverse diseases remains elusive, but several hypotheses seek to clarify it.

### *Structural Model*

The "structural hypothesis" posits that *LMNA* mutations alter the nuclear lamina structure and organization, increasing nuclear fragility and mechanosensitivity. The physical continuum existing between the surrounding environment and nuclear interior through lamins highlights its role in signal mechanotransduction. Hence, cells carrying LMNA mutations within load-bearing tissues, such as striated muscles, bone, cartilage, and cardiovascular tissue, may be unable to properly respond to mechanical cues from the microenvironment. The "structural hypothesis" offers a valuable perspective for explaining striated muscle diseases. It is based on the premise that the striated muscles are constantly subjected to mechanical forces and that mutations in "nucleocytoskeletal" support elements make them susceptible to damage from recurrent stress. Several studies support the structural model as a key player in the pathogenesis of CardioLaminopathy. An investigation using mouse embryonic fibroblasts revealed that the absence of both lamins A and C leads to disrupted nuclear shape, reduced nuclear stiffness, and decreased cell viability under strain. In contrast, cells lacking lamins B1 maintained proper nuclear mechanics, despite an increased frequency of nuclear blebbing [55]. Several investigations support the disruption of biochemical properties and nuclear mechanics in laminopathies [77, 101–106]. These studies highlight the role of mechanotransduction defects in laminopathies, though they alone cannot fully explain the wide range of phenotypes and the complex pathogenesis of these disorders.

## *Signaling Model*

The signaling model is based on the regulation of signaling pathways by nuclear lamins. Indeed, changes in nuclear lamina associated with lamin mutations could modulate signaling pathways through the physical continuum between the cytoplasm and the nucleoplasm, thanks to the LINC complex.

Several studies have previously shown the dysregulation of three distinct pathways, which are extracellular signal-regulated kinase 1/2 (ERK1/2), protein kinase B/mammalian target of rapamycin complex 1 (AKT/mTORC1), and WNT/β-catenin signaling. In laminopathies, especially in dilated cardiomyopathy, both from patients and *Lmna*$^{H222P/H222P}$ mouse model, lamin mutation appears sufficient to perturb the nuclear structure. This disruption leads to hyperactivation of mitogen-activated protein (MAP) kinases associated with hyperphosphorylation and nuclear sequestration of its downstream target ERK1/2 [107]. The ERK1/2 hyperphosphorylation event is preceded by an activation of transforming growth factor β1 (TGF-β1) and its Smad effectors [PMID: 15276457, 27131347]. In a hyperphosphorylated state, ERK1/2 phosphorylates cofilin-1, which in turn disassembles actin sarcomeres resulting in alterations of left ventricular cardiac contractions [108]. The phosphorylated cofilin-1 leads to a sequestration of the myocardin-related transcription factor A (MRTF-A) in the cytoplasm, decreasing the serum response factor (SRF) downstream activity in the nucleus. Therefore, the level of α-tubulin is decreased, associated with a mislocalization of the cardiac connexin 43, important for intercalated disks in cardiomyocytes, leading to cardiac conduction defects [9, 109]. These studies then demonstrate a cytoskeleton and signaling pathway interplay in cardiomyopathy pathogenesis. This phenotype occurs before the initiation of cardiac impairments, demonstrating that MAP kinase pathway hyperactivation is a causative event in CardioLaminopathy.

Additionally, the nuclear sequestration of hyperphosphorylated ERK1/2 leads to the hyperactivation of the AKT/mTORC1 signaling pathway through the activation of an ERK1/2-specific phosphatase, resulting in a reduced lifespan associated with a decrease of autophagy [110]. The third pathway impacted by *Lmna* mutation is the WNT/β-catenin signaling pathway, which is involved in proliferation, differentiation, and apoptosis. In this disease context, Wnt and β-catenin both decrease [6, 111], correlated with a mislocalization and a decline of the expression of cardiac connexin 43, which is involved in the intercalated disk stabilization in cardiomyocytes [111–113].

## *Chromatin Hypothesis*

A recently identified function of nuclear lamins is their role in spatial genome organization, which subsequently regulates gene expression [114, 115]. This crucial role is facilitated by the interaction between lamins and chromatin. Additionally, the physical continuum between the extracellular environment and the nuclear interior seems to play a crucial role in modulating the genome, influencing gene transcription based on the external signals perceived by cells. The "chromatin hypothesis" postulates that changes in genome organization, specifically in LADs, could lead to an aberrant gene expression in lamin-related diseases.

The genome's organization extends beyond a linear arrangement, intricately shaping a three-dimensional (3D) chromatin configuration (Fig. 1.4).

This configuration exhibits tissue-specific patterns within mammalian nuclei, facilitating the appropriate control of gene expression profiles specific to each cell type. Various architectural domains operating across different genomic scales are crucial for densely packing all chromosomes within the limited confines of the nucleus, ensuring the maintenance of genome organization and stability. The mammalian genome is organized in chromatin, whose basic building block is the nucleosome. Each nucleosome consists of a histone octamer around which 145–147 DNA base pairs are wound [116]. Nucleosomes are grouped into clusters, and each contains approximately 1–2 kilobases (kb) of DNA [117]. These clusters then collectively constitute chromatin nanodomains (CNDs), each spanning roughly 100 kilobases (kb) in size. CNDs serve both structural and functional purposes. They effectively bring enhancers and promoters into closer spatial proximity, thereby promoting intra-chromosomal (cis) DNA interactions at both short and long ranges [118]. Disrupting these enhancer-promoter contacts can alter gene expression and lead to developmental defects and disease, including congenital disorders and cancer [119]. On a broader scale, CNDs and CCCTC-binding factor (CTCF)-cohesin-dependent chromatin loops form topological associated domains (TADs) ranging from 100 kb to 1 Mb. These TADs engage in interactions with one another that are influenced by their epigenetic state [120]. TADs are known to be highly preserved and remain mostly constant across various cell types. In contrast, the specific chromatin interactions or loops within these TADs exhibit greater tissue-specific variations [121]. At the scale-up of hundreds of megabases, chromatin is subdivided into two A and B chromosomal compartments [122, 123]. The A compartment contains chromatin regions marked by an abundance of active chromatin markers, where genes are highly transcribed and centrally located in the nucleus. Oppositely, the B compartment is characterized by enrichment in repressive chromatin marks and poor gene content at the nuclear periphery. The few genes in the B compartment are typically silenced or weakly transcribed [124]. At the highest chromatin scale, interphase chromosomes segregate into distinct nuclear regions known as chromosome territories [125]. Parts of these territories can merge through a variety of chromatin interactions, whether they originate from the same chromosome or different chromosomes, resulting in the formation of hubs sharing similar functions [118]. In mammalian cells, some large chromatin domains can be associated with distinct nuclear components. For instance, lamina-associated domains (LADs) link to the nuclear lamina, pericentromere-associated domains (PADs) connect with the pericentromeric heterochromatin, and the nucleolus-associated domains (NADs) are affiliated to the nucleolus. During mitosis or pathogenesis, these large chromatin

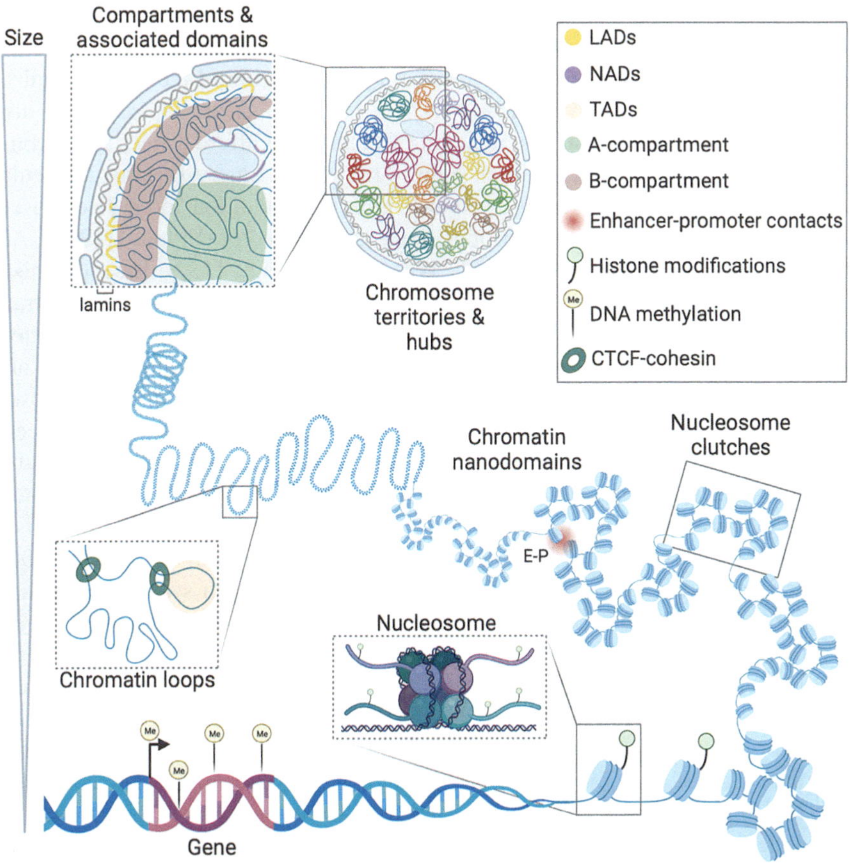


**Fig. 1.4** Three-dimensional genome organization and epigenetic modifications in mammalian cells. Double-stranded DNA formed a helix. Gene can be methylated in their promoter or gene body. DNA is wrapped around nucleosomes to form chromatin. A nucleosome is composed of a histone octamer containing two pairs of each histone H2A, H2B, H3, and H4. Nucleosome histone tails can be epigenetically modified by specific enzymes. Nucleosomes are associated in nucleosome clutches, which are contained in a bigger structure called chromatin nanodomains, favoring contacts between enhancers and promoters. Chromatin nanodomains form several chromatin loops containing TADs thanks to cohesin-CTCF factors. Chromatin is separated into two A and B compartments. Chromatin domains in interactions with the nucleolus are called NADs, and those associated with the nuclear lamina are called LADs. At the larger scale, each chromosome occupied its own territory. Me, methylation; E-P, enhancer-promoter interaction; CTCF, CCCTC-binding factor; LADs, lamina-associated domains; NADs, nucleolus-associated domains. Created with BioRender

domains can be changed in composition and/or reorganized [126–131]. This reorganization may result in a compartment change, thereby influencing gene expression.

The interaction between A-type lamins and chromatin has led to dysregulated cell type-specific gene expression emerging as a plausible mechanism to partially elucidate the pathogenesis of CardioLaminopathy. Firstly, evidence indicates that lamins play a crucial role in organizing chromatin through LADs and, consequently, in regulating gene expression [114, 126, 132, 133]. Then, the "chromatin hypothesis" is supported by several studies suggesting an alteration of the LAD organization in CardioLaminopathy [5, 134]. These LAD organization changes were also observed in other laminopathies, including Hutchinson-Gilford progeria syndrome (HGPS) associated with premature aging [135–137], Emery-Dreifuss muscular dystrophy (EDMD) involving the skeletal muscle in addition to cardiac muscle [127], and familial partial lipodystrophy of the Dunnigan type (FPLD2) [137]. LAD repositioning seems to be specific to the type of laminopathy [137]. The reorganization of LADs in CardioLaminopathy has been demonstrated to correlate with transcriptional changes, notably affecting the expression of transcription factors associated with biological processes such as cell cycle, cell death/survival, and metabolism in explanted human hearts from patients with DCM [5]. Similar observations were made in other studies, further highlighting the character of the cell-specific gene expression modulation in disease. Indeed, the study was conducted in cardiomyocytes, adipocytes, and hepatocytes derived from induced pluripotent stem cells (iPSCs) provided by patients with CardioLaminopathy. Only cardiomyocytes (iPSC-CMs) exhibited LAD reorganization and transcriptional changes [134]. Transcriptional changes resulting from the compartmentation switch of LADs in CardioLaminopathy are coupled with a decrease of LADs associated with A-type lamins and an increase of open chromatin marks in the promoter of LAD genes [138]. Therefore, this study demonstrated that *LMNA* p.K117f mutation can induce local changes in LADs correlated with transcriptional activation. Similarly, iPSC-CMs carrying *LMNA* p.K219T mutation exhibit conduction defects associated with the repression of *SCN5A* gene, a gene encoding a sodium channel, orchestrated through an epigenetic regulatory mechanism [139]. One possible scenario is that this gene was originally situated within a LAD submitted to a switch from the nuclear interior to the nuclear periphery, leading to its repression [140]. Only ~1% of the genome seems to be linked to a transition between A and B compartments, leading to changes in cell-specific gene expression in CardioLaminopathy [141]. However, these findings indicate that even a subtle alteration in chromatin compartment organization in mutant cells can still contribute to the phenotype observed in CardioLaminopathy. Hence, chromatin compartmentation may not be the primary pathogenic mechanism in CardioLaminopathy but could contribute to the development and expansion of the disease. A-type lamins are predominantly present at the nuclear periphery, but a fraction of these proteins can also be found in the nuclear interior. Therefore, open chromatin (i.e., euchromatin) can also interact with nucleoplasmic lamins [142] and is relocated in DCM [143, 144]. Additionally to LADs, other structures of the 3D genome can be remodulated in laminopathies. For instance, changes in chromosome territory positioning have been previously reported in these diseases, resulting in abnormal gene expression [145, 146].

## Conclusion and Future Perspectives

The role of nuclear lamins in chromatin organization and regulation in cardiac muscle cells remains poorly understood. While studies have uncovered alterations in chromatin regulation causing some of the symptoms in CardioLaminopathy, these findings are largely restricted to experimental models (in vitro and in vivo), each with inherent technical and theoretical limitations. No single technique can comprehensively capture the full spectrum of chromatin dysregulation or provide a systemic understanding of its complexity. Integrating multiple data sets on chromatin organization promises significantly enhancing our knowledge by increasing informational depth and enabling a more holistic, system-level perspective. However, much work remains to be done, and advancing this field will be critical for breakthroughs in both the diagnosis and treatment of CardioLaminopathy.

**Acknowledgments** This work was supported by the Association Française contre les Myopathies, Institut National de la Santé et de la Recherche Médicale, and Sorbonne Université. MK received funding from the Région Occitanie.

**Compliance with Ethical Standards** The authors declare that the research was conducted in the absence of any commercial or financial relationships that could be construed as a potential conflict of interest.

## References

1. Crasto S, My I, Di Pasquale E (2020) The broad spectrum of LMNA cardiac diseases: from molecular mechanisms to clinical phenotype. Front Physiol 11:761
2. Bonne G, Barletta MRD, Varnous S et al (1999) Mutations in the gene encoding lamin A/C cause autosomal dominant Emery-Dreifuss muscular dystrophy. Nat Genet 21:285–288
3. Muchir A, Worman HJ (2019) Emery-Dreifuss muscular dystrophy: focal point nuclear envelope. Curr Opin Neurol 32(5):728–734
4. Heymans S, Lakdawala NK, Tschöpe C, Klingel K (2023) Dilated cardiomyopathy: causes, mechanisms, and current and future treatment approaches. Lancet 402:998–1011
5. Cheedipudi SM, Matkovich SJ, Coarfa C et al (2019) Genomic reorganization of lamin-associated domains in cardiac myocytes is associated with differential gene expression and DNA methylation in human dilated cardiomyopathy. Circ Res 124:1198–1213
6. Muchir A, Wu W, Choi JC et al (2012) Abnormal p38 mitogen-activated protein kinase signaling in dilated cardiomyopathy caused by lamin A/C gene mutation. Hum Mol Genet 21:4325–4333
7. Rouhi L, Auguste G, Zhou Q et al (2022) Deletion of the Lmna gene in fibroblasts causes senescence-associated dilated cardiomyopathy by activating the double-stranded DNA damage response and induction of senescence-associated secretory phenotype. J Cardiovasc Aging 2:30
8. Nikolova V, Leimena C, McMahon AC et al (2004) Defects in nuclear structure and function promote dilated cardiomyopathy in lamin A/C–deficient mice. J Clin Invest 113:357–369

9. Le Dour C, Chatzifrangkeskou M, Macquart C et al (2022) Actin-microtubule cytoskeletal interplay mediated by MRTF-A/SRF signaling promotes dilated cardiomyopathy caused by LMNA mutations. Nat Commun 13:7886
10. Muchir A (2000) Identification of mutations in the gene encoding lamins A/C in autosomal dominant limb girdle muscular dystrophy with atrioventricular conduction disturbances (LGMD1B). Hum Mol Genet 9:1453–1459
11. Muchir A, Van Engelen BG, Lammens M et al (2003) Nuclear envelope alterations in fibroblasts from LGMD1B patients carrying nonsense Y259X heterozygous or homozygous mutation in lamin A/C gene. Exp Cell Res 291:352–362
12. Ben Yaou R, Yun P, Dabaj I et al (2021) International retrospective natural history study of *LMNA*-related congenital muscular dystrophy. Brain. Communications 3:fcab075
13. Renou L, Stora S, Yaou RB et al (2008) Heart-hand syndrome of Slovenian type: a new kind of laminopathy. J Med Genet 45:666–671
14. Şilfeler DB, Karateke A, Keskin Kurt R et al (2014) Malouf syndrome with hypergonadotropic hypogonadism and cardiomyopathy: two-case report and literature review. Case Rep Obstet Gynecol 2014:1–3
15. Gersak K, Strgulc M, Gorjup V et al (2013) Dilated cardiomyopathy and ovarian dysgenesis in a patient with Malouf syndrome: a case report. Mol Med Rep 8:1311–1314
16. Clayton P, Fischer B, Mann A et al (2010) Mutations causing Greenberg dysplasia but not Pelger anomaly uncouple enzymatic from structural functions of a nuclear membrane protein. Nucleus 1:354–366
17. Sandre-Giovannoli AD, Chaouch M, Kozlov S et al (2002) Homozygous defects in LMNA, encoding Lamin A/C nuclear-envelope proteins, cause autosomal recessive axonal neuropathy in human (charcot-marie-tooth disorder Type 2) and mouse. Am J Hum Genet 70:726
18. Padiath QS, Fu Y-H (2010) Autosomal dominant leukodystrophy caused by Lamin B1 duplications: a clinical and molecular case study of altered nuclear function and disease. In: Methods in cell biology. Elsevier, pp 337–357
19. Raininko R, Gosky M, Padiath QS LMNB1-related autosomal dominant leukodystrophy
20. Neri I, Ramazzotti G, Mongiorgi S et al (2023) Understanding the ultra-rare disease autosomal dominant leukodystrophy: an updated review on morpho-functional alterations found in experimental models. Mol Neurobiol 60:6362–6372
21. Damiano JA, Afawi Z, Bahlo M et al (2015) Mutation of the nuclear lamin gene *LMNB2* in progressive myoclonus epilepsy with early ataxia. Hum Mol Genet 24:4483–4490
22. Barascu A, Le Chalony C, Pennarun G et al (2012) Oxidative stress induces an ATM-independent senescence pathway through p38 MAPK-mediated lamin B1 accumulation: Senescence through lamin B1 accumulation. EMBO J 31:1080–1094
23. Cristofoli F, Moss T, Moore HW et al (2020) De Novo variants in LMNB1 cause pronounced syndromic microcephaly and disruption of nuclear envelope integrity. Am J Hum Genet 107:753–762
24. Parry DA, Martin C-A, Greene P et al (2021) Heterozygous lamin B1 and lamin B2 variants cause primary microcephaly and define a novel laminopathy. Genet Med 23:408–414
25. De Castro SCP, Malhas A, Leung K-Y et al (2012) Lamin B1 polymorphism influences morphology of the nuclear envelope, cell cycle progression, and risk of neural tube defects in mice. PLoS Genet 8:e1003059
26. Frost B, Hemberg M, Lewis J, Feany MB (2014) Tau promotes neurodegeneration through global chromatin relaxation. Nat Neurosci 17:357–366
27. Frost B, Bardai FH, Feany MB (2016) Lamin dysfunction mediates neurodegeneration in tauopathies. Curr Biol 26:129–136
28. Oosterwijk JC (2003) Congenital abnormalities reported in Pelger-Huet homozygosity as compared to Greenberg/HEM dysplasia: highly variable expression of allelic phenotypes. J Med Genet 40:937–941
29. Fernandez-Pombo A, Diaz-Lopez EJ, Castro AI et al (2023) Clinical spectrum of LMNA-associated Type 2 familial partial lipodystrophy: a systematic review. Cells 12:725

30. Hegele RA, Cao H, Liu DM et al (2006) Sequencing of the reannotated LMNB2 gene reveals novel mutations in patients with acquired partial lipodystrophy. Am J Hum Genet 79:383–389
31. Simsek-Kiper PO, Roach E, Utine GE, Boduroglu K (2014) Barraquer–Simons syndrome: a rare clinical entity. American J of Med Genetics Pt A 164:1756–1760
32. Gonzalo S, Kreienkamp R, Askjaer P (2017) Hutchinson-Gilford Progeria syndrome: a premature aging disease caused by LMNA gene mutations. Ageing Res Rev 33:18–29
33. Sears RM, Roux KJ (2022) Mechanisms of A-Type lamin targeting to nuclear ruptures are disrupted in lMNA- and BANF1-associated progerias. Cells 11:865
34. Oshima J, Hisama FM (2014) Search and insights into novel genetic alterations leading to classical and atypical Werner syndrome. Gerontology 60:239–246
35. Novelli G, Muchir A, Sangiuolo F et al (2002) Mandibuloacral Dysplasia is caused by a mutation in LMNA-encoding Lamin A/C. Am J Hum Genet 71:426–431
36. Agarwal AK (2003) Zinc metalloproteinase, ZMPSTE24, is mutated in mandibuloacral dysplasia. Hum Mol Genet 12:1995–2001
37. Moulson CL, Go G, Gardner JM et al (2005) Homozygous and compound heterozygous mutations in ZMPSTE24 cause the laminopathy restrictive dermopathy. J Invest Dermatol 125:913–919
38. Navarro CL, Cadiñanos J, Sandre-Giovannoli AD et al (2005) Loss of ZMPSTE24 (FACE-1) causes autosomal recessive restrictive dermopathy and accumulation of Lamin A precursors. Hum Mol Genet 14:1503–1513
39. Fatkin D, MacRae C, Sasaki T et al (1999) Missense mutations in the rod domain of the Lamin A/C gene as causes of dilated cardiomyopathy and conduction-system disease. N Engl J Med 341:1715–1724
40. Tesson F, Saj M, Uvaize MM et al (2014) Lamin A/C mutations in dilated cardiomyopathy. Cardiol J 21:331–342
41. Taylor MRG, Fain PR, Sinagra G et al (2003) Natural history of dilated cardiomyopathy due to lamin A/C gene mutations. J Am Coll Cardiol 41:771–780
42. Meune C, Van Berlo JH, Anselme F et al (2006) Primary prevention of sudden death in patients with Lamin A/C gene mutations. N Engl J Med 354:209–210
43. Schultheiss H-P, Fairweather D, Caforio ALP et al (2019) Dilated cardiomyopathy. Nat Rev Dis Primers 5:32
44. de Leeuw R, Gruenbaum Y, Medalia O (2018) Nuclear Lamins: thin filaments with major functions. Trends Cell Biol 28:34–45
45. Dittmer TA, Misteli T (2011) The lamin protein family. Genome Biol 12:222
46. Gerace L, Blobel G (1980) The nuclear envelope lamina is reversibly depolymerized during mitosis. Cell 19:277–287
47. Broers JLV, Machiels BM, Kuijpers HJH et al (1997) A- and B-type lamins are differentially expressed in normal human tissues. Histochem Cell Biol 107:505–517
48. DeBoy E, Puttaraju M, Jailwala P et al (2017) Identification of novel RNA isoforms of *LMNA*. Nucleus 8:573–582
49. Furukawa K, Hotta Y (1993) cDNA cloning of a germ cell specific lamin B3 from mouse spermatocytes and analysis of its function by ectopic expression in somatic cells. EMBO J 12:97–106
50. Evangelisti C, Rusciano I, Mongiorgi S et al (2022) The wide and growing range of lamin B-related diseases: from laminopathies to cancer. Cell Mol Life Sci 79:126
51. Worman HJ, Lazaridis I, Georgatos SD (1988) Nuclear lamina heterogeneity in mammalian cells. Differential expression of the major lamins and variations in lamin B phosphorylation. J Biol Chem 263:12135–12141
52. Georgatos SD, Stournaras C, Blobel G (1988) Heterotypic and homotypic associations between the nuclear lamins: site-specificity and control by phosphorylation. Proc Natl Acad Sci USA 85:4325–4329

53. Nmezi B, Xu J, Fu R et al (2019) Concentric organization of A- and B-type lamins predicts their distinct roles in the spatial organization and stability of the nuclear lamina. Proc Natl Acad Sci USA 116:4307–4315
54. Wintner O, Hirsch-Attas N, Schlossberg M et al (2020) A unified linear viscoelastic model of the cell nucleus defines the mechanical contributions of lamins and chromatin. Adv Sci 7:1901222
55. Lammerding J, Fong LG, Ji JY et al (2006) Lamins A and C but Not Lamin B1 regulate nuclear mechanics. J Biol Chem 281:25768–25780
56. Chu C-T, Chen Y-H, Chiu W-T, Chen H-C (2021) Tyrosine phosphorylation of lamin A by Src promotes disassembly of nuclear lamina in interphase. Life Sci Alliance 4:e202101120
57. Hocevar BA, Burns DJ, Fields AP (1993) Identification of protein kinase C (PKC) phosphorylation sites on human lamin B. Potential role of PKC in nuclear lamina structural dynamics. J Biol Chem 268:7545–7552
58. Heald R, McKeon F (1990) Mutations of phosphorylation sites in lamin A that prevent nuclear lamina disassembly in mitosis. Cell 61:579–589
59. Peter M, Nakagawa J, Dorée M et al (1990) In vitro disassembly of the nuclear lamina and M phase-specific phosphorylation of lamins by cdc2 kinase. Cell 61:591–602
60. Cano-Monreal GL, Wylie KM, Cao F et al (2009) Herpes simplex virus 2 UL13 protein kinase disrupts nuclear lamins. Virology 392:137–147
61. Caruso LB, Guo R, Keith K et al (2022) The nuclear lamina binds the EBV genome during latency and regulates viral gene expression. PLoS Pathog 18:e1010400
62. Wu S, Pan S, Zhang L et al (2016) Herpes Simplex Virus 1 induces phosphorylation and reorganization of Lamin A/C through the $\gamma_1$ 34.5 protein that facilitates nuclear egress. J Virol 90:10414–10422
63. Speese SD, Ashley J, Jokhi V et al (2012) Nuclear envelope budding enables large ribonucleoprotein particle export during synaptic Wnt signaling. Cell 149:832–846
64. Velez-Aguilera G, Nkombo Nkoula S, Ossareh-Nazari B et al (2020) PLK-1 promotes the merger of the parental genome into a single nucleus by triggering lamina disassembly. eLife 9:e59510
65. Murray-Nerger LA, Justice JL, Rekapalli P et al (2021) Lamin B1 acetylation slows the G1 to S cell cycle transition through inhibition of DNA repair. Nucleic Acids Res 49:2044–2064
66. Karoutas A, Szymanski W, Rausch T et al (2019) The NSL complex maintains nuclear architecture stability via lamin A/C acetylation. Nat Cell Biol 21:1248–1260
67. Moriuchi T, Kuroda M, Kusumoto F et al (2016) Lamin A reassembly at the end of mitosis is regulated by its SUMO-interacting motif. Exp Cell Res 342:83–94
68. Li Y, Jiang X, Zhang Y et al (2019) Nuclear accumulation of UBC9 contributes to SUMOylation of lamin A/C and nucleophagy in response to DNA damage. J Exp Clin Cancer Res 38:67
69. Zhang Y-Q, Sarge KD (2008) Sumoylation regulates lamin A function and is lost in lamin A mutants associated with familial cardiomyopathies. J Cell Biol 182:35–39
70. Simon DN, Domaradzki T, Hofmann WA, Wilson KL (2013) Lamin A tail modification by SUMO1 is disrupted by familial partial lipodystrophy–causing mutations. MBoC 24:342–350
71. Liczmanska M, Tatham MH, Mojsa B et al (2023) SUMO protease SENP6 protects the nucleus from hyperSUMOylation-induced laminopathy-like alterations. Cell Rep 42:112960
72. Krishnamoorthy V, Khanna R, Parnaik VK (2018) E3 ubiquitin ligase HECW2 targets PCNA and lamin B1. Biochimica et Biophysica Acta (BBA). Mol Cell Res 1865:1088–1104
73. Khanna R, Krishnamoorthy V, Parnaik VK (2018) E3 ubiquitin ligase RNF 123 targets lamin B1 and lamin-binding proteins. FEBS J 285:2243–2262
74. Simon D, Wriston A, Fan Q et al (2018) OGT (O-GlcNAc transferase) selectively modifies multiple residues unique to Lamin A. Cells 7:44
75. Kim S-J, Yoo BC, Uhm C-S, Lee S-W (2011) Posttranslational arginine methylation of lamin A/C during myoblast fusion. Biochimica et Biophysica Acta (BBA)—Proteins and Proteomics 1814:308–317

76. Broers JLV, Peeters EAG, Kuijpers HJH et al (2004) Decreased mechanical stiffness in LMNA−/− cells is caused by defective nucleo-cytoskeletal integrity: implications for the development of laminopathies. Hum Mol Genet 13:2567–2580
77. Lammerding J, Schulze PC, Takahashi T et al (2004) Lamin A/C deficiency causes defective nuclear mechanics and mechanotransduction. J Clin Invest 113:370–378
78. Cho S, Vashisth M, Abbas A et al (2019) Mechanosensing by the Lamina protects against nuclear rupture, DNA damage, and cell-cycle arrest. Dev Cell 49:920–935.e5
79. Urciuoli E, Peruzzi B (2022) The Paradox of nuclear Lamins in pathologies: apparently controversial roles explained by tissue-specific mechanobiology. Cells 11:2194
80. Swift J, Ivanovska IL, Buxboim A et al (2013) Nuclear Lamin-A scales with tissue stiffness and enhances matrix-directed differentiation. Science 341:1240104
81. Buxboim A, Swift J, Irianto J et al (2014) Matrix elasticity regulates Lamin-A,C phosphorylation and turnover with feedback to actomyosin. Curr Biol 24:1909–1917
82. Kovacs MT, Vallette M, Wiertsema P et al (2023) DNA damage induces nuclear envelope rupture through ATR-mediated phosphorylation of lamin A/C. Mol Cell 83:3659–3668.e10
83. Denais CM, Gilbert RM, Isermann P et al (2016) Nuclear envelope rupture and repair during cancer cell migration. Science 352:353–358
84. Gibbs-Seymour I, Markiewicz E, Bekker-Jensen S et al (2015) Lamin A/C-dependent interaction with 53BP1 promotes cellular responses to DNA damage. Aging Cell 14:162–169
85. Etourneaud L, Moussa A, Rass E et al (2021) Lamin B1 sequesters 53BP1 to control its recruitment to DNA damage. Sci Adv 7:eabb3799
86. Maynard S, Keijzers G, Akbari M et al (2019) Lamin A/C promotes DNA base excision repair. Nucleic Acids Res gkz912:11709
87. Singh M, Hunt CR, Pandita RK et al (2013) Lamin A/C depletion enhances DNA damage-induced stalled replication fork arrest. Mol Cell Biol 33:1210–1222
88. Chen SN, Lombardi R, Karmouch J et al (2019) DNA damage response/TP53 pathway is activated and contributes to the pathogenesis of dilated cardiomyopathy associated with LMNA (Lamin A/C) mutations. Circ Res 124:856–873
89. Cheedipudi SM, Asghar S, Marian AJ (2022) Genetic ablation of the DNA damage response pathway attenuates lamin-associated dilated cardiomyopathy in mice. JACC: basic to translational. Science 7:1232–1245
90. Wang Y, Jiang J, He L et al (2019) Effect of lamin-A expression on migration and nuclear stability of ovarian cancer cells. Gynecol Oncol 152:166–176
91. Gonzalez-Suarez I, Redwood AB, Perkins SM et al (2009) Novel roles for A-type lamins in telomere biology and the DNA damage response pathway. EMBO J 28:2414–2427
92. Pennarun G, Picotto J, Etourneaud L et al (2021) Increase in lamin B1 promotes telomere instability by disrupting the shelterin complex in human cells. Nucleic Acids Res 49:9886–9905
93. Matias I, Diniz LP, Damico IV et al (2022) Loss of lamin-B1 and defective nuclear morphology are hallmarks of astrocyte senescence in vitro and in the aging human hippocampus. Aging Cell 21:e13521
94. Decker ML, Chavez E, Vulto I, Lansdorp PM (2009) Telomere length in Hutchinson-Gilford Progeria syndrome. Mech Ageing Dev 130:377–383
95. Shumaker DK, Solimando L, Sengupta K et al (2008) The highly conserved nuclear lamin Ig-fold binds to PCNA: its role in DNA replication. J Cell Biol 181:269–280
96. Moir RD, Spann TP, Herrmann H, Goldman RD (2000) Disruption of nuclear lamin organization blocks the elongation phase of DNA replication. J Cell Biol 149:1179–1192
97. Spann TP, Moir RD, Goldman AE et al (1997) Disruption of nuclear lamin organization alters the distribution of replication factors and inhibits DNA synthesis. J Cell Biol 136:1201–1212
98. Kumaran RI, Muralikrishna B, Parnaik VK (2002) Lamin A/C speckles mediate spatial organization of splicing factor compartments and RNA polymerase II transcription. J Cell Biol 159:783–793

99. Legartová S, Fagherazzi P, Stixová L et al (2021) The SC-35 splicing factor interacts with RNA Pol II and A-Type Lamin depletion weakens this interaction. Cells 10:297
100. Ito K, Patel PN, Gorham JM et al (2017) Identification of pathogenic gene mutations in *LMNA* and *MYBPC3* that alter RNA splicing. Proc Natl Acad Sci USA 114:7689–7694
101. Dahl KN, Scaffidi P, Islam MF et al (2006) Distinct structural and mechanical properties of the nuclear lamina in Hutchinson–Gilford progeria syndrome. Proc Natl Acad Sci USA 103:10271–10276
102. Osmanagic-Myers S, Kiss A, Manakanatas C et al (2018) Endothelial progerin expression causes cardiovascular pathology through an impaired mechanoresponse. J Clin Invest 129:531–545
103. Bertrand AT, Ziaei S, Ehret C et al (2014) Cellular micro-environments reveal defective mechanosensing responses and elevated YAP signaling in LMNA-mutated muscle precursors. J Cell Sci jcs.144907:2873
104. Zwerger M, Jaalouk DE, Lombardi ML et al (2013) Myopathic lamin mutations impair nuclear stability in cells and tissue and disrupt nucleo-cytoskeletal coupling. Hum Mol Genet 22:2335–2349
105. Lammerding J, Hsiao J, Schulze PC et al (2005) Abnormal nuclear shape and impaired mechanotransduction in emerin-deficient cells. J Cell Biol 170:781–791
106. Owens D, Messéant J, Moog S et al (2020) Lamin-related congenital muscular dystrophy alters mechanical signaling and skeletal muscle growth. IJMS 22:306
107. Muchir A, Pavlidis P, Decostre V et al (2007) Activation of MAPK pathways links LMNA mutations to cardiomyopathy in Emery-Dreifuss muscular dystrophy. J Clin Invest 117:1282–1293
108. Chatzifrangkeskou M, Yadin D, Marais T et al (2018) Cofilin-1 phosphorylation catalyzed by ERK1/2 alters cardiac actin dynamics in dilated cardiomyopathy caused by lamin A/C gene mutation. Hum Mol Genet 27:3060–3078
109. Macquart C, Jüttner R, Morales Rodriguez B et al (2018) Microtubule cytoskeleton regulates Connexin 43 localization and cardiac conduction in cardiomyopathy caused by mutation in A-type lamins gene. Hum Mol Genet 28:4043
110. Choi JC, Wu W, Phillips E et al (2018) Elevated dual specificity protein phosphatase 4 in cardiomyopathy caused by lamin A/C gene mutation is primarily ERK1/2-dependent and its depletion improves cardiac function and survival. Hum Mol Genet 27:2290–2305
111. Le Dour C, Macquart C, Sera F et al (2017) Decreased WNT/β-catenin signalling contributes to the pathogenesis of dilated cardiomyopathy caused by mutations in the lamin a/C gene. Hum Mol Genet ddw389:ddw389
112. Mounkes LC, Kozlov SV, Rottman JN, Stewart CL (2005) Expression of an LMNA-N195K variant of A-type lamins results in cardiac conduction defects and death in mice. Hum Mol Genet 14:2167–2180
113. Chen SC, Kennedy BK, Lampe PD (2013) Phosphorylation of connexin43 on S279/282 may contribute to laminopathy-associated conduction defects. Exp Cell Res 319:888–896
114. Shimi T, Pfleghaar K, Kojima S et al (2008) The A- and B-type nuclear lamin networks: microdomains involved in chromatin organization and transcription. Genes Dev 22:3409–3421
115. Kim Y, Zheng X, Zheng Y (2019) Role of lamins in 3D genome organization and global gene expression. Nucleus 10:33–41
116. Zhou K, Gaullier G, Luger K (2019) Nucleosome structure and dynamics are coming of age. Nat Struct Mol Biol 26:3–13
117. Ricci MA, Manzo C, García-Parajo MF et al (2015) Chromatin fibers are formed by heterogeneous groups of nucleosomes in vivo. Cell 160:1145–1158
118. Jerkovic I, Cavalli G (2021) Understanding 3D genome organization by multidisciplinary methods. Nat Rev Mol Cell Biol 22:511–528
119. Kempfer R, Pombo A (2020) Methods for mapping 3D chromosome architecture. Nat Rev Genet 21:207–226

120. Rocha PP, Raviram R, Bonneau R, Skok JA (2015) Breaking TADs: insights into hierarchical genome organization. Epigenomics 7:523–526
121. Achinger-Kawecka J, Taberlay PC, Clark SJ (2016) Alterations in three-dimensional organization of the cancer genome and epigenome. Cold Spring Harb Symp Quant Biol 81:41–51
122. Simonis M, Klous P, Splinter E et al (2006) Nuclear organization of active and inactive chromatin domains uncovered by chromosome conformation capture–on-chip (4C). Nat Genet 38:1348–1354
123. Lieberman-Aiden E, van Berkum NL, Williams L et al (2009) Comprehensive mapping of long-range interactions reveals folding principles of the human genome. Science 326:289–293
124. Jung N, Kim T-K (2021) Advances in higher-order chromatin architecture: the move towards 4D genome. BMB Rep 54:233–245
125. Cremer T, Cremer M (2010) Chromosome territories. Cold Spring Harb Perspect Biol 2:a003889–a003889
126. Solovei I, Wang AS, Thanisch K et al (2013) LBR and Lamin A/C sequentially tether peripheral heterochromatin and inversely regulate differentiation. Cell 152:584–598
127. Perovanovic J, Dell'Orso S, Gnochi VF et al (2016) Laminopathies disrupt epigenomic developmental programs and cell fate. Sci Transl Med 8:335ra58
128. Wijchers PJ, Geeven G, Eyres M et al (2015) Characterization and dynamics of pericentromere-associated domains in mice. Genome Res 25:958–969
129. Krigerts J, Salmina K, Freivalds T et al (2021) Differentiating cancer cells reveal early large-scale genome regulation by pericentric domains. Biophys J 120:711–724
130. Peng T, Hou Y, Meng H et al (2023) Mapping nucleolus-associated chromatin interactions using nucleolus Hi-C reveals pattern of heterochromatin interactions. Nat Commun 14:350
131. Bersaglieri C, Santoro R (2019) Genome Organization in and around the Nucleolus. Cells 8:579
132. Guelen L, Pagie L, Brasset E et al (2008) Domain organization of human chromosomes revealed by mapping of nuclear lamina interactions. Nature 453:948–951
133. Zullo JM, Demarco IA, Piqué-Regi R et al (2012) DNA sequence-dependent compartmentalization and silencing of chromatin at the nuclear lamina. Cell 149:1474–1487
134. Shah PP, Lv W, Rhoades JH et al (2021) Pathogenic LMNA variants disrupt cardiac lamina-chromatin interactions and de-repress alternative fate genes. Cell Stem Cell 28:938–954.e9
135. Ikegami K, Secchia S, Almakki O et al (2020) Phosphorylated Lamin A/C in the nuclear interior binds active enhancers associated with abnormal transcription in progeria. Dev Cell 52:699–713.e11
136. Köhler F, Bormann F, Raddatz G et al (2020) Epigenetic deregulation of lamina-associated domains in Hutchinson-Gilford progeria syndrome. Genome Med 12:46
137. Paulsen J (2017) Chrom3D: three-dimensional genome modeling from Hi-C and nuclear lamin-genome contacts. 15
138. Lee J, Termglinchan V, Diecke S et al (2019) Activation of PDGF pathway links LMNA mutation to dilated cardiomyopathy. Nature 572:335–340
139. Salvarani N, Crasto S, Miragoli M et al (2019) The K219T-Lamin mutation induces conduction defects through epigenetic inhibition of SCN5A in human cardiac laminopathy. Nat Commun 10:2267
140. Lund E, Oldenburg AR, Delbarre E et al (2013) Lamin A/C-promoter interactions specify chromatin state–dependent transcription outcomes. Genome Res 23:1580–1589
141. Bertero A, Fields PA, Smith AST et al (2019) Chromatin compartment dynamics in a haploinsufficient model of cardiac laminopathy. J Cell Biol 218:2919–2944
142. Gesson K, Vidak S, Foisner R (2014) Lamina-associated polypeptide (LAP)2α and nucleoplasmic lamins in adult stem cell regulation and disease. Semin Cell Dev Biol 29:116–124
143. Zhang X, Shao X, Zhang R et al (2021) Integrated analysis reveals the alterations that LMNA interacts with euchromatin in LMNA mutation-associated dilated cardiomyopathy. Clin Epigenetics 13:3

144. Feng Y, Cai L, Hong W et al (2022) Rewiring of 3D chromatin topology orchestrates transcriptional reprogramming and the development of human dilated cardiomyopathy. Circulation 145:1663–1683
145. Meaburn KJ, Cabuy E, Bonne G et al (2007) Primary laminopathy fibroblasts display altered genome organization and apoptosis. Aging Cell 6:139–153
146. Mewborn SK, Puckelwartz MJ, Abuisneineh F et al (2010) Altered chromosomal positioning, compaction, and gene expression with a Lamin A/C gene mutation. PLoS One 5:e14342

# Chapter 2
# Role of Nuclear Envelope Proteins in the Structure and Function of the Neuromuscular Junction: Focus on Subsynaptic Nuclei

**Céline Douarre, Bruno Cadot, Antoine Muchir, and Stéphanie Bauché**

**Abstract** The role of the nuclear envelope (NE) in skeletal muscle function was first recognized 30 years ago when mutations in NE-associated genes, such as SUN-1, Syne1, Syne2, and LMNA, were linked to muscular dystrophies, also known as nuclear envelopathies. These findings underscored the critical role of NE components in maintaining muscle fiber integrity and function. NE proteins, including lamin A/C, SUN-1/2, nesprins, and LAP1, play a key role in anchoring and positioning nuclei within muscle fibers. In particular, the correct positioning of subsynaptic nuclei (SSNs) beneath the neuromuscular junction (NMJ) is essential for their differentiation and functional specialization, ensuring efficient neuromuscular transmission. The crucial role of the NE in SSNs and NMJ function has been further emphasized by its association with an atypical form of congenital myasthenic syndrome (CMS). Despite significant advances in understanding NMJ formation and motor neuron-myofiber communication, the mechanisms governing gene regulation in SSNs remain largely unexplored.

**Keywords** Nuclear envelope · Neuromuscular junction · Subsynaptic nuclei · Emery-dreifuss muscular dystrophy · Congenital myasthenic syndrome

---

C. Douarre · B. Cadot · A. Muchir · S. Bauché (✉)
Sorbonne Université, INSERM U974, Institute of Myology, Center of Research in Myology, Paris, France
e-mail: bruno.cadot@inserm.fr; a.muchir@inserm.fr; stephanie.godard-bauche@sorbonne-universite.fr

© The Author(s), under exclusive license to Springer Nature Switzerland AG 2025 
K. Sengupta, A. Muchir (eds.), *Lamins in Genome Organization and Diseases*, Subcellular Biochemistry 115, https://doi.org/10.1007/978-3-032-00537-3_2

# Introduction

Neuromuscular junctions (NMJ) are key sites of communication between motor neurons and muscle fibers, essential for muscle contraction. The precise regulation of NMJ structure and function largely relies on the activity of subsynaptic nuclei, a subset of muscle nuclei located beneath the synapse. These nuclei play a central role in gene expression specific to NMJ maintenance. Nuclear envelope (NE) proteins, which structure and organize the interface between the nucleus and the cytoskeleton, have emerged as key players in these processes. In addition to their role in nuclear organization, NE proteins are involved in the regulation of subsynaptic signaling, though this remains poorly understood. A deeper understanding of their functions within subsynaptic nuclei could shed light on the pathological mechanisms underlying neuromuscular diseases associated with mutations in these proteins, such as certain myopathies or congenital myasthenic syndromes (CMS).

# Nuclei, Nuclear Envelope Proteins, and Muscular Disorders

## *Nuclei and Nuclear Envelope (NE) Proteins*

In higher eukaryotic cells, the nucleus is surrounded by a complex structure called the nuclear envelope (NE), composed of three distinct components: the inner and outer nuclear membranes, the nuclear pore complexes, and the nuclear lamina [1]. This NE serves as a physical barrier, isolating the chromosomes from the cytoplasm and playing a central role in the spatial organization and protection of genetic material [2]. The NE is also composed of a unique set of membrane proteins, including lamin A/C, nesprin-1 and nesprin-2, emerin, and SUN-1, all of which interact with the nuclear lamina to maintain nuclear structure and regulate various nuclear functions [3–5].

In the specialized context of mammalian skeletal muscle fibers (myofibers), the NE acquires additional importance, partly due to the presence of hundreds of nuclei within a single cell, but also because it acquired, during the process of differentiation, microtubules organizing properties previously devoted to the centrosome [6]. In mature myofibers, these nuclei are strategically positioned and evenly spaced along the periphery, just below the plasma membrane. This precise nuclear arrangement of nuclei is essential for the proper functioning of the muscle fiber, ensuring the efficient coordination of gene expression and cellular processes across the extensive and multi-nucleated landscape of the myofiber [7].

## *NE Diseases*

Mutations in genes encoding NE proteins, including lamin A/C, nesprin-1 and nesprin-2, and SUN-1, lead to a range of disorders collectively known as "laminopathies" or "nuclear envelopathies." [8, 9]

Laminopathies and nuclear envelopathies encompass a wide range of diseases that can affect several tissues and organs, including muscles and heart. The manifestations of these diseases are diverse and reflect the essential roles played by NE proteins in different cellular contexts including lamin A/C, emerin, nesprins, and SUN proteins [10]. Mutations in *LMNA* gene, which codes for lamins A and C, lead to diseases such as Emery-Dreifuss muscular dystrophy, characterized by muscle weakness associated with dilated cardiomyopathy [11, 12]. Mutations in other genes coding for proteins such as nesprins and SUN proteins, all of which contribute to the structural and functional integrity of the nuclear envelope, lead to disorders with muscle weakness and dystrophy frequently associated with cardiomyopathy [13, 14].

## NE Proteins and Neuromuscular Junction (NMJ)

Skeletal muscle contraction is initiated at the neuromuscular junction (NMJ), a highly specialized synapse connecting the motor neuron endings directly to the myofiber [15, 16]. This critical site of communication ensures that the electrical signals from the nervous system are effectively translated into mechanical contractions of the muscle. Within the myofiber, which contains hundreds of nuclei, a distinct subset of these nuclei – typically between four and eight – is strategically positioned beneath the NMJ [17, 18]. These subsynaptic nuclei (SSN) possess a unique transcriptomic profile that is essential for the specialized functions of neuromuscular transmission [19] (Fig. 2.1).

The precise positioning of SSNs during muscle differentiation is crucial for both the formation and maintenance of the NMJ [18]. This positioning is controlled by a complex interplay between the cytoskeleton and various NE proteins, which are key players in nuclear positioning, chromatin organization, and gene regulation [20, 21]. NE proteins such as lamins, nesprin, and SUN proteins, among others, have well-established roles in maintaining nuclear architecture and regulating gene expression [22–24] (Fig. 2.2). However, despite significant advances in understanding the general functions of these NE proteins, their specific roles in the SSNs at the NMJ remain poorly understood. This gap in knowledge highlights the need for further research to elucidate how NE proteins contribute to the specialized functions of SSNs and their impact on the overall efficiency and stability of neuromuscular transmission.

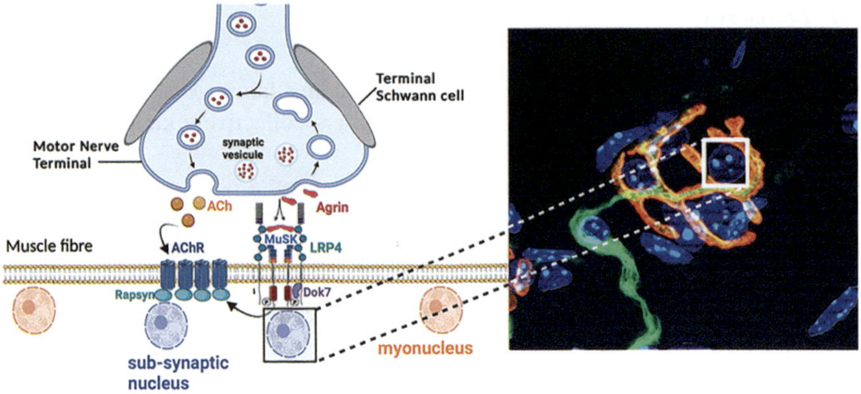


**Fig. 2.1** Schematic and representative structure of the subsynaptic and extra-synaptic nuclei at the NMJ. Confocal image of a mouse NMJ with staining of AChR (red), terminal axon (green), and nuclei (blue). Scale bar = 10 μm. (Created with BioRender)

## Neuromuscular Junction (NMJ) Mechanism and Function

### *NMJ Formation and Maintenance*

The NMJ is a neurochemical synapse whose highly specialized anatomical and molecular structure enables neuromuscular transmission [25]. It is composed of three distinct cell types that are essential to its proper functioning: a motor neuron nerve ending that synthesizes the neurotransmitter (acetylcholine, ACh), a terminal Schwann cell that caps the nerve ending, and, finally, the skeletal muscle fiber [26–28]. A basal lamina surrounds and connects these three components, creating a microenvironment that facilitates the exchange of molecular signals and controls the formation, maturation, and maintenance of this peripheral synapse. The NMJ is the most extensively studied neurochemical synapse in the vertebrate nervous system, with a highly organized structure for chemical neurotransmission. The presynaptic nerve terminal contains a molecular apparatus for the evoked release of the neurotransmitter (acetylcholine), which binds to its specific receptor (acetylcholine receptor, AChR) in the postsynaptic muscle fiber containing all the components responsible for the change in membrane potential when the neurotransmitter is released [15, 16]. Electrical activity activates transcription of synaptic protein genes – including AChR subunits – in SSNs and also affects all myonuclei outside the NMJ, repressing gene expression of postsynaptic proteins [29, 30]. Adequate AChR density at the NMJ involves two mechanisms. The first corresponds to the formation of the NMJ with the clustering and stabilization of AChR molecules under the sarcolemma through the release of a neurotrophic factor, neural agrin [31], and a complex postsynaptic protein scaffold composed of Lrp4 [32, 33] and its specific co-receptor MuSK [34, 35]. The second, which corresponds to the maintenance of the NMJ, results from the specialization of (some) subsynaptic nuclei to

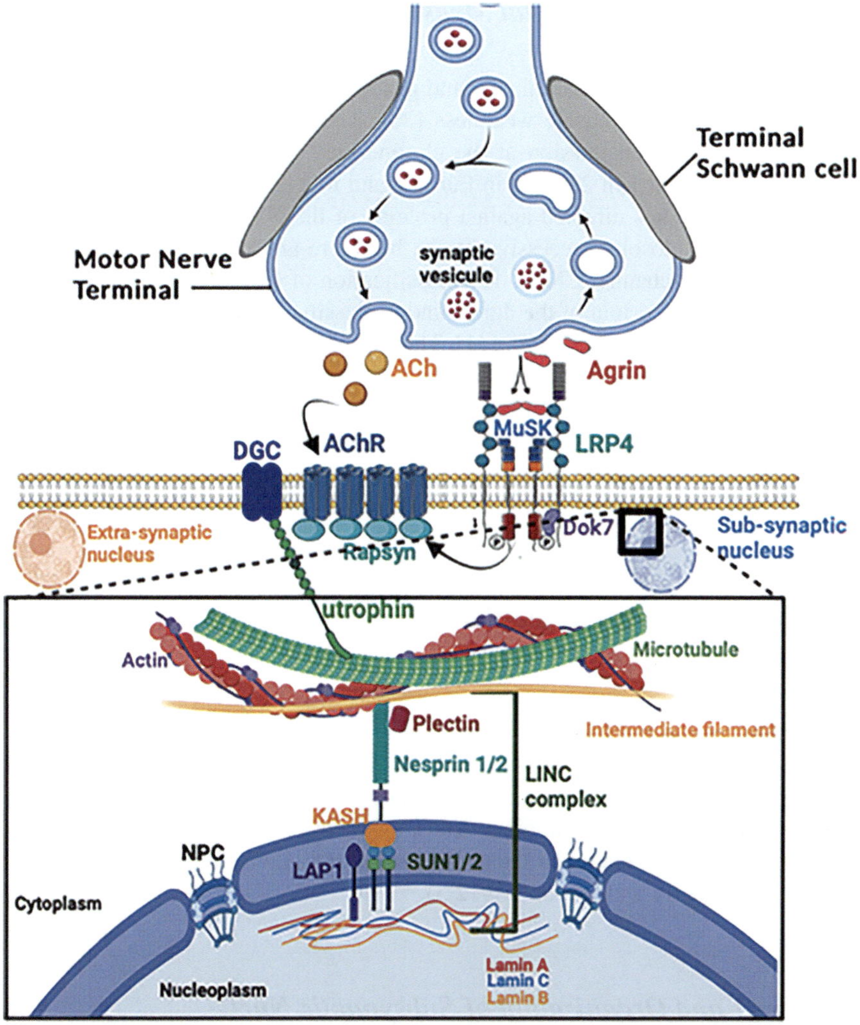


**Fig. 2.2** Schematic representation of interactions with postsynaptic proteins at the NMJ and NE proteins in subsynaptic nuclei. At the NMJ, the subsynaptic nuclei are specialized beneath the nerve terminal by a complex interacting network of proteins linked to the dystrophin-glycoprotein complex (DGC) via the dystroglycan. Utrophin binds to the DGC and the actin cytoskeleton. Nesprin-1/nesprin-2 binds to the actin cytoskeleton and the NE via its KASH domain and forms the LINC complex with SUN-1/2, LAP1, and the lamins. (Created with BioRender)

allow the expression of synaptic genes encoding the components of the NMJ, including the AChR subunits [18, 36, 37]. NMJ dysfunction mainly affects the function of acetylcholine receptors (AChR) and can result in NMJ disorders or premature death.

### *NMJ Pathology: Congenital Myasthenic Syndrome (CMS)*

CMSs form a group of rare, clinically and genetically heterogeneous diseases characterized by fatigable muscle weakness [38]. They are inherited and caused by defective synaptic transmission at the cholinergic NMJ [39]. The prevalence of CMS is estimated at 1 in 250,000 in Europe, and it is rarer than myasthenia gravis due to autoantibodies directed against proteins of the NMJ [40]. Most CMSs are inherited in an autosomal recessive mode, but there are a few cases of dominant inheritance (slow-channel CMS). The classification of CMS is pathophysiological and depends on the origin of the defect and is classified into presynaptic, synaptic, postsynaptic, and ubiquitous CMS [41]. The number of genes known to cause CMS when mutated is currently 35, and the relationship between fatigable muscle weakness and defective function is fairly well understood for many of them [41–43]. However, the frontier forms between CMS and some congenital myopathies highlight the poorly understood interaction between synaptic and extra-synaptic molecules in the neuromuscular system.

## Role of NE Proteins in the Formation and Maintenance of NMJ

NE proteins play a crucial role in a number of well-studied biological mechanisms such as the maintenance and integrity of the nucleus, the chromatin organization, and the regulation of signaling pathways that control gene expression [12, 44, 45]. There are distinct nuclear subtypes: extra-synaptic (myonuclei) and subsynaptic nuclei (SSN) specialized under the nerve ending and in which the roles of NE proteins are still poorly understood (Fig. 2.3).

### *Nesprin-1 and Organization of Subsynaptic Nuclei*

Nesprins are conserved proteins, characterized by a central stem of spectrin repeats (SR), which are anchored to the outer nuclear membrane by a C-terminal KASH (Klarsicht/ANC-1/Syne homology) domain [4]. Nesprin-1 is encoded by the *Syne1* gene and is a key component of the linker of nucleoskeleton and cytoskeleton (LINC) complex, which connects the nuclear envelope to the cytoskeleton [46]. Several isoforms can exist in different tissues, and the shortest isoform, lacking the actin binding domain, is specific to striated muscles. The GIANT isoform is correlated with actin filaments and facilitates the precise positioning of nuclei within muscle fibers [23]. The role of nesprin-1, the largest of the four mammalian nesprins, has been demonstrated in central and peripheral nervous systems [13, 23]. The importance of nesprin-1 in neurons was illustrated by the identification of mutations

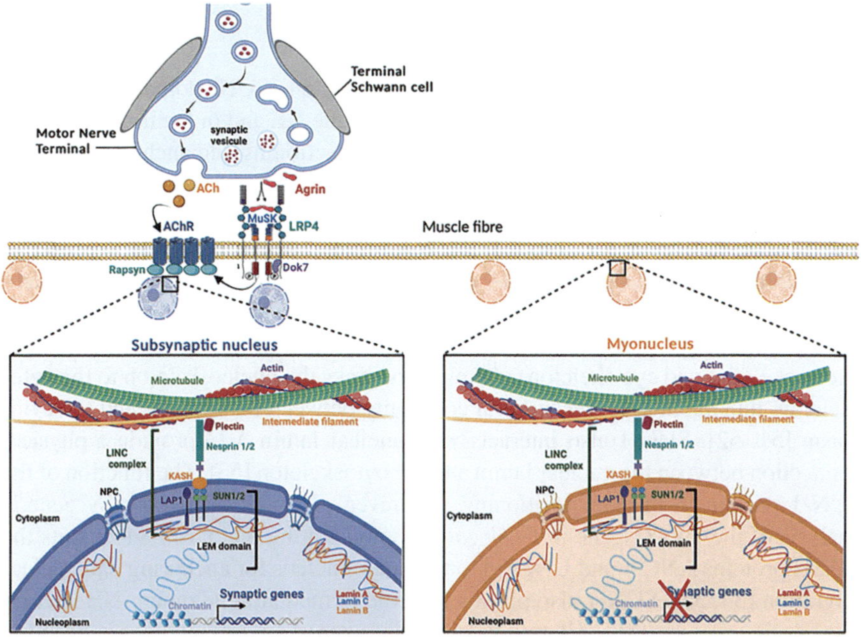


**Fig. 2.3** Schematic representation of the transcriptional regulation of synaptic genes in subsynaptic and extra-synaptic nuclei involving NE proteins. In subsynaptic nuclei, transcriptional regulation is initiated by the release of neuronal agrin, which binds to LRP4 to activate the Lrp4/MuSK complex and induce, via LINK complex proteins, the activation of the LEM domain linked to the chromatin system and the transcription of synaptic genes. Ach, acetylcholine; AChR, acetylcholine receptor; KASH, Klarsicht/ANC-1/Syne homology; LAP1, lamin-associated protein 1; NPC, nuclear pore complex. (Created with BioRender)

leading to deletion of the KASH domain of nesprin-1 in patients with the autosomal recessive cerebellar ataxia [47]. Similar mutations leading to disconnection of nesprin-1 from the LINC complex have been identified in patients with Emery-Dreifuss muscular dystrophy (EDMD) [13, 48]. One of the characteristics of nesprin-1 at the NMJ is its accumulation in synaptic nuclei after birth and which appears to be enriched in the SSN after denervation [49]. Studies have also shown that the localization and structure of nesprin-1 strongly suggest that it is involved in the migration of myonuclei into myotubes and/or their anchorage in the postsynaptic apparatus [21, 49]. In addition, studies in mouse models have shown that the association of nesprin-1 with the NE is necessary for the migration and maintenance of SSNs, which are essential for muscle function [49]. Nesprin-1 is the first NE protein found to be selectively associated with SSN [50].

## *SUN-1 and Synaptic Nuclear Anchorage*

The SUN-1 protein is a member of the Sad1/UNC-84 (SUN) domain family and plays a crucial role in maintaining the integrity of the NE and in facilitating various cellular processes such as migration and nuclear positioning and anchoring. SUN-1 is encoded by *SUN-1* gene and is widely expressed in various tissues and cell types, reflecting its essential role in maintaining nuclear envelope integrity and mediating nuclear-cytoskeletal interactions [5]. The N-terminal domain of SUN-1 is located in the nucleoplasm with domains that anchor the protein to the inner nuclear membrane. Its C-terminal domain, located in the perinuclear space, is essential for binding to the KASH domain proteins of the nesprins making up the LINC (linker of nucleoskeleton and cytoskeleton) complex that links the nucleoskeleton to the cytoskeleton, thus facilitating mechanical continuity between the nucleus and the cytoplasm [51, 52]. SUN-1 also interacts with nuclear lamin A to provide a physical connection between the nuclear lamin and the cytoskeleton [53]. The function of the SUN-1 protein in nuclear positioning was revealed for the first time by genetic analysis of the unc-84 gene in *C. elegans*, demonstrating that UNC-84 recruits the KASH proteins ANC-1 and UNC-83 to the outer nucleus for anchoring and nuclear migration [54, 55]. SUN-1 also plays a key role in mediating telomere-NE interactions during meiosis [56]. Alterations in SUN-1 expression or function are linked to various diseases, including muscular dystrophies, neuropathies, and a number of types of cancers, highlighting its role in cell homeostasis and function [14, 57, 58].

SUN-1 has been shown to play an important role in synaptic nuclear anchoring in Sun1 knockout mice, with disruption of Syne-1 localization [24]. These studies clearly show that SUN-1 functions critically for Syne-1 localization at the NE, which is essential for correct positioning of SSNs and myonuclei.

## *The Role of Lamin A/C in NMJ Maintenance*

The nuclear lamina is a thin layer of protein beneath the NE, composed mainly of type A and B lamins, members of the type V intermediate filament family [3, 11, 59]. The *LMNA* gene encodes type A and C lamins, which are involved in a number of cellular, molecular, and biophysical processes such as maintenance of nuclear structure via the LINC complex, DNA replication, and chromatin organization [60, 61]. Lamin A/C is ubiquitously expressed in all differentiated cells with a high level in skeletal muscle cells [22]. Lamin A/C mutations have been associated with a wide range of diseases, commonly referred to as laminopathies [8]. Studies in mouse models of laminopathy suggest that mutations in the *LMNA* gene alter the structural organization of the nuclear envelope in the subsynaptic nuclei, leading to a recruitment defect and poor positioning of these nuclei below the synapse [62]. Other recent studies reveal the role of lamin A/C in the maintenance of the NMJ and suggest that nuclear dysfunction or deficiency may contribute to NMJ deficits in aged muscle [63]. These studies also revealed an important role for lamin A/C in the muscle but no impairment of motor neurons in mouse models. In addition, the role of rapsyn and Dok-7, proteins essential for the clustering and stabilization of AChRs, in the pathophysiology has been demonstrated in mice model invalidated for the *LMNA* gene [63, 64].

## *LAP1 Protein and Its Involvement in Neurotransmission*

Lamin-1-associated protein (LAP1), encoded by the *TOR1AIP1* gene, is a ubiquitously expressed nuclear protein located on the inner nuclear membrane [65, 66]. Several isoforms of LAP1 have been reported in various species, but only two have been identified in humans, LAP1B and LAP1C, which have been associated with various biological processes and human diseases including muscular dystrophy, cardiomyopathy, and multisystemic disease [67, 67]. LAP1B is expressed at higher levels in the brain, heart, and liver, whereas LAP1C is more abundant in the lung, kidney, and spleen [68, 69]. The crucial role of NE proteins in specific SSNs and in the function of the NMJ has recently been highlighted by the description of mutations in the *TOR1AIP1* gene associated with an atypical form of congenital myasthenic syndrome [70, 71]. Patient biopsies revealed nuclear abnormalities reminiscent of those observed in patients with other TOR1AIP1-related disorders associated with functional defects of the NMJ. In addition, LAP1-deficient mice develop progressive muscle weakness and susceptibility to fatigue associated with morphological abnormalities of NMJs, including increased numbers of SSNs and abnormal expression of NMJ proteins [70]. Although direct evidence for a role for LAP1 in synaptic transmission and plasticity in the MNJ is limited, its involvement in NE integrity and gene regulation suggests that it may influence the expression of genes essential for synaptic function.

## Conclusion

The role of NE in skeletal muscle function came out into the open 30 years ago when mutations in genes encoding proteins of the NE, such as SUN-1 (encoded by *SUN-1* gene), nesprin-1/nesprin-2 (encoded by the *Syne1* and *Syne2* genes), and lamin A/C (encoded by the *LMNA* gene), have been implicated in muscular dystrophies (i.e., nuclear envelopathies), underscoring the critical role of NE components in maintaining muscle fiber integrity and function. NE proteins, such as lamin A/C, SUN-1/2, nesprins, and LAP1, are involved in anchoring and positioning the nuclei within the muscle fiber. Proper positioning, especially of SSNs beneath the NMJ, is crucial for the differentiation and functional specialization necessary for effective neuromuscular transmission. The crucial role of NE in specific SSNs and in the function of the NMJ has recently been highlighted and associated with an atypical form of congenital myasthenic syndrome (CMS). Although the formation of the NMJ and the communication between motor neurons and myofiber are well described, the mechanisms coordinating gene regulation in SSNs are still poorly understood [19].

## Compliance with Ethical Standards

To uphold the fundamental ethical standards required, all experimental data presented in this chapter are derived from studies conducted in full compliance with international ethical guidelines. These include the Declaration of Helsinki for research involving human subjects and the three R principles (Reduction, Refinement, Replacement) for animal experimentation. Furthermore, the use of preexisting work strictly adheres to established rules for citation and proper acknowledgment of the contributions of previous authors. This commitment ensures the integrity, transparency, and respect that are essential for advancing scientific knowledge.

## References


1. Senior A, Gerace L (1988) Integral membrane proteins specific to the inner nuclear membrane and associated with the nuclear lamina. J Cell Biol 107:2029–2036
2. Wong X, Melendez-Perez AJ, Reddy KL (2022) The nuclear lamina. Cold Spring Harb Perspect Biol 14:a040113
3. Wilson KL (2000) The nuclear envelope, muscular dystrophy and gene expression. Trends Cell Biol 10:125–129
4. Apel ED, Lewis RM, Grady RM, Sanes JR (2000) Syne-1, A dystrophin- and Klarsicht-related protein associated with synaptic nuclei at the neuromuscular junction. J Biol Chem 275:31986–31995
5. Liu Q et al (2007) Functional association of Sun1 with nuclear pore complexes. J Cell Biol 178:785–798
6. Gimpel P et al (2017) Nesprin-1α-dependent microtubule nucleation from the nuclear envelope via Akap450 is necessary for nuclear positioning in muscle cells. Curr Biol 27:2999–3009.e9
7. Liu J et al (2020) Regulation of myonuclear positioning and muscle function by the skeletal muscle-specific CIP protein. Proc Natl Acad Sci 117:19254–19265
8. Worman HJ, Bonne G (2007) "Laminopathies": a wide spectrum of human diseases. Exp Cell Res 313:2121–2133
9. Worman HJ, Ostlund C, Wang Y (2010) Diseases of the nuclear envelope. Cold Spring Harb Perspect Biol 2:a000760–a000760
10. Janin A, Bauer D, Ratti F, Millat G, Méjat A (2017) Nuclear envelopathies: a complex LINC between nuclear envelope and pathology. Orphanet J Rare Dis 12:147
11. Mattout A, Dechat T, Adam SA, Goldman RD, Gruenbaum Y (2006) Nuclear lamins, diseases and aging. Curr Opin Cell Biol 18:335–341
12. Worman HJ (2012) Nuclear lamins and laminopathies. J Pathol 226:316–325
13. Zhang Q et al (2007) Nesprin-1 and -2 are involved in the pathogenesis of Emery–Dreifuss muscular dystrophy and are critical for nuclear envelope integrity. Hum Mol Genet 16:2816–2833

14. Li P, Meinke P, Huong LTT, Wehnert M, Noegel AA (2014) Contribution of SUN1 mutations to the pathomechanism in muscular dystrophies. Hum Mutat 35:452–461
15. Slater CR (2008) Structural factors influencing the efficacy of neuromuscular transmission. Ann N Y Acad Sci 1132:1–12
16. Slater CR (2015) The functional organization of motor nerve terminals. Prog Neurobiol 134:55–103
17. Schaeffer L, De Kerchove d'Exaerde A, Changeux J-P (2001) Targeting transcription to the neuromuscular synapse. Neuron 31:15–22
18. Méjat A, Ravel-Chapuis A, Vandromme M, Schaeffer L (2003) Synapse-specific gene expression at the neuromuscular junction. Ann N Y Acad Sci 998:53–65
19. Belotti E, Schaeffer L (2020) Regulation of gene expression at the neuromuscular junction. Neurosci Lett 735:135163
20. Burke B (2019) Chain reaction: LINC complexes and nuclear positioning. F1000Res 8:136
21. Ruegg MA (2005) Organization of synaptic myonuclei by Syne proteins and their role during the formation of the nerve–muscle synapse. Proc Natl Acad Sci 102:5643–5644
22. Frock RL et al (2006) Lamin A/C and emerin are critical for skeletal muscle satellite cell differentiation. Genes Dev 20:486–500
23. Zhang J et al (2010) Nesprin 1 is critical for nuclear positioning and anchorage. Hum Mol Genet 19:329–341
24. Lei K et al (2009) SUN1 and SUN2 play critical but partially redundant roles in anchoring nuclei in skeletal muscle cells in mice. Proc Natl Acad Sci 106:10207–10212
25. Wessler I (1988) Neurotransmission at the NMJ. Trends Pharmacol Sci 9:125–126
26. Reddy LV, Koirala S, Sugiura Y, Herrera AA, Ko C-P (2003) Glial cells maintain synaptic structure and function and promote development of the neuromuscular junction in vivo. Neuron 40:563–580
27. Peng HB et al (2003) Differential effects of neurotrophins and Schwann cell-derived signals on neuronal survival/growth and synaptogenesis. J Neurosci 23:5050–5060
28. Robitaille R (1995) Purinergic receptors and their activation by endogenous purines at perisynaptic glial cells of the frog neuromuscular junction. J Neurosci 15:7121–7131
29. Changeux JP et al (1990) The acetylcholine receptor: functional architecture and regulation. Adv Second Messenger Phosphoprotein Res 24:15–19
30. Wood SJ, Slater R, C. (2001) Safety factor at the neuromuscular junction. Prog Neurobiol 64:393–429
31. Bezakova G, Ruegg MA (2003) New insights into the roles of agrin. Nat Rev Mol Cell Biol 4:295–309
32. Zhang B et al (2008) LRP4 serves as a coreceptor of Agrin. Neuron 60:285–297
33. Kim N et al (2008) Lrp4 is a receptor for Agrin and forms a complex with MuSK. Cell 135:334–342
34. Glass DJ et al (1996) Agrin acts via a MuSK receptor complex. Cell 85:513–523
35. Watty A et al (2000) The in vitro and in vivo phosphotyrosine map of activated MuSK. Proc Natl Acad Sci 97:4585–4590
36. Wallace BG, Zhican Q, Richard HL (1991) Agrin induces phosphorylation of the nicotinic acetylcholine receptor. Neuron 6:869–878
37. Ferns M, Deiner M, Hall Z (1996) Agrin-induced acetylcholine receptor clustering in mammalian muscle requires tyrosine phosphorylation. J Cell Biol 132:937–944
38. Vanhaesebrouck AE, Beeson D (2019) The congenital myasthenic syndromes: expanding genetic and phenotypic spectrums and refining treatment strategies. Curr Opin Neurol 32:696–703
39. Stojkovic T et al (2022) Diagnostic yield of a practical electrodiagnostic protocol discriminating between different congenital myasthenic syndromes. Neuromuscul Disord 32:870–878
40. Binks S, Vincent A, Palace J (2016) Myasthenia gravis: a clinical-immunological update. J Neurol 263:826–834

41. Nicole S et al (2017) Congenital myasthenic syndromes or inherited disorders of neuromuscular transmission: recent discoveries and open questions. J Neuromuscul Dis 4:269–284
42. Ramdas S, Beeson D (2021) Congenital myasthenic syndromes: where do we go from here? Neuromuscul Disord 31:943–954
43. Henehan L, Beeson D, Palace J (2024) Congenital myasthenic syndromes. Pract Neurol 24:185–187
44. Broers JLV, Ramaekers FCS, Bonne G, Yaou RB, Hutchison CJ (2006) Nuclear Lamins: laminopathies and their role in premature ageing. Physiol Rev 86:967–1008
45. Muchir A et al (2012) Treatment with selumetinib preserves cardiac function and improves survival in cardiomyopathy caused by mutation in the Lamin A/C gene. Cardiovasc Res 93:311–319
46. Mislow JMK et al (2002) Nesprin-1α self-associates and binds directly to emerin and Lamin A in vitro. FEBS Lett 525:135–140
47. Indelicato E et al (2019) SYNE1-ataxia: novel genotypic and phenotypic findings. Parkinsonism Relat Disord 62:210–214
48. Zhou C et al (2017) Novel nesprin-1 mutations associated with dilated cardiomyopathy cause nuclear envelope disruption and defects in myogenesis. Hum Mol Genet 26:2258–2276
49. Grady RM, Starr DA, Ackerman GL, Sanes JR, Han M (2005) Syne proteins anchor muscle nuclei at the neuromuscular junction. Proc Natl Acad Sci 102:4359–4364
50. Morel V, Lepicard S, Rey N, A., Parmentier, M.-L. & Schaeffer, L. (2014) Drosophila Nesprin-1 controls glutamate receptor density at neuromuscular junctions. Cell Mol Life Sci 71:3363–3379
51. Tzur YB, Wilson KL, Gruenbaum Y (2006) SUN-domain proteins: 'Velcro' that links the nucleoskeleton to the cytoskeleton. Nat Rev Mol Cell Biol 7:782–788
52. Worman HJ, Gundersen GG (2006) Here come the SUNs: a nucleocytoskeletal missing link. Trends Cell Biol 16:67–69
53. Haque F et al (2006) SUN1 interacts with nuclear Lamin A and cytoplasmic Nesprins to provide a physical connection between the nuclear Lamina and the cytoskeleton. Mol Cell Biol 26:3738–3751
54. Malone CJ, Fixsen WD, Horvitz HR, Han M (1999) UNC-84 localizes to the nuclear envelope and is required for nuclear migration and anchoring during C. elegans development. Development 126:3171–3181
55. Starr DA et al (2001) unc-83 encodes a novel component of the nuclear envelope and is essential for proper nuclear migration. Development 128:5039–5050
56. Ding X et al (2007) SUN1 is required for telomere attachment to nuclear envelope and gametogenesis in mice. Dev Cell 12:863–872
57. Wang J-Y et al (2015) Sun1 deficiency leads to cerebellar ataxia in mice. Dis Model Mech dmm.019240
58. Matsumoto A, Matsuura N, Hieda M (2018) Detection of SUN1 splicing variants at the mRNA and protein levels in cancer. In: Gundersen GG, Worman HJ (eds) The LINC complex, vol 1840. Springer, New York, NY, pp 307–319
59. Burke B, Stewart CL (2002) Life at the edge: the nuclear envelope and human disease. Nat Rev Mol Cell Biol 3:575–585
60. Mounkes L, Kozlov S, Burke B, Stewart CL (2003) The laminopathies: nuclear structure meets disease. Curr Opin Genet Dev 13:223–230
61. Burke B, Stewart CL (2006) The laminopathies: the functional architecture of the nucleus and its contribution to disease. Annu Rev Genomics Hum Genet 7:369–405
62. Méjat A et al (2009) Lamin A/C–mediated neuromuscular junction defects in Emery-Dreifuss muscular dystrophy. J Cell Biol 184:31–44
63. Gao N et al (2020) A role of Lamin A/C in preventing neuromuscular junction decline in mice. J Neurosci 40:7203–7215
64. Arimura S et al (2014) DOK7 gene therapy benefits mouse models of diseases characterized by defects in the neuromuscular junction. Science 345:1505–1508

65. Lornage X et al (2021) Selective loss of a LAP1 isoform causes a muscle-specific nuclear envelopathy. Neurogenetics 22:33–41
66. Mackels L, Liu X, Bonne G, Servais L (2023) TOR1AIP1-associated nuclear envelopathies. Int J Mol Sci 24:6911
67. Fichtman B et al (2019) Combined loss of LAP1B and LAP1C results in an early onset multi-systemic nuclear envelopathy. Nat Commun 10:605
68. Shin J-Y et al (2013) Lamina-associated polypeptide-1 interacts with the muscular dystrophy protein Emerin and is essential for skeletal muscle maintenance. Dev Cell 26:591–603
69. Shin J-Y, Dauer WT, Worman HJ (2014) Lamina-associated polypeptide 1: protein interactions and tissue-selective functions. Semin Cell Dev Biol 29:164–168
70. Cossins J et al (2020) Congenital myasthenic syndrome due to a TOR1AIP1 mutation: a new disease pathway for impaired synaptic transmission. Brain Commun 2:fcaa174
71. Malfatti E et al (2022) A *TOR1AIP1* variant segregating with an early onset limb girdle myasthenia—support for the role of LAP1 in NMJ function and disease. Neuropathol Appl Neurobiol 48:e12743

# Chapter 3
# Tubular Lamin- and Emerin-Lined Nuclear Envelope Invaginations Isolate Chromatin Domains for Gene Expression Regulation

**Qing Qin Ji, Zewen Ding, Jingxian Duan, Jiahui Nie, Chenli Wu, Sutherland K. Maciver, Thomas Theil, A. Mark Evans, and Eric C. Schirmer**

**Abstract** Lamins provide physical support to the nucleus and maintain nuclear shape. In many disease states linked to lamin and other nuclear envelope (NE) protein mutations, altered nuclear shape is viewed as a reflection of aberrant lamin and nuclear functions; however, there are also cell types with dramatically invaginated nuclei made with wild-type functional lamins. Closer inspection suggests four classes of NE invagination (NEI): large NE indentations or clefts, tubular NE invaginations that span from one side of the nucleus to the other, blind tubular NE invaginations that project variable distances into the nucleus, and blind invaginations of the inner nuclear membrane alone. These NEIs are likely formed by a combination of NE interactions with the cytoskeleton and with chromatin. Specific functions of

Q. Q. Ji · E. C. Schirmer (✉)
Institute of Cell Biology, University of Edinburgh, Max Born Crescent, Edinburgh, UK
e-mail: e.schirmer@ed.ac.uk

Z. Ding · C. Wu · S. K. Maciver · T. Theil
Centre for Discovery Brain Sciences, Edinburgh Medical School, Biomedical Sciences, University of Edinburgh, Edinburgh, UK

J. Duan
Centre for Discovery Brain Sciences, Edinburgh Medical School, Biomedical Sciences, University of Edinburgh, Edinburgh, UK

Centre for Cardiovascular Sciences, College of Medicine and Veterinary Medicine, University of Edinburgh, Edinburgh, UK

Institute of Biomedical and Health Engineering, Shenzhen Institute of Advanced Technology, Chinese Academy of Sciences, Shenzhen, China

J. Nie · A. M. Evans
Centre for Discovery Brain Sciences, Edinburgh Medical School, Biomedical Sciences, University of Edinburgh, Edinburgh, UK

Centre for Cardiovascular Sciences, College of Medicine and Veterinary Medicine, University of Edinburgh, Edinburgh, UK

© The Author(s), under exclusive license to Springer Nature Switzerland AG 2025 
K. Sengupta, A. Muchir (eds.), *Lamins in Genome Organization and Diseases*, Subcellular Biochemistry 115, https://doi.org/10.1007/978-3-032-00537-3_3

NEIs are still being investigated, but their increasing relative nuclear surface area could increase NE/lamin functions in genome organization and regulation. Indeed, NEIs have recently been shown to hold a unique epigenetic signature compared to other NE regions in the same cell. Furthermore, studies have shown that tubular NEIs contain a $Ca^{2+}$ store and that each NEI demarcates a cytoplasmic nanocourse, where calcium appears to be exchanged across the ONM between the NEI lumen and thus isolated cytoplasmic nanocourses. Therefore, $Ca^{2+}$ flux at NEIs could additionally provide a more direct/rapid route to genome regulation compared to traditional signalling cascades. Here we discuss what is currently known about NEIs and the various possibilities of how they may contribute to genome regulation.

**Keywords** Nucleus · Nuclear envelope invagination · Nucleoplasmic reticulum · Lamin · Emerin · Calcium · Chromatin · Genome organization

## Introduction

The nuclear envelope (NE) is a complex structure composed of two phospholipid bilayers: the outer (ONM) and inner nuclear membranes (INM). The ONM connects with the sarcoplasmic/endoplasmic reticulum (S/ER), making the NE lumen continuous with the S/ER lumen [115]. This lumen, also called the perinuclear space (PNS), is approximately 50 nm wide from INM to ONM in metazoan cells and 30 nm in yeast [107]. INM/ONM merges where nuclear pore complexes (NPCs), the gateways for macromolecule selective transport [125], are inserted (Fig. 3.1). Underlying the INM is the nuclear lamina, an intermediate filament network that is principally composed of separate A-type and B-type lamin polymers [132, 154] and provides physical/mechanical support to the NE [69]. Both ONM and INM have also hundreds of NE transmembrane proteins (NETs), the majority of which are tissue-specific [64]. Many INM NETs bind both lamins and chromatin [85, 151, 156], but the functions of most NETs are as yet unknown. The Sad1/UNC84 (SUN) family of INM proteins binds the lamin polymer on the nucleoplasmic side and on the lumenal side binds the NE spectrin repeat protein (nesprin) family of ONM proteins [117]. The cytoplasmic face of nesprins further connects to all major cytoskeletal networks (microfilaments, microtubules, and intermediate filaments), thus providing a mechanism for mechanosignal transduction to the nucleus while also contributing to nuclear stability.

Historically, when just lamins and NPCs were the only NE proteins known, the consensus view was that the NE was principally a barrier that also provided some structural support and that all NEs were essentially the same. This view ignored the

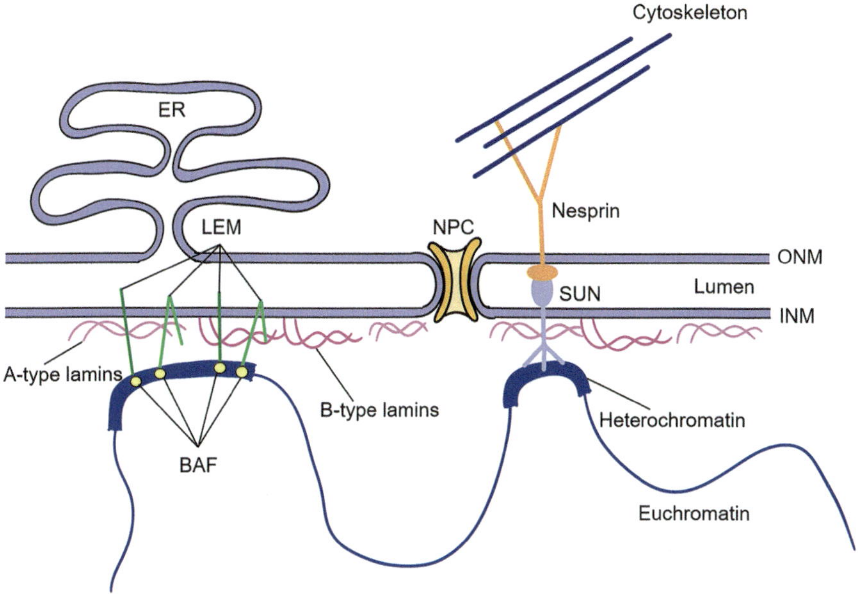


**Fig. 3.1** Schematic of the NE. The NE is a double-membrane system comprised of two distinct lipid bilayers—the inner (INM) and outer nuclear membranes (ONM)—together with a panoply of transmembrane proteins (e.g., NPC proteins, nesprins, SUN proteins, LEM-domain proteins, and others), the lamin polymer that underlies the INM, and the contents of the NE lumen. The NE is continuous with the ER, and their lumens are also continuous. Several ONM proteins connect to the cytoskeleton, while several INM proteins bind to chromatin, principally heterochromatin and chromatin proteins such as BAF

very large elephant in the room that electron microscopy (EM) atlases for many years [37] had shown many distinct and characteristic nuclear shapes and peripheral chromatin distributions for different cell types in different tissues. Liver hepatocytes, for example, typically have spheroid nuclei with clusters of heterochromatin and euchromatin, while liver Kupffer cells tend to have oblong and highly variable nuclear shapes with heterochromatin throughout the nuclear periphery. Blood cells have likely the widest range of nuclear shapes, from classical spherical lymphocyte nuclei to the bi-lobed eosinophils and five-lobed neutrophils with each lobe connected by thin NE-limited chromatin sheets. Moreover, resting lymphocytes have enormous concentrations of heterochromatin throughout the nuclear periphery, which largely dissipates upon activation [51]. More recent studies have shown that even the spherical lymphocyte nuclei dramatically change shape and reposition within the cell when the cells move toward targets and yet again when establishing the synapse with the target [152]. Another facet affecting nuclear shape is invaginations of the NE that can penetrate deep into the nucleus (Fig. 3.2a).

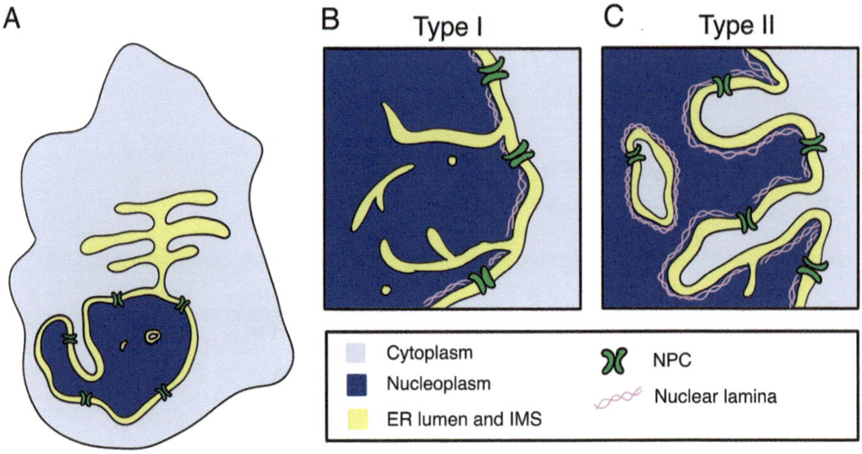


**Fig. 3.2** Schematic of two types of NEIs. Type I NEIs (B) contain just INM, whereas type II NEIs (C) contain both INM and ONM and accordingly form cytoplasmic pockets. NEIs appear as holes in the nucleus when seen in the cross section, but serial sections reveal them to continue to the top of bottom of the nucleus. In some cases, NEIs completely traverse the entire nucleus

## Nuclear Envelope Invaginations (NEIs)

It is hard to track the first clear observations of NEIs since they would have been caught in images from when EM first began to be used on cells and tissue sections. A good reference point is a review from Jos Broers and colleagues in which many early observations are recorded [137]. One early study from nearly 60 years ago was focused just on NEIs, following contracting smooth muscle cells from mouse jejunum [70]. They found some invaginations that completely traversed the nucleus as well as branching of invaginations. These NEIs are approximately 100–200 nm in width; however, it is hard to identify whether the NEIs are tubular or cleft-like from 2D EM. Shortly after this study, NEIs were also noted in skeletal muscle cells [40]. These deep invaginations formed of both ONM and INM are 100–200 nm across. The first NEI that may be tubular as suggested from 2D EM was reported in 1979 [8]: the authors showed a ring-like structure in the flat section of a TG cell derived from a tumor of a human oviduct, which would be a tubular NEI if the ring-like structures observed were continuous, but this was not tested. One of the most striking types of smooth muscle cells for NEIs is primary pulmonary arterial smooth muscle cells (PASMCs; Fig. 3.3a, b). In addition to muscle, NEIs have been observed to characteristically occur in certain types of brain regions and specific neuronal cell types. For example, in hamster pyramidal neurons of the frontal cortex, large NEIs with significant spacing of often hundreds of nanometer across the penetrating cytoplasmic pockets were noted during development and aging [11]. NEIs that are 300–500 nm in width characteristically crossing one third to half the nucleus were observed in the cat medial geniculate body projecting into the auditory cortex [150].

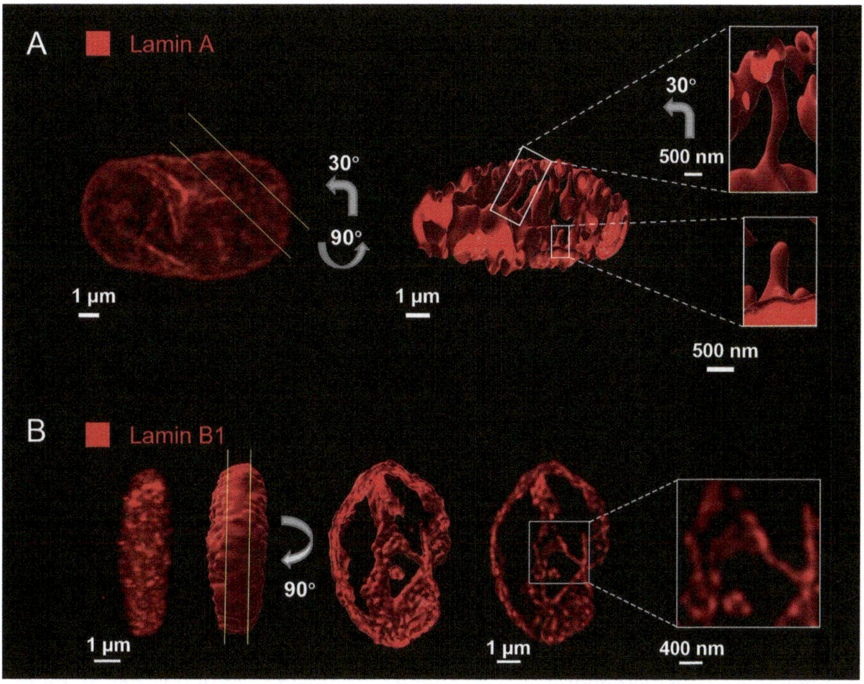


**Fig. 3.3** NEIs observed in different cells. (**a**) Rat PASMC NE labeled with lamin A and 3D reconstructions. (**b**) Human cortical organoid neuron labeled with lamin B1

David Vaux's lab led the way in not only characterizing these structures in a quantitative manner but also in demonstrating that specific types of NEI formed highly organized tubular networks [41]. He used the term nucleoplasmic reticulum that is synonymous with NEI and classified several subtypes of invaginations [79]. The two core types depend on whether the ONM is part of an invagination: type I invaginations are those with only INM inserting into the nucleoplasm (Fig. 3.2b), while type II invaginations consist of both INM and ONM, forming a cytoplasmic core on the NE surface that penetrates deep within the outermost nuclear shell (Fig. 3.2c). Type II NEI subdivides into clefts and, critically, two types of tubular NEI, blind and transnuclear.

While the earliest studies of NEIs relied just on EM, the subsequent discovery of the lamin polymer underlying the NE and finding that lamin antibody staining could visualize NEIs by immunofluorescence microscopy has led to many new discoveries of NEIs. For example, ventricular-subventricular zone neural stem cells, particularly those differentiating into astrocytes, have nuclei with smaller lobules that connect to the larger nucleus with NE-limited chromatin sheets [17], just like those that connect neutrophil lobules. This observation of NEIs in stem cells lends to introducing another typical niche for finding NEIs—embryonic development. In an early study, the LaVelle lab looked at hamster facial motor neurons at 14-day fetus, newborn, and several different days postnatally. They observed that both nuclear shape and size changed over this period but also that there were no NEIs at early stages that then became induced with 300–400 nm in width during the developmental period and, interestingly, were limited to just one side of the nucleus [20]. Striking lamin B-delimited invaginations that in cross section made the nucleus look like Swiss cheese were observed at the zygote stage of rabbit embryonic development that slowly dissipated in number until major genome remodeling that occurs around the 8–16-cell stage [114]. Tubular NEIs 300–400 nm wide were also observed in bovine pre-implantation development where it was noted that the timing of NEI appearance corresponded with the switch from maternal to embryonic mRNA production and a drop in nuclear size, suggesting that they form to accommodate the increased ratio of NE area to nuclear size as nuclear size decreases [113]. NEIs are also found in neurons of human cortical organoids, and an example of these labeled with lamin B1 is shown in Fig. 3.3b. Finally, one other tissue type, reproductive cells, has had frequent reports of NEIs, including oocytes, endometrium, and sperm [25, 90, 95]. Similarly, there have been reports of NEIs in a number of cancer tissue culture cell lines [21], but because of differences in the behavior of immortalized cells and 2D tissue culture, we will similarly not further discuss this literature. Of note, NEIs in all the above systems tended to have diameters of many hundreds of nanometer with the exception of those observed in the bovine pre-implantation and muscle systems, which were very thin (<200 nm in cytoplasmic cross section). Whether differences in the diameter have functional significance remains to be demonstrated; however, studies discussed later in this review on the thin NEIs of PASMCs (Fig. 3.3a) being used as $Ca^{2+}$ stores may indicate unique functions based on diameter such that diameter should perhaps be considered as another classification parameter.

NEIs have also been found in a wide range of organisms in addition to mammals, suggesting they may reflect a more fundamental mechanism used by cells rather than just an aspect of higher order differentiation in organisms with many complex tissues. Interestingly, similarly thin NEIs as the PASMCs were also observed in onion nuclei [23], and extensive invaginations were observed in Cercozoa [16] and dinoflagellates [44], spreading these observations beyond the Animal kingdom into the Plant and Protista kingdoms.

## Roles of Lamins, NETs, and Cytoskeleton in Nuclear Shape and Size

As was noted above, lamin B1 is often used in immunofluorescence staining to identify NEIs, as well as nuclear shape and size. PASMC NEIs are often lined by A-type lamins, though the smaller and tubular NEIs appear to always be lined by A-type but not lined by B-type lamins. A-type lamins are the ones most supported experimentally for functioning in nuclear stiffness and stability [33, 69, 140]. It is a common misconception, though, to think of lamins as building blocks. As intermediate filaments, the lamin polymer is far more tensile than tubulin and actin polymers [56]. In fact, isolated nuclei are capable of enormous stretch under micropipette aspiration without breaking (AC [123]). Lamins can stretch under conditions where tubulin and actin polymers break; so where microtubules were viewed somewhat like a medieval buttress in the tensegrity model [54], the lamin polymer might be viewed more like a trampoline net in response to compression forces. Thus, while the nuclear lamina physically supports the NE and keeps its shape, it does so more by enabling it to both buffer and respond to internal forces from chromosomes (the largest molecule in the cell) or external forces from the cytoskeleton without breaking than by forming a rigid shell. When the lamina is mostly buffering these forces and they are reasonably balanced, a nucleus would be expected to be mostly spherical or ovoid; however, when cytoskeletal forces predominate, the lamina would likely respond by forming NEIs. This is currently just hypothetical, but it is consistent with other studies on nuclear shape and size.

The human leukemic HL-60 cell line is an excellent model to study nuclear shape changes. In the undifferentiated basal state of the cells, the nuclei are largely spherical. However, after retinoic acid treatment, which induces granulopoiesis, the nucleus differentiates to be highly lobulated, first forming the bi-lobed nucleus like eosinophils and then eventually a four- to five-lobed nucleus like neutrophils (Fig. 3.4, top). In support of the above idea, the Olins lab found that cytoskeletal polarization occurred during differentiation [99] and that disrupting cytoskeletal elements interfered with the nuclear shape changes [101]. However, the cytoskeleton was not operating in isolation. They also found that as HL-60 cells differentiated into granulocytes, levels dropped for lamin A/C, B1, and LINC proteins nesprin 1 and 2 and SUN1 [97, 100]. At the same time, lamin B receptor (LBR) levels increased [101]. This could just relate to LBR interactions with B-type lamins, but notably LBR also binds to heterochromatin protein 1 (HP1) [156], thus potentially bringing forces from the genome into the dynamic. Compellingly, in an LBR knockdown HL-60 cell line, the addition of retinoic acid still induced granulopoiesis and resulted in a lamin A/C level drop, but the HL-60 cells retained a more ovoid nuclear shape [98]. At the same time, the lamin A drop is also important for the nuclear shape changes because lamin A overexpression in the HL-60 cell line exhibited reduced LBR levels and reduced lobulation [124] (Fig. 3.4, bottom).

The lamin A effects are consistent with other reports of A-type lamins being more important than B-type lamins for nuclear stiffness [68, 69] and findings that the lamina was more soluble when lamin A was overexpressed in HL-60 cells, which differentiated into the granulocyte lineage as opposed to a macrophage-like lineage that can be induced with phorbol esters [129]. Notably, the latter study also indicated that lamin A and lamin B1 are more likely to self-assemble than to form an integrated polymer, while lamin B2 likely mediates interactions between the two polymers. Subsequent studies using super-resolution microscopy have now confirmed the separation of lamin A and lamin B1 networks [132, 155], though the different studies and techniques differed on a number of other aspects of the networks. Thus, differences in lamina composition could yield considerably different shapes and mechanical properties to nuclei in different cell types, a conclusion

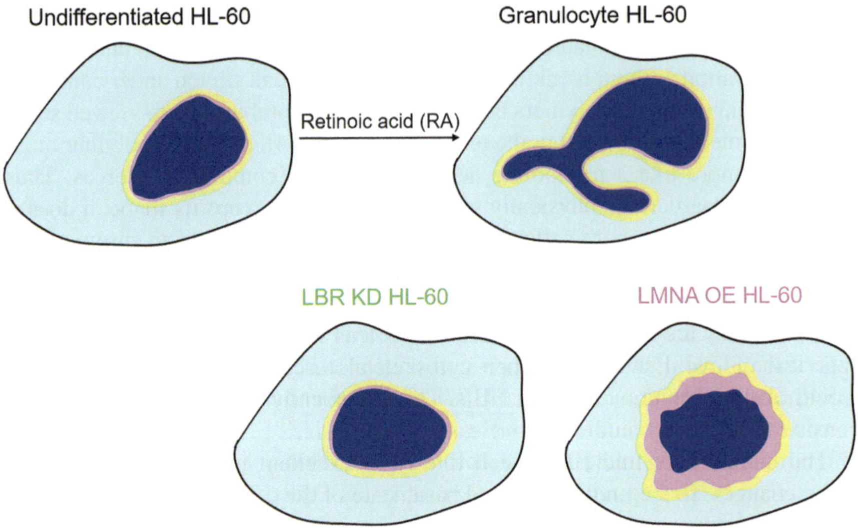


**Fig. 3.4** Schematic of nucleus shape changes in HL-60 differentiation. HL-60 cells provide a unique model system in which dramatic changes in nuclear shape and architectural components occur during induced differentiation that can be manipulated to test the importance of different proteins to the process of inducing shape changes. Upper cells: when treated with retinoic acid (RA), the cells differentiate in the granulocyte lineages, and their round nucleus undergoes massive lobulation, first typically generating a bi-lobed nucleus that eventually develops to a five-lobed nucleus characteristic of neutrophils. Treatment with phorbol esters causes different changes as the cells become more macrophage-like. Lower cells: knockdown of INM protein LBR or overexpression of lamin A results in a failure to achieve characteristic shape changes upon RA induction of the granulocytic lineage

consistent with observations that different cell types have characteristic differences in the levels of the four major lamin subtypes A, C, B1, and B2 [10]. Moreover, there are several additional lamin splice variants that occur in specific cell types such as sperm [2, 131]. These results altogether indicate that lamins play crucial roles in regulating the shape of the nucleus, but they do so working in concert with other NE proteins and their connections to both the cytoskeleton and chromatin.

Nuclear size also varies between different species and cell types, and research shows that lamins also play a role in determining nuclear size. Two related *Xenopus* species were used for these studies: the pseudotetraploid *Xenopus laevis* and the diploid *Xenopus tropicalis*, which in addition to being diploid also has smaller nuclei as the DNA content typically correlates with nuclear size. Levels of two transport factors, importin α and Ntf2, differed between the egg extracts of the *Xenopus* species. Both factors modulate lamin B3 import, suggesting that lamin B3 modulates nuclear size [73], and this is consistent with B3 depletion blocking nuclear growth in egg extracts [92]. Also, mRNA microinjection of importin α, which promotes import rate and therefore induced lamin B3 import, led to increased nuclear size in early embryos. However, there are other factors that could contribute to nuclear size, particularly the more than doubling that occurs between when the nucleus reforms in telophase of one mitosis and after DNA replication but before the next cell division [38, 60]. Therefore, nucleo-cytoplasmic transport factors altered in these experiments could have many indirect effects and even simply be limiting for nuclear growth because of all the proteins that must be imported to support DNA replication (ploidy also being related to nuclear size). However, subsequent studies from the Levy lab found more direct evidence for lamins as a major structural element of the NE to be limiting for nuclear size increases that occur during interphase. They altered the concentration of different types of lamins (lamins A, B1, B2, and B3) in cultured mammalian cell lines and *Xenopus laevis* eggs and embryos and found increasing nuclear size with low lamin concentration and decreasing size with high concentration [57].

Many studies of nuclear size have been more focused on the idea of scaling because cytoplasmic and nuclear size tend to increase together, maintaining a particular ratio often called the karyoplasmic ratio. Mechanisms for maintaining this ratio are conserved across evolution and can be altered in disease states [15, 34, 60, 91]. One particularly relevant disease state is cancer where characteristic nuclear size changes independent of ploidy are associated with changes to increasing cancer grade and metastasis in at least 19 cancer types [26, 160]. However, each cancer type is different in that the degree and directionality of nuclear size changes are characteristic for each type of cancer. For example, in breast cancer, nearly twofold nuclear size increases correlate with increased metastasis [12], but in small cell squamous lung cancer, smaller nuclear size changes correlate with increased metastasis [67]. This led to the postulate that disruption of the balance of nucleo-cytoskeletal connections with nuclear size changes could affect cell motility and the hypothesis that reversing the nuclear size changes could inhibit cell motility associated with metastasis [130]. Indeed, a screen for drugs that would reverse nuclear size changes in three different cancer types identified many drugs with such effects and found that correcting the nuclear size change also inhibited cell motility in wound healing assays and invasion assays [144]. Moreover, for each cancer type tested, a largely different and non-overlapping set of compounds was identified. Interestingly, while these compounds ranged from beta-blockers to anti-helminths and had many known protein targets, a chemical genomic analysis indicated that the most common target of the different drugs was lamin A [144]. However, it is important to note that many other targets including NETs were also identified that can also affect nuclear size [120, 130].

## Possible Cytoskeletal Involvement in NEI Formation

The exact mechanism of NEI formation is unclear. NEIs are likely to be formed by forces acting on the NE (as supported by the HL-60 lobulation studies), and different hypotheses argue whether the forces come from the chromatin side or the cytoskeleton side. There are also some hypotheses postulating that the NEIs are formed by affinity or favorable chemical environments. In cell types such as human breast epithelial cells and the HMT-3522 cell line, irregular shapes of nuclei are possibly related to the physical stress applied on the NE [59, 72]. In some cancer cells, the formation of NEIs might also result from extranuclear cytoskeletal stresses. For example, in pancreatic carcinoma cells, there are intermediate filament bundles consisting of vimentin and keratins around the nucleus. However, in other cancer cells like adrenal adenocarcinoma cells, the intermediate filament bundles are absent, but NEIs are still formed. These results together indicate that the cytoskeleton may contribute in NEI formation in some cells, but the mechanism is not universal.

Linker of the nucleoskeleton and cytoskeleton (LINC) complexes consist of SUN domain containing proteins in the INM, which interact with nesprins in the ONM. LINC proteins form a bridge across the NE, and they associate with cytoskeletal components actin, microtubules, and intermediate filaments, therefore enabling them to transmit forces and mechanical signalling across the nuclear boundary and perhaps determine nuclear shape [9, 135, 136]. Accordingly, the thin NEIs in onion nuclei were lined by actin, further suggesting cytoskeletal-LINC interactions being involved in formation/maintenance of NEIs [23], and the actin-LINC nexus is likely also involved in the thicker NEIs as well because thicker NEIs of connective tissue fibroblasts accumulated alpha smooth muscle actin [138]. Interestingly, in the dinoflagellates, there were two types of NEIs, one with an extensive branching network and the other with very straight grooves containing microtubules [44], consistent with findings reported for PASMC [32]. In dinoflagellates, this network likely has a more specific function in supporting cell division as it was only observed in mitotic cells. By contrast, in PASMC, such networks are clearly present in contractile myocytes but disassemble upon cell proliferation in tissue culture. This gets to the most interesting observation—that different proteins involved in NE-cytoskeletal interactions can achieve similar ends or specialize based on alternate partners in different cell types. An elegant example of this was a study from Iris Meier's group in which they investigated the CRWN1-SUN-WIP-WIT-myosin XI-i nexus—the plant equivalent nexus to lamin-SUN-nesprin-motor protein-actin. They found that the lamin and LINC equivalents both contribute independently to nuclear shape, that different SUN family members contribute unequally to nuclear shape, and that the roles of the SUN and WIT paralogs in nuclear shape diverged between different plant tissues [159].

While some NEIs presumably form from pre-existing NE that needs somewhere to fold in as nuclear volume decreases without a reduction in NE area such as in bovine pre-implantation (J. [113]), in other cases, de novo lipid synthesis occurs to increase NE area during NEI formation. A nice study from the Vaux group followed the formation of not only shallow cleft-like but also tubular NEIs in endometrial cells in response to hormone treatments [116]. They found an increase in nucleoplasmic reticulum/NEIs over time after treating with estrogen with NEIs first appearing within 3 h of treatment and large seemingly connected networks having formed by 48 h. Since phosphatidylcholine production is induced by estrogen, they deuterium-labeled a precursor, pulsed the cells, and found its accumulation in the induced NEIs using NanoSIMS [116]. By contrast, in pulmonary arterial myocytes, dissolution of NEIs is observed during serum-induced proliferation [32].

## NEI Dynamicity

The endometrial experimental system mentioned above highlights another aspect of some NEIs—their dynamicity. Changes in endometrial NEIs were noted many years ago from in vitro fertilization studies using GnRH-a, FSH, and hCG. This treatment resulted in induction of possibly tubular NEIs (called NCS for nuclear channel system in the study at hand), and these NEIs are approximately 100–200 nm in width, but these subsequently dissipated [95]. Hormones also could induce changes in non-gonadal cell types, with estrogen inducing cleft-like NEI (~500 nm) changes also in rat hypothalamic ventromedial neurons [58].

While it is unclear how dynamic NEIs are in general and if different subtypes of NEIs are more or less dynamic, they do in general appear to be responsive to signalling or environmental changes. Much of the literature on changing NEIs is related to toxicity because such changes in NE morphology were viewed (perhaps wrongly from what we know now) as indicating defects or sickness. For example, it was observed that there was a large increase in shallow NEIs in rat liver nuclei (normally quite spherical) when the rats were given anticoagulant rodenticides [48]. Another example is a recent study investigating possible links between aluminum and brain disorders such as dementia. Here the authors found that treatment of rats with aluminum silicate induced deep and approximately 500-nm-wide NEIs in the Purkinje cells as well as nuclear shape abnormalities, and this could be largely reversed by treatment of the rats with extracts that bees collect from plants [102]. Even common environmental changes such as hyperthermia could impact NEIs as one study heated cells to 45.5 °C and found this induced possibly tubular NEIs based on lamin staining and these dissipated when cells recovered [36]. This was viewed as reflecting damage to the nucleus, but it could also reflect a response to adapt to the heat, better safekeeping the DNA.

Studying normal states and dynamicity of NEIs as well as changes in disease states should be useful in understanding both the mechanisms of their formation and their function as it has been proposed by the Vaux lab based on their extensive observations of both that the underlying mechanisms are the same (Marek Mateusz [30]). Another possible research area where studying such changes may be informative is in genome regulation that could potentially occur through changes in NEIs. Early studies showed induction of NEIs associated with radiation treatments that might be due to changes in NE-chromatin interactions associated with DNA damage [141]. Even more excitingly for this, in the PASMC system, the extensive NEI network is lost when acutely isolated cells are maintained in culture and transition from differentiated to proliferating myocytes in culture [32]. The complete loss of NEIs is accompanied by global activation of genes that facilitate proliferation and repression of those that oppose it [147], suggesting potential NEI roles in gene regulation.

## Chromatin Interactions with the NE

In addition to revealing NEIs, the old EM atlases and early studies of NEIs by EM showed that heterochromatin tends to be more concentrated at the nuclear periphery and the more active euchromatin in the nuclear interior [126, 158]. However, it is only in the past quarter century that specific NE functions in gene regulation have been revealed. These include but are not restricted to transcription factor sequestration, transcriptional repressor recruitment, affinity tethering of heterochromatin, epigenetic silencing of chromatin, DNA damage repair, and gene expression regulation through control of genome organization.

1. *Transcription factor sequestration*—Lamins were shown to bind transcriptional regulators pRb [104] and cFos [55], thus preventing their acting on target genes in the active, euchromatic regions of the nucleus. Similarly, the NET MAN1 binds to Smad transcription factors, thus keeping them at the NE away from target genes [103, 105]. The NET emerin does the same thing, binding transcription factor Lmo7, but interestingly, in this case, since Lmo7 acts on the *emerin* gene, the sequestration functions as part of a feedback loop so that it only sequesters the transcription factor when there are high levels of emerin [53].
2. *Transcriptional repressor recruitment*—The transcriptional repressor germ cell-less binds to both the NETs lamina-associated protein 2 (LAP2) [94] and emerin [52], presumably to better repress genes at the NE.
3. *Affinity tethering of heterochromatin*—Early studies from the Worman lab found that the NET LBR binds HP1, thus explaining in part the tendency for heterochromatin to accumulate at the nuclear periphery [156]. Many other affinity interactions that contribute to heterochromatin tethering have also been shown such as the NETs LAP2, emerin, and MAN1 (all of which bind lamins) associating with the chromatin-bound barrier-to-autointegration factor (BAF) [27, 62, 122, 158, 161].
4. *Epigenetic silencing of chromatin*—The NET LAP2 binds the histone deacetylase HDAC3, thus promoting the addition of epigenetic silencing marks to chromatin once at the NE [134]. A similar interaction was also demonstrated between lamin A and HDAC2 [84].
5. *DNA damage repair*—Genome instability from weakened NE interactions was one mechanism suggested to explain pathologies in some of the laminopathies (NE-linked disorders), particularly the premature aging disorder Hutchison-Guilford progeria syndrome. Much work has thus focused on NE interactions with the DNA damage repair machinery covering a wide range of functions including NE contacts stabilizing damaged regions to promote repair [119], NE recruitment of factors such as STING, and changes in lamin interactions with the progerin mutation [50, 66]. These are nicely reviewed by Graziano et al. [46].
6. *Genome organization*—Before the advent of high-throughput genomics, several studies had shown one gene at a time that important developmental genes moved between the nuclear periphery and nuclear interior concomitant with their repression or activation (the nuclear periphery being generally considered as a

repressive environment). For example, the *IgH* locus is released from the NE when *v-d-j* recombination is activated [65] and the *Mash1* locus is released from the NE during neural development [149]. It is important to note, however, that while the general tendency is for the NE to be a silencing environment, there are also examples of genes being activated at the NE such as the gene encoding proteolipid protein in oligodendrocytes [93]. Thus, the NE is perhaps better described as a regulatory environment rather than as a repressive environment.

The subsequent development of high-throughput techniques such as DamID—where DNA at the NE can be identified by fusing lamin B1 to a bacterial methylase so it confers a unique methylation that can then be isolated and sequenced—revealed that genome regions in contact with the nuclear lamina called lamina-associated domains (LADs) tend to be gene-poor and contain repressive epigenetic marks [109, 110]. Moreover, using this technique, the van Steensel lab showed that during neural cell development, many specific genes are released or recruited to the NE concomitant with their activation/repression [108]. Much of LAD formation is thought to be driven by heterochromatin interactions with the NE; however, tissue-specific NETs also clearly mediate many of these changes in LADs and thus tissue-specific patterns of genome organization [24, 43, 63, 121, 161].

To date, the role of NEIs in genome regulation has not been to our knowledge investigated. In theory, NEIs should increase the relative amount of nuclear surface area to total nuclear volume and thus to the extent that the NE is a silencing environment should increase silencing. However, another possibility was recently proposed by the Evans lab [32] that NEIs would create subnuclear domains that would isolate groups of genes, enhancers, and other regulatory factors. This could reduce the area that regulatory factors need to sample to find genes to act on and also reduce the area that genes and enhancers released from the NE at NEIs need to sample to find one another. This adds a further layer to modeling studies on the release of genes for expression from the peripheral NE into the wider nucleoplasm, which indicated that for two loci to find each other, sampling a 1–1.5-μm-diameter space by Brownian motion can easily take 24 h to interact, but if the space is reduced to 0.5–0.8 μm diameter, then they can find one another within an hour [28], and we found that genes and enhancers many Mb apart in the linear genome and kept far apart at the NE in resting lymphocytes interact upon their release with lymphocyte activation [122]. Thus, NEIs could add an additional level of fine-tuned regulation to many of the aforementioned modes of gene regulation.

## $Ca^{2+}$ Stores in NEIs

$Ca^{2+}$ stores are normally distributed throughout the S/ER from its origin at the ONM to the plasma membrane (PM) [112], and different regions of this network can deliver $Ca^{2+}$ signals of different forms and functions. In 1999, the Iino lab had elegantly identified a variety of different $Ca^{2+}$ signals in smooth muscles in situ in intact sections of the rat tail artery [3]. Subsequently, studies demonstrated that different regions of S/ER contained different pumps and transporters and that these different S/ER regions could deliver spatially segregated $Ca^{2+}$ signals [19, 32, 61] (Fig. 3.5a), despite the fact that the lumen of the NE is continuous with that of the S/ER [45, 139, 146].

Nevertheless and regardless of cell type, the general consensus was that propagating global $Ca^{2+}$ waves gained entry into the nucleoplasm across the NE and NEIs, and it was suggested that different spatiotemporal patterns of $Ca^{2+}$ release could coordinate gene expression and cell plasticity [7, 35]. Indeed early electrophysiological studies on nuclei isolated from *Xenopus* oocytes indicated that $Ca^{2+}$ release channels were widely distributed across the ONM (DO [78]; Don-On Daniel [77]), while $Ca^{2+}$ imaging suggested that agonist-evoked activation of inositol 1,4,5-trisphosphate receptors ($IP_3Rs$) and ryanodine receptors (RyRs) on such nuclei resulted in marked increases in $Ca^{2+}$ within the nucleoplasm [14]; accordingly, the presence of $Ca^{2+}$ channels and other ion transporters was later confirmed by a number of proteomic studies [63, 64, 128, 148]. Moreover, more recent studies on acutely isolated ventricular myocytes appeared to support the consensus that $Ca^{2+}$ release from the NE and NEI entered the nucleoplasm [111]. This viewpoint was supported by confocal imaging studies on a variety of cell lines that confirmed the release of $Ca^{2+}$ into the nucleoplasm from $Ca^{2+}$ stores within the NE lumen. For example, sarco−/endoplasmic reticulum $Ca^{2+}$-ATPase (SERCA) pumps and $IP_3Rs$ calcium release channels were observed on the NE at large (>0.5 μm) cleft-like NEIs on HeLa cells [74, 75, 89], C6 glioma cells [76], H1299 cells [89], and MDCK cells [22]. Also RyRs were observed on the NE at cleft-like NEIs on C2C12 cells [83] and primary cultures of neonatal cardiomyocytes [47, 71, 133]. In addition, the activation of $IP_3Rs$ and RyRs was shown to trigger the release of calcium into the nucleoplasm of nuclei isolated from primary cultures of human aortic smooth muscle cells (hVSMC), the nuclei of which incorporated large (>1 μm) lamin-positive cleft-like and/or tubular NEI [4].

However, an earlier study suggested that a nuclear buffer barrier may block direct calcium entry into the nucleoplasm in differentiated neurons [1]. Adding to this, a recent study on PASMC demonstrated that when compared to the perinuclear SR, NEIs hold discrete subtypes of SERCA pump (SERCA1) and $Ca^{2+}$ release channel (RyR1), while the peripheral NE does not. More strikingly still, it was demonstrated that these NEIs represent epigenetically distinct units of the NE that may provide for discrete mechanisms of gene expression regulation [32] (Fig. 3.5b). Moreover, application of the selective RyR1 agonist maurocalcine and a physiological stimulus (angiotensin II) confirmed this, by selectively inducing calcium release into cytoplasmic nanocourses demarcated by NEIs [31], and changes in calcium flux at NEI were also shown to impact gene expression [31].

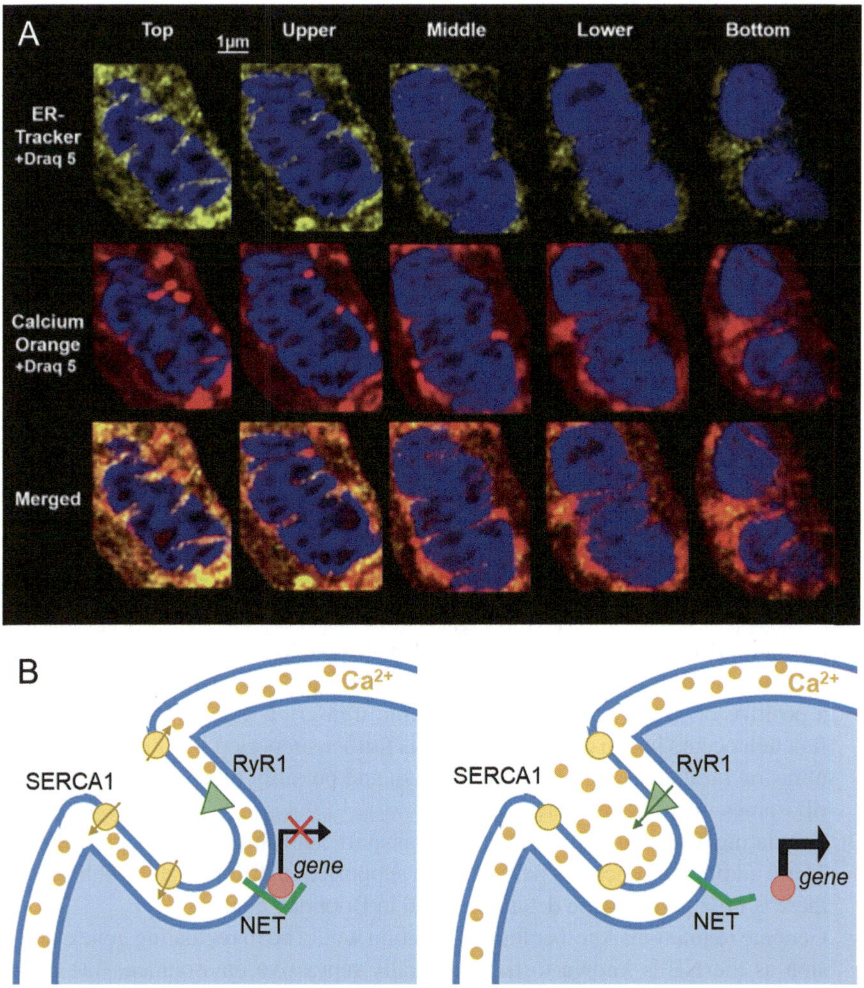


**Fig. 3.5** NEIs can function as a calcium store. (**a**) A series of confocal Z sections through a rat PASMC; the SR and NE are labeled with ER tracker (yellow), calcium was identified by Calcium Orange (red), and the nucleoplasm was identified by Draq5 (blue) [31]. (**b**) Schematic of the hypothesized mechanism of calcium coordinating gene expression. In this particular scenario, when $Ca^{2+}$ ions are in the lumen and bound to the NE, the conformation transduced across the INM promotes nucleoplasmic tethering of a gene. When $Ca^{2+}$ is pumped out of the lumen into the cytoplasmic stores of the NEIs, then ion binding is lost, and a different protein conformation is achieved for the NET that results in a loss of the gene tethering and activation of the released gene in the nucleoplasm

## NEI Functions

The functions of NEIs at this point are mostly just supposition, but one general idea is that they might support NE functions in deeper nuclear regions remote from the outermost nuclear shell (ONS) [79]. A reason that may contribute to why functions of NEIs are still not fully understood may be that NEIs have many disparate roles so that attempts to find parallels besides the crudely structural would frequently be unsuccessful. There is some support for a few very different functions listed below.

1. Fast nuclear energy: Many EM images of skeletal muscle cells show large, cleft-like invaginations that are not deep, but reasonably frequently contain mitochondria [5, 40]. Thus, it might make sense for these NEIs to help direct ATP rapidly into the nucleus for rapid changes in gene expression and genome organization induced by signals coming at the neuromuscular junction where nuclei tend to accumulate in these cells.
2. Translation: Another function that would make sense would be to support more rapid protein translation as soon as mRNAs are exported in some cell types (notably this could work together with point 1 above). Although the dark ribosomal spots that are observed in the RER and ONM are not observed on most NEIs, in the nuclei of trophoblast giant cells, the cleft-like and/or large tubular (>1 μm) NEIs are lined with such ribosomal spots by EM and stain with Unr, a cytoplasmic RNA-binding protein that regulates RNA turnover and can be both a positive or negative regulator of specific transcripts [127]. That this type of function occurs just in certain cell types is further supported in *Drosophila* where ultrastructural analysis indicated cleft-like and possibly tubular NEIs lined with ribosomes in polytene nuclei [106].
3. Regulating $Ca^{2+}$ signalling, material transport, and nuclear lipid metabolism: These additional functions have all been supported in different studies; however, there is little mechanistic detail reviewed in Bootman et al. [7].
4. Genome regulation: Another logical function would be in regulating gene expression as the NE is known to be a generally repressive environment due to the accumulation of epigenetic silencing marks and enzymes there. While this makes sense later in development, it is unlikely to be the case for the many changes in NEIs in early embryonic development because it was noted that chromatin was not in contact with the NEIs in the early embryonic stages where they transiently occur [113].

Another possibility for PASMC NEIs is that the NEI provides a regulatory environment closer to the interior of the nucleus, which has higher transcriptional activity, thus shortening the distance the gene needs to travel upon activation or inhibition, and allows faster gene expression regulation. Two genes were picked to test this hypothesis: DNA mismatch repair protein MutL homolog 1 (Mlh1), which can be repressed through interaction with H3K9me2 [18], and S100 calcium-binding protein A9 (S100a9) that associates with barrier-to-autointegration factor (BAF), a chromatin protein [142]. In previous research, these two genes showed reduced expression levels when the calcium flow from the NEI store into the cytoplasmic lumen was blocked by tetracaine (TTC) treatment in rat PASMCs [32]. However, the expression of only one, *S100a9*, was abolished upon dissolution of NEIs in proliferating cells.

To visualize the location of these genes, immuno-FISH was performed with co-staining for lamin A in untreated rat PASMCs and PASMCs treated with either TTC or maurocalcine (MCa), which, respectively, block or promote $Ca^{2+}$ flux. *S100a9* was often associated with or very close to the NE/NEIs for cells in their basal state, while *Mlh1* was typically a clear distance from the NE/NEIs. However, both genes moved toward the NE/NEIs upon treatments with TTC, while they moved further away from the NE with MCa treatment (Fig. 3.6). This result indicated that $Ca^{2+}$ flux is likely to play a direct role in the regulation of gene positioning and expression.

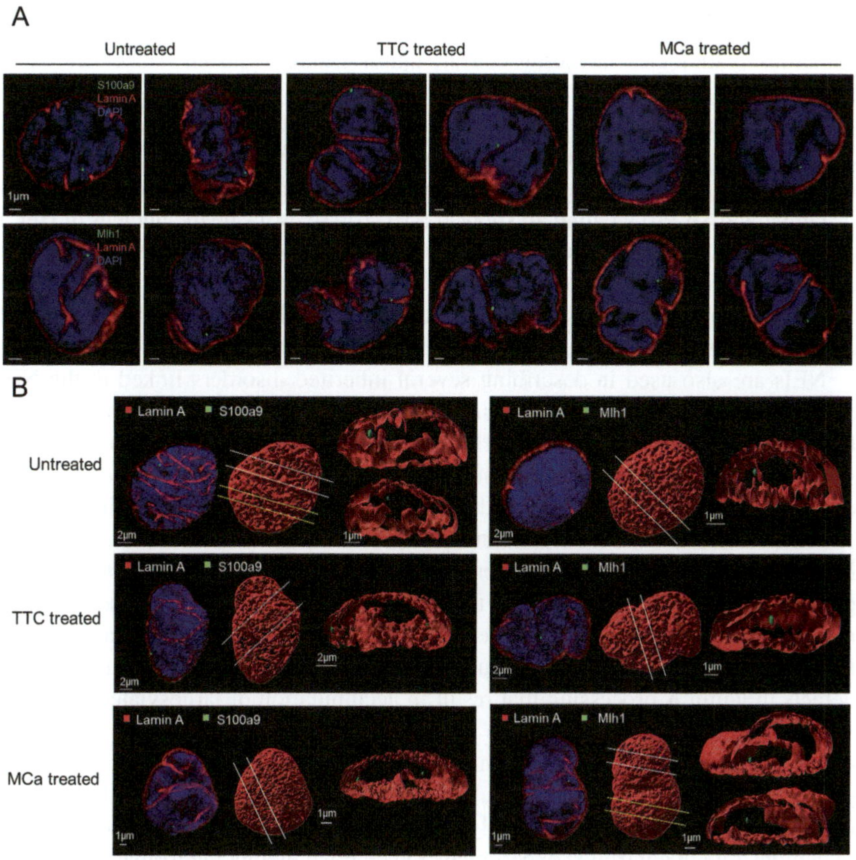


**Fig. 3.6** In rat PASMCs, S100a9 and Mlh1 were moving when the calcium flux was blocked or induced. (**a**) FISH showing the location of the genes (green) in the nucleus with NE identified by lamin A (red). (**b**) 3D reconstruction of the genes and NE identified by lamin A

Nonetheless, the mechanism by which $Ca^{2+}$ flux regulates gene positioning and expression remains unclear. Thus, much more experimentation will be required before the many possible mechanisms can be distinguished. However, it is already clear that $Ca^{2+}$ flux occurs at NEIs, which correlates with changes in gene positioning and gene expression, and that drug-induced $Ca^{2+}$ flux results in similar gene repositioning and gene expression changes.

## NEIs in Disease

Just as prion conformational changes were initially viewed as disease pathology but subsequently found to also be a wide-ranging regulatory mechanism in cellular health [49], so it is with NEIs. Before the recent findings that NEIs are similarly used for cell regulatory functions, NEI abundance was initially associated with numerous pathologies including cancer, laminopathies, poison, Alzheimer's disease, and many others.

As noted earlier, nuclear shape and size changes are commonly used in cancer diagnosis and prognosis. This has been the case for over 150 years [6]—ever since microscopy gained ~1 μm resolution in the 1830s. Over a quarter century ago, Con A, lamin, and Yo-Pro labeling all showed NEI structures inside the nucleus of HeLa cells [41], and today, NEIs are widely included in the cancer diagnostic matrix (Ashraf N [80]). NEIs visualized by staining for lamins and NETs are as commonly used as nuclear size changes for diagnosis/prognosis in breast, prostate, and thyroid cancers [13, 39, 118].

NEIs are also used in describing several inherited disorders linked to the NE, collectively termed "laminopathies." The first described of these was Emery-Dreifuss muscular dystrophy (EDMD) for which many studies demonstrated massively misshapen and invaginated nuclei in the affected muscle both when it was caused by mutations in lamin A and when it was caused by mutations in the NET emerin [82, 96]. Similar nuclear changes were subsequently observed in many other NE-linked disorders [81]. In the premature-aging laminopathy Hutchison-Guilford progeria syndrome (HGPS), mutant farnesylated lamin A accumulates and results in more abundant NEIs compared to healthy cells [86]. Interestingly, similar NEIs are formed in naturally aging cells due to disrupted cellular homeostasis yielding defects in lamin A processing that result in accumulation of farnesylated prelamin A, which functions similarly to the mutant lamin A in HGPS [87]. As mentioned above, the Vaux lab developed a method for deuterium labeling of phospholipid precursors that detected the accumulation of nascent phospholipids in the INM of NEIs with NanoSIMS [116]. Notably, they had previously used the same approach to demonstrate in controlled experiments the de novo membrane formation in NEIs induced by expression of the HGPS mutant version of lamin A [29], further demonstrating a role for lamins in NEI formation and a potential role for NEIs in development of pathology. This process appears to require the lipid moiety attached to lamin A as treating HGPS cells with farnesyltransferase or geranylgeranyltransferase I inhibitors resulted in a decrease in the number of NEIs [145].

Other diseases not traditionally linked to the NE also yield NEIs as part of their pathology. Alzheimer's disease (AD) is a neurodegenerative disease that despite many years of research still has many unknowns. In 2016, Frost reported reduced lamin B1 but not lamin A levels alongside significantly increased NEI numbers both in human AD brains and in a model system involving transgenic human tau in *Drosophila* neurons [42]. These were very similar to what has been described in laminopathies. Thus, the range of disorders where NEIs accumulate goes well beyond the laminopathies.

Likewise, there is some indication of $Ca^{2+}$ flux playing a role also in laminopathies. EDMD characteristically has conduction defects leading to cardiomyopathy associated with it [153]. Notably, ~21% of EDMD cases are caused by mutations in lamin A and ~ 19% to mutations in emerin, and roughly 15% are caused by mutations in muscle-specific NETs including the $Ca^{2+}$ channel regulator Tmem38a, also referred to as TRIC-A for trimeric intracellular cation-selective channel A [88, 143].

Finally, the potential link between NEI and disease is underscored by the findings noted above in PASMCs. This relates to group III pulmonary hypertension (high blood pressure in the lung) that is associated with sleep apnea, interstitial lung disease, and chronic obstructive pulmonary disease (COPD). Group III pulmonary hypertension is progressive, due to the prolonged action of hypoxia and other vasoactive inputs that together elicit maladaptive gene expression changes that select for proliferative/secretory myocyte phenotypes that, in part, drive further deleterious remodeling of the pulmonary vasculature [157]. It is possible, therefore, that in these abnormally proliferating PASMCs, NEIs are lost and/or reconfigure, similar to changes observed with proliferation of acutely isolated PASMC in culture [31].

## Conclusions

There is clear evidence of NEIs being induced or depleted in association with different stages in development and tissue differentiation. There is also clear evidence that NEIs change in a number of disease states. Whether NEIs in disease reflect an attempt of the cell to reconfigure itself to respond to aspects of disease pathology or are a reflection of disease pathology remains to be discerned. However, the observations of many distinct functions for different types of NEIs in a variety of cell types provide multiple ways that cells could try to adapt or respond to different diseases by inducing NEIs to change nuclear architecture and function. Whether using NEIs for focused protein production in ribosomes lining them, altering $Ca^{2+}$ signalling networks, fine-tuning changes in gene expression, or massively altering gene expression to change cellular phenotype, these amazing invaginations of the NE are clearly important and worthy of further study.

**Author Contributions** All authors contributed to the conception of this review. QJ and ECS wrote the manuscript. AME gave detailed feedback on calcium signalling, epigenetic segregation, and gene expression at NEI. TH and SM gave further feedback on this manuscript. QJ studies gene movement at NEI with ECS and AME. JN studies calcium signalling at NEI and LINC complex organization in pulmonary arterial myocytes with AME. ZD studies calcium signalling and NEI in neurons with TH and AME. CW studies calcium signalling and NEI function in *Dictyostelium* with SM and AME.

**Compliance with Ethical Standards** This study was performed in line with the legal requirements or guidelines in the UK for the care and use of animals.

## References

1. al-Mohanna FA, Caddy KW, Bolsover SR (1994) The nucleus is insulated from large cytosolic calcium ion changes. Nature 367(6465):745–750
2. Alsheimer M, Benavente R (1996) Change of karyoskeleton during mammalian spermatogenesis: expression pattern of nuclear Lamin C2 and its regulation. Exp Cell Res 228(2):181–188
3. Asada Y, Yamazawa T, Hirose K, Takasaka T, Iino M (1999) Dynamic Ca2+ signalling in rat arterial smooth muscle cells under the control of local renin–angiotensin system. J Physiol 521(Pt 2):497
4. Avedanian L, Jacques D, Bkaily G (2011) Presence of tubular and reticular structures in the nucleus of human vascular smooth muscle cells. J Mol Cell Cardiol 50(1):175–186
5. Baumann M, Steichen-Gersdorf E, Krabichler B, Petersen B-S, Weber U, Schmidt WM et al (2017) Homozygous SYNE1 mutation causes congenital onset of muscular weakness with distal arthrogryposis: a genotype–phenotype correlation. Eur J Hum Genet 25(2):262–266
6. Beale L (1860) Results of the chemical and microscopical examination of solid organs and secretions. Examination of sputum from a case of cancer of the pharynx and the adjacent parts. Arch Med 2:44
7. Bootman MD, Fearnley C, Smyrnias I, MacDonald F, Roderick HL (2009) An update on nuclear calcium signalling. J Cell Sci 122(14):2337–2350
8. Bourgeois C, Hemon D, Bouteille M (1979) Structural relationship between the nucleolus and the nuclear envelope. J Ultrastruct Res 68(3):328–340
9. Bouzid T, Kim E, Riehl BD, Esfahani AM, Rosenbohm J, Yang R et al (2019) The LINC complex, mechanotransduction, and mesenchymal stem cell function and fate. J Biol Eng 13(1):1–12
10. Broers JL, Machiels BM, Kuijpers HJ, Smedts F, Van Den Kieboom R, Raymond Y, Ramaekers FC (1997) A-and B-type lamins are differentially expressed in normal human tissues. Histochem Cell Biol 107:505–517
11. Buschmann MT, Geoffroy JS, LaVelle A (1981) Nuclear envelope invaginations in hamster pyramidal cells during development and aging. Neurobiol Aging 2(1):27–32
12. Bussolati G, Maletta F, Asioli S, Annaratone L, Sapino A, Marchiò C (2014) "To be or not to be in a good shape": diagnostic and clinical value of nuclear shape irregularities in thyroid and breast cancer. Recent Advances May Elucidate Past Paradoxes, Cancer Biology and the Nuclear Envelope, pp 101–121
13. Bussolati G, Marchiò C, Gaetano L, Lupo R, Sapino A (2008) Pleomorphism of the nuclear envelope in breast cancer: a new approach to an old problem. J Cell Mol Med 12(1):209–218
14. Cancela JM, Van Coppenolle F, Galione A, Tepikin AV, Petersen OH (2002) Transformation of local Ca2+ spikes to global Ca2+ transients: the combinatorial roles of multiple Ca2+ releasing messengers. EMBO J 21:909

15. Cavalier-Smith T (2005) Economy, speed and size matter: evolutionary forces driving nuclear genome miniaturization and expansion. Ann Bot 95(1):147–175
16. Cavalier-Smith T, Lewis R, Chao EE, Oates B, Bass D (2009) Helkesimastix marina n. sp. (Cercozoa: Sainouroidea superfam. N.) a gliding zooflagellate of novel ultrastructure and unusual ciliary behaviour. Protist 160(3):452–479
17. Cebrian-Silla A, Alfaro-Cervello C, Herranz-Perez V, Kaneko N, Park DH, Sawamoto K et al (2017) Unique organization of the nuclear envelope in the post-natal quiescent neural stem cells. Stem Cell Reports 9(1):203–216
18. Chen H, Yan Y, Davidson TL, Shinkai Y, Costa M (2006) Hypoxic stress induces dimethylated histone H3 lysine 9 through histone methyltransferase G9a in mammalian cells. Cancer Res 66(18):9009–9016
19. Clark JH, Kinnear NP, Kalujnaia S, Cramb G, Fleischer S, Jeyakumar LH et al (2010) Identification of functionally segregated sarcoplasmic reticulum calcium stores in pulmonary arterial smooth muscle. J Biol Chem 285(18):13542–13549
20. Clark P, Jones KJ, LaVelle A (1990) Ultrastructural and morphometric analysis of nucleolar and nuclear changes during the early growth period in hamster facial neurons. J Comp Neurol 302(4):749–760
21. Clubb B, Locke M (1998) 3T3 cells have nuclear invaginations containing F-actin. Tissue Cell 30(6):684–691
22. Collado-Hilly M, Shirvani H, Jaillard D, Mauger JP (2010) Differential redistribution of Ca2+-handling proteins during polarisation of MDCK cells: effects on Ca2+ signalling. Cell Calcium 48(4):215–224
23. Collings DA, Carter CN, Rink JC, Scott AC, Wyatt SE, Allen NS (2000) Plant nuclei can contain extensive grooves and invaginations. Plant Cell 12(12):2425–2439
24. Czapiewski R, Batrakou DG, de Las Heras JI, Carter RN, Sivakumar A, Sliwinska M et al (2022) Genomic loci mispositioning in Tmem120a knockout mice yields latent lipodystrophy. Nat Commun 13(1):321
25. Dabiké M, Preller A (1999) Cytoarchitecture of Caudiverbera caudiverbera stage VI oocytes: a light and electron microscope study. Anat Embryol 199:489–497
26. de Las Heras JI, Schirmer EC (2014) The nuclear envelope and cancer: a diagnostic perspective and historical overview. Recent Advances May Elucidate Past Paradoxes, Cancer Biology and the Nuclear Envelope, pp 5–26
27. De Oca RM, Lee KK, Wilson KL (2005) Binding of barrier to autointegration factor (BAF) to histone H3 and selected linker histones including H1. 1. J Biol Chem 280(51):42252–42262
28. Dekker J, Mirny L (2016) The 3D genome as moderator of chromosomal communication. Cell 164(6):1110–1121
29. Drozdz MM, Jiang H, Pytowski L, Grovenor C, Vaux DJ (2017) Formation of a nucleoplasmic reticulum requires de novo assembly of nascent phospholipids and shows preferential incorporation of nascent lamins. Sci Rep 7(1):7454
30. Drozdz MM, Vaux DJ (2017) Shared mechanisms in physiological and pathological nucleoplasmic reticulum formation. Nucleus 8(1):34–45
31. Duan, J. (2018). Cell-wide web of cytoplasmic nanocourses coordinates calcium signalling
32. Duan J, Navarro-Dorado J, Clark JH, Kinnear NP, Meinke P, Schirmer EC, Evans AM (2019) The cell-wide web coordinates cellular processes by directing site-specific Ca2+ flux across cytoplasmic nanocourses. Nat Commun 10(1):2299
33. Dutta S, Bhattacharyya M, Sengupta K (2016) Implications and assessment of the elastic behavior of lamins in laminopathies. Cells 5(4):37
34. Edens LJ, White KH, Jevtic P, Li X, Levy DL (2013) Nuclear size regulation: from single cells to development and disease. Trends Cell Biol 23(4):151–159
35. Eder A, Bading H (2007) Calcium signals can freely cross the nuclear envelope in hippocampal neurons: somatic calcium increases generate nuclear calcium transients. BMC Neurosci 8(1):1–11

36. Falloon EA, Dynlacht JR (2002) Reversible changes in the nuclear lamina induced by hyperthermia. J Cell Biochem 86(3):451–460
37. Fawcett DW (1981) The cell. W.B. Saunders Philadelphia, Philadelphia
38. Fidorra J, Mielke T, Booz J, Feinendegen L (1981) Cellular and nuclear volume of human cells during the cell cycle. Radiat Environ Biophys 19:205–214
39. Fischer AH (2014) The diagnostic pathology of the nuclear envelope in human cancers. Cancer Biology and the Nuclear Envelope: Recent Advances May Elucidate Past Paradoxes, pp 49–75
40. Franke WW, Schinko W (1969) Nuclear shape in muscle cells. J Cell Biol 42(1):326
41. Fricker M, Hollinshead M, White N, Vaux D (1997) Interphase nuclei of many mammalian cell types contain deep, dynamic, tubular membrane-bound invaginations of the nuclear envelope. J Cell Biol 136(3):531–544
42. Frost B, Bardai FH, Feany MB (2016) Lamin dysfunction mediates neurodegeneration in tauopathies. Curr Biol 26(1):129–136
43. Gatticchi L, de Las Heras JI, Sivakumar A, Zuleger N, Roberti R, Schirmer EC (2020) Tm7sf2 disruption alters radial gene positioning in mouse liver leading to metabolic defects and diabetes characteristics. Front Cell Develop Biol 8:592573
44. Gavelis GS, Herranz M, Wakeman KC, Ripken C, Mitarai S, Gile GH et al (2019) Dinoflagellate nucleus contains an extensive endomembrane network, the nuclear net. Sci Rep 9(1):839
45. Gordienko D, Greenwood I, Bolton T (2001) Direct visualization of sarcoplasmic reticulum regions discharging Ca2+ sparks in vascular myocytes. Cell Calcium 29(1):13–28
46. Graziano S, Kreienkamp R, Coll-Bonfill N, Gonzalo S (2018) Causes and consequences of genomic instability in laminopathies: replication stress and interferon response. Nucleus 9(1):289–306
47. Guatimosim S, Amaya MJ, Guerra MT, Aguiar CJ, Goes AM, Gómez-Viquez NL et al (2008) Nuclear Ca2+ regulates cardiomyocyte function. Cell Calcium 44(2):230–242
48. Gül N, Yiğit N, Saygılı F, Demirel E, Geniş C (2016) Comparison of the effects of difenacoum and brodifacoum on the ultrastructure of rat liver cells. Arch Ind Hyg Toxicol 67(3):204–209
49. Halfmann R, Jarosz DF, Jones SK, Chang A, Lancaster AK, Lindquist S (2012) Prions are a common mechanism for phenotypic inheritance in wild yeasts. Nature 482(7385):363–368
50. Hilton BA, Liu J, Cartwright BM, Liu Y, Breitman M, Wang Y et al (2017) Progerin sequestration of PCNA promotes replication fork collapse and mislocalization of XPA in laminopathy-related progeroid syndromes. FASEB J 31(9):3882
51. Hirschhorn R, Decsy MI, Troll W (1971) The effect of PHA stimulation of human peripheral blood lymphocytes upon cellular content of euchromatin and heterochromatin. Cell Immunol 2(6):696–701
52. Holaska JM, Lee KK, Kowalski AK, Wilson KL (2003) Transcriptional repressor germ cell-less (GCL) and barrier to autointegration factor (BAF) compete for binding to emerin in vitro. J Biol Chem 278(9):6969–6975
53. Holaska JM, Rais-Bahrami S, Wilson KL (2006) Lmo7 is an emerin-binding protein that regulates the transcription of emerin and many other muscle-relevant genes. Hum Mol Genet 15(23):3459–3472
54. Ingber DE (2018) From mechanobiology to developmentally inspired engineering. Philosophical Transactions of the Royal Society B: Biological Sciences 373(1759):20170323
55. Ivorra C, Kubicek M, González JM, Sanz-González SM, Álvarez-Barrientos A, O'Connor J-E et al (2006) A mechanism of AP-1 suppression through interaction of c-Fos with Lamin A/C. Genes Dev 20(3):307–320
56. Janmey PA, Euteneuer U, Traub P, Schliwa M (1991) Viscoelastic properties of vimentin compared with other filamentous biopolymer networks. J Cell Biol 113(1):155–160
57. Jevtić P, Edens LJ, Li X, Nguyen T, Chen P, Levy DL (2015) Concentration-dependent effects of nuclear lamins on nuclear size in Xenopus and mammalian cells. J Biol Chem 290(46):27557–27571

58. Jones KJ, Pfaff DW, McEwen BS (1985) Early estrogen-induced nuclear changes in rat hypothalamic ventromedial neurons: an ultrastructural and morphometric analysis. J Comp Neurol 239(3):255–266
59. Jorgens DM, Inman JL, Wojcik M, Robertson C, Palsdottir H, Tsai W-T et al (2017) Deep nuclear invaginations are linked to cytoskeletal filaments–integrated bioimaging of epithelial cells in 3D culture. J Cell Sci 130(1):177–189
60. Jorgensen P, Edgington NP, Schneider BL, Rupeš I, Tyers M, Futcher B (2007) The size of the nucleus increases as yeast cells grow. Mol Biol Cell 18(9):3523–3532
61. Kinnear NP, Wyatt CN, Clark JH, Calcraft PJ, Fleischer S, Jeyakumar LH et al (2008) Lysosomes co-localize with ryanodine receptor subtype 3 to form a trigger zone for calcium signalling by NAADP in rat pulmonary arterial smooth muscle. Cell Calcium 44(2):190–201
62. Koch AJ, Holaska JM (2014) Emerin in health and disease. Semin Cell Dev Biol. Elsevier 29:95–106
63. Korfali N, Wilkie GS, Swanson SK, Srsen V, Batrakou DG, Fairley EA et al (2010) The leukocyte nuclear envelope proteome varies with cell activation and contains novel transmembrane proteins that affect genome architecture. Mol Cell Proteomics 9(12):2571–2585
64. Korfali N, Wilkie GS, Swanson SK, Srsen V, de Las Heras J, Batrakou DG et al (2012) The nuclear envelope proteome differs notably between tissues. Nucleus 3(6):552–564
65. Kosak ST, Skok JA, Medina KL, Riblet R, Le Beau MM, Fisher AG, Singh H (2002) Subnuclear compartmentalization of immunoglobulin loci during lymphocyte development. Science 296(5565):158–162
66. Kubben N, Voncken JW, Demmers J, Calis C, van Almen G, Pinto YM, Misteli T (2010) Identification of differential protein interactors of Lamin A and progerin. Nucleus 1(6):513–525
67. Ladekarl M, Bœk-Hansen T, Henrik-Nielsen R, Mouritzen C, Henriques U, Sørensen FB (1995) Objective malignancy grading of squamous cell carcinoma of the lung. Stereologic estimates of mean nuclear size are of prognostic value, independent of clinical stage of disease. Cancer 76(5):797–802
68. Lammerding J, Fong LG, Ji JY, Reue K, Stewart CL, Young SG, Lee RT (2006) Lamins A and C but not Lamin B1 regulate nuclear mechanics. J Biol Chem 281(35):25768–25780
69. Lammerding J, Schulze PC, Takahashi T, Kozlov S, Sullivan T, Kamm RD et al (2004) Lamin A/C deficiency causes defective nuclear mechanics and mechanotransduction. J Clin Invest 113(3):370–378
70. Lane BP (1965) Alterations in the cytologic detail of intestinal smooth muscle cells in various stages of contraction. J Cell Biol 27(1):199–213
71. Lee SH, Hadipour-Lakmehsari S, Miyake T, Gramolini AO (2018) Three-dimensional imaging reveals endo(sarco)plasmic reticulum-containing invaginations within the nucleoplasm of muscle. Am J Physiol Cell Physiol 314(3):C257–C267
72. Lele TP, Dickinson RB, Gundersen GG (2018) Mechanical principles of nuclear shaping and positioning. J Cell Biol 217(10):3330–3342
73. Levy DL, Heald R (2010) Nuclear size is regulated by importin α and Ntf2 in Xenopus. Cell 143(2):288–298
74. Lui PP, Chan FL, Suen Y, Kwok T, Kong S (2003) The nucleus of HeLa cells contains tubular structures for Ca2+ signaling with the involvement of mitochondria. Biochem Biophys Res Commun 308(4):826–833
75. Lui PP, Kong SK, Kwok TT, Lee CY (1998a) The nucleus of HeLa cell contains tubular structures for Ca2+ signalling. Biochem Biophys Res Commun 247(1):88–93
76. Lui PP, Lee CY, Tsang D, Kong SK (1998b) Ca2+ is released from the nuclear tubular structure into nucleoplasm in C6 glioma cells after stimulation with phorbol ester. FEBS Lett 432(1–2):82–87
77. Mak D-OD, Foskett JK (1997) Single-channel kinetics, inactivation, and spatial distribution of inositol trisphosphate (IP3) receptors in Xenopus oocyte nucleus. J Gen Physiol 109(5):571–587

78. Mak D, Foskett JK (1994) Single-channel inositol 1, 4, 5-trisphosphate receptor currents revealed by patch clamp of isolated Xenopus oocyte nuclei. J Biol Chem 269(47):29375–29378
79. Malhas A, Goulbourne C, Vaux DJ (2011) The nucleoplasmic reticulum: form and function. Trends Cell Biol 21(6):362–373
80. Malhas AN, Vaux DJ (2014) Nuclear envelope invaginations and cancer. Cancer Biology and the Nuclear Envelope: Recent Advances May Elucidate Past Paradoxes, pp 523–535
81. Maraldi NM, Lattanzi G, Capanni C, Columbaro M, Mattioli E, Sabatelli P et al (2006) Laminopathies: a chromatin affair. Adv Enzym Regul 46(1):33–49
82. Maraldi NM, Squarzoni S, Sabatelli P, Lattanzi G, Ognibene A, Manzoli FA (2002) Emery–Dreifuss muscular dystrophy, nuclear cell signaling and chromatin remodeling. Adv Enzym Regul 42:1–18
83. Marius P, Guerra MT, Nathanson MH, Ehrlich BE, Leite MF (2006) Calcium release from ryanodine receptors in the nucleoplasmic reticulum. Cell Calcium 39(1):65–73
84. Mattioli E, Andrenacci D, Garofalo C, Prencipe S, Scotlandi K, Remondini D et al (2018) Altered modulation of Lamin A/C-HDAC2 interaction and p21 expression during oxidative stress response in HGPS. Aging Cell 17(5):e12824
85. Mattout A, Cabianca DS, Gasser SM (2015) Chromatin states and nuclear organization in development—a view from the nuclear lamina. Genome Biol 16:1–15
86. McClintock D, Gordon LB, Djabali K (2006) Hutchinson–Gilford progeria mutant lamin A primarily targets human vascular cells as detected by an anti-Lamin A G608G antibody. Proc Natl Acad Sci 103(7):2154–2159
87. McClintock D, Ratner D, Lokuge M, Owens DM, Gordon LB, Collins FS, Djabali K (2007) The mutant form of Lamin A that causes Hutchinson-Gilford progeria is a biomarker of cellular aging in human skin. PLoS One 2(12):e1269
88. Meinke P, Kerr AR, Czapiewski R, de Las Heras JI, Dixon CR, Harris E et al (2020) A multistage sequencing strategy pinpoints novel candidate alleles for Emery-Dreifuss muscular dystrophy and supports gene misregulation as its pathomechanism. EBioMedicine 51:102587
89. Nalaskowski MM, Fliegert R, Ernst O, Brehm MA, Fanick W, Windhorst S et al (2011) Human inositol 1,4,5-trisphosphate 3-kinase isoform B (IP3KB) is a nucleocytoplasmic shuttling protein specifically enriched at cortical actin filaments and at invaginations of the nuclear envelope. J Biol Chem 286(6):4500–4510
90. Ndiaye M, Mattei X (1993) Process of nuclear envelope reduction in spermiogenesis of a mosquito, Culex tigripes. Mol Reprod Develop 34(4):416–419
91. Neumann FR, Nurse P (2007) Nuclear size control in fission yeast. J Cell Biol 179(4):593–600
92. Newport JW, Wilson KL, Dunphy WG (1990) A Lamin-independent pathway for nuclear envelope assembly. J Cell Biol 111(6):2247–2259
93. Nielsen JA, Hudson LD, Armstrong RC (2002) Nuclear organization in differentiating oligodendrocytes. J Cell Sci 115(21):4071–4079
94. Nili E, Cojocaru GS, Kalma Y, Ginsberg D, Copeland NG, Gilbert DJ et al (2001) Nuclear membrane protein LAP2β mediates transcriptional repression alone and together with its binding partner GCL (germ-cell-less). J Cell Sci 114(18):3297–3307
95. Novotný R, Malínský J, Oborná I, Dostál J (1999) Nuclear channel system (NCS) in normal endometrium and after hormonal stimulation. Biomedical Papers
96. Ognibene A, Sabatelli P, Petrini S, Squarzoni S, Riccio M, Santi S et al (1999) Nuclear changes in a case of X-linked Emery-Dreifuss muscular dystrophy. Muscle Nerve 22(7):864–869
97. Olins AL, Buendia B, Herrmann H, Lichter P, Olins DE (1998) Retinoic acid induction of nuclear envelope-limited chromatin sheets in HL-60. Exp Cell Res 245(1):91–104
98. Olins AL, Ernst A, Zwerger M, Herrmann H, Olins DE (2010) An in vitro model for Pelger-Huet anomaly: stable knockdown of Lamin B receptor in HL-60 cells. Nucleus 1(6):506–512
99. Olins AL, Herrmann H, Lichter P, Olins DE (2000) Retinoic acid differentiation of HL-60 cells promotes cytoskeletal polarization. Exp Cell Res 254(1):130–142

100. Olins AL, Hoang TV, Zwerger M, Herrmann H, Zentgraf H, Noegel AA et al (2009) The LINC-less granulocyte nucleus. Eur J Cell Biol 88(4):203–214
101. Olins AL, Olins DE (2004) Cytoskeletal influences on nuclear shape in granulocytic HL-60 cells. BMC Cell Biol 5(1):1–18
102. Omar NA, Abu-Almaaty AH, Abd El-Aziz YM, Abdeen AM, Mohamed F, Hashem MMM, Hammad S (2019) Impacts of Egyptian propolis extract on rat cerebellum intoxicated by aluminum silicate: histopathological studies. Environ Sci Pollut Res Int 26(21):22061–22068
103. Osada S-I, Ohmori S-Y, Taira M (2003) XMAN1, an inner nuclear membrane protein, antagonizes BMP signaling by interacting with Smad1. in Xenopus embryos 130:1783
104. Ozaki T, Saijo M, Murakami K, Enomoto H, Taya Y, Sakiyama S (1994) Complex formation between Lamin A and the retinoblastoma gene product: identification of the domain on Lamin A required for its interaction. Oncogene 9(9):2649–2653
105. Pan D, Estévez-Salmerón LD, Stroschein SL, Zhu X, He J, Zhou S, Luo K (2005) The integral inner nuclear membrane protein MAN1 physically interacts with the R-Smad proteins to repress signaling by the transforming growth factor-β superfamily of cytokines. J Biol Chem 280(16):15992–16001
106. Park PC, De Boni U (1992) Nuclear membrane modifications in polytene nuclei of Drosophila melanogaster: serial reconstruction and cytochemistry. Anat Rec 234(1):15–26
107. Peeters BW, Piët AC, Fornerod M (2022) Generating membrane curvature at the nuclear pore: A lipid point of view. Cells 11(3):469
108. Peric-Hupkes D, Meuleman W, Pagie L, Bruggeman SW, Solovei I, Brugman W et al (2010) Molecular maps of the reorganization of genome-nuclear lamina interactions during differentiation. Mol Cell 38(4):603–613
109. Pickersgill H, Kalverda B, De Wit E, Talhout W, Fornerod M, Van Steensel B (2006) Characterization of the Drosophila melanogaster genome at the nuclear lamina. Nat Genet 38(9):1005–1014
110. Pindyurin AV, Moorman C, de Wit E, Belyakin SN, Belyaeva ES, Christophides GK et al (2007) SUUR joins separate subsets of PcG, HP1 and B-type Lamin targets in Drosophila. J Cell Sci 120(14):2344–2351
111. Plackic J, Preissl S, Nikonova Y, Pluteanu F, Hein L, Kockskamper J (2016) Enhanced nucleoplasmic Ca(2+) signaling in ventricular myocytes from young hypertensive rats. J Mol Cell Cardiol 101:58–68
112. Plattner H, Verkhratsky A (2015) Evolution of calcium signalling. Cell Calcium 57:121–122
113. Popken J, Graf A, Krebs S, Blum H, Schmid VJ, Strauss A et al (2015) Remodeling of the nuclear envelope and lamina during bovine preimplantation development and its functional implications. PLoS One 10(5):e0124619
114. Popken J, Schmid VJ, Strauss A, Guengoer T, Wolf E, Zakhartchenko V (2016) Stage-dependent remodeling of the nuclear envelope and lamina during rabbit early embryonic development. J Reprod Dev 62(2):127–135
115. Prunuske AJ, Ullman KS (2006) The nuclear envelope: form and reformation. Curr Opin Cell Biol 18(1):108–116
116. Pytowski L, Drozdz MM, Jiang H, Hernandez Z, Kumar K, Knott E, Vaux DJ (2019) Nucleoplasmic reticulum formation in human endometrial cells is steroid hormone responsive and recruits nascent components. Int J Mol Sci 20(23)
117. Razafsky D, Hodzic D (2009) Bringing KASH under the SUN: the many faces of nucleo-cytoskeletal connections. J Cell Biol 186(4):461–472
118. Recupero D, Annaratone L, Maletta F, Bussolati G (2010) Nuclear shape in papillary thyroid carcinoma: a role for Lamin B receptor. Romanian J Morphol Embryol 51(4):615–620
119. Redwood AB, Perkins SM, Vanderwaal RP, Feng Z, Biehl KJ, Gonzalez-Suarez I et al (2011) A dual role for A-type lamins in DNA double-strand break repair. Cell Cycle 10(15):2549–2560
120. Rizzotto A, Tollis S, Pham NT, Zheng Y, Abad MA, Wildenhain J et al (2023) Reduction in nuclear size by DHRS7 in prostate cancer cells and by estradiol propionate in DHRS7-depleted cells. Cells 13(1):57

121. Robson MI, de Las Heras JI, Czapiewski R, Lê Thành P, Booth DG, Kelly DA et al (2016) Tissue-specific gene repositioning by muscle nuclear membrane proteins enhances repression of critical developmental genes during myogenesis. Mol Cell 62(6):834–847
122. Robson MI, de Las Heras JI, Czapiewski R, Sivakumar A, Kerr AR, Schirmer EC (2017) Constrained release of lamina-associated enhancers and genes from the nuclear envelope during T-cell activation facilitates their association in chromosome compartments. Genome Res 27(7):1126–1138
123. Rowat A, Lammerding J, Ipsen JH (2006) Mechanical properties of the cell nucleus and the effect of emerin deficiency. Biophys J 91(12):4649–4664
124. Rowat AC, Jaalouk DE, Zwerger M, Ung WL, Eydelnant IA, Olins DE et al (2013) Nuclear envelope composition determines the ability of neutrophil-type cells to passage through micron-scale constrictions. J Biol Chem 288(12):8610–8618
125. Rush C, Jiang Z, Tingey M, Feng F, Yang W (2023) Unveiling the complexity: assessing models describing the structure and function of the nuclear pore complex. Front Cell Develop Biol 11
126. Rynearson AL, Sussman CR (2011) Nuclear structure, organization, and oncogenesis. J Gastrointest Cancer 42:112–117
127. Saltel F, Giese A, Azzi L, Elatmani H, Costet P, Ezzoukhry Z et al (2017) Unr defines a novel class of nucleoplasmic reticulum involved in mRNA translation. J Cell Sci 130(10):1796–1808
128. Schirmer EC, Florens L, Guan T, Yates JR III, Gerace L (2003) Nuclear membrane proteins with potential disease links found by subtractive proteomics. Science 301(5638):1380–1382
129. Schirmer EC, Gerace L (2004) The stability of the nuclear lamina polymer changes with the composition of Lamin subtypes according to their individual binding strengths. J Biol Chem 279(41):42811–42817
130. Schirmer EC, Latonen L, Tollis S (2022) Nuclear size rectification: A potential new therapeutic approach to reduce metastasis in cancer. Front Cell Develop Biol 10:1022723
131. Schütz W, Alsheimer M, Öllinger R, Benavente R (2005) Nuclear envelope remodeling during mouse spermiogenesis: postmeiotic expression and redistribution of germline Lamin B3. Exp Cell Res 307(2):285–291
132. Shimi T, Kittisopikul M, Tran J, Goldman AE, Adam SA, Zheng Y et al (2015) Structural organization of nuclear lamins A, C, B1, and B2 revealed by superresolution microscopy. Mol Biol Cell 26(22):4075–4086
133. Shimojima M, Yuasa S, Motoda C, Yozu G, Nagai T, Ito S et al (2017) Emerin plays a crucial role in nuclear invagination and in the nuclear calcium transient. Sci Rep 7:44312
134. Somech R, Shaklai S, Geller O, Amariglio N, Simon AJ, Rechavi G, Gal-Yam EN (2005) The nuclear-envelope protein and transcriptional repressor LAP2β interacts with HDAC3 at the nuclear periphery, and induces histone H4 deacetylation. J Cell Sci 118(17):4017–4025
135. Sosa BA, Kutay U, Schwartz TU (2013) Structural insights into LINC complexes. Curr Opin Struct Biol 23(2):285–291
136. Starr DA, Fridolfsson HN (2010) Interactions between nuclei and the cytoskeleton are mediated by SUN-KASH nuclear-envelope bridges. Annu Rev Cell Dev Biol 26:421–444
137. Stiekema M, Houben F, Verheyen F, Borgers M, Menzel J, Meschkat M et al (2022) The role of Lamins in the Nucleoplasmic reticulum, a pleiomorphic organelle that enhances nucleo-cytoplasmic interplay. Front Cell Dev Biol 10:914286
138. Storch KN, Taatjes DJ, Bouffard NA, Locknar S, Bishop NM, Langevin HM (2007) Alpha smooth muscle actin distribution in cytoplasm and nuclear invaginations of connective tissue fibroblasts. Histochem Cell Biol 127(5):523–530
139. Subramanian K, Meyer T (1997) Calcium-induced restructuring of nuclear envelope and endoplasmic reticulum calcium stores. Cell 89(6):963–971
140. Swift J, Ivanovska IL, Buxboim A, Harada T, Dingal PDP, Pinter J et al (2013) Nuclear Lamin-A scales with tissue stiffness and enhances matrix-directed differentiation. Science 341(6149):1240104
141. Szekely J, Copps T, Morash B (1980) Radiation-induced invagination of the nuclear envelope. Radiat Res 83(3):621–632

142. Takama H, Sugiura K, Ogawa Y, Muro Y, Akiyama M (2013) Possible roles of barrier-to-autointegration factor 1 in regulation of keratinocyte differentiation and proliferation. J Dermatol Sci 71(2):100–106
143. Takeshima H, Nishimura S, Matsumoto T, Ishida H, Kangawa K, Minamino N et al (1989) Primary structure and expression from complementary DNA of skeletal muscle ryanodine receptor. Nature 339(6224):439–445
144. Tollis S, Rizzotto A, Pham NT, Koivukoski S, Sivakumar A, Shave S et al (2022) Chemical interrogation of nuclear size identifies compounds with cancer cell line-specific effects on migration and invasion. ACS Chem Biol 17(3):680–700
145. Varela I, Pereira S, Ugalde AP, Navarro CL, Suárez MF, Cau P et al (2008) Combined treatment with statins and aminobisphosphonates extends longevity in a mouse model of human premature aging. Nat Med 14(7):767–772
146. Verkhratsky A (2005) Physiology and pathophysiology of the calcium store in the endoplasmic reticulum of neurons. Physiol Rev 85(1):201–279
147. Wang G, Jacquet L, Karamariti E, Xu Q (2015) Origin and differentiation of vascular smooth muscle cells. J Physiol 593(14):3013–3030
148. Wilkie GS, Korfali N, Swanson SK, Malik P, Srsen V, Batrakou DG et al (2011) Several novel nuclear envelope transmembrane proteins identified in skeletal muscle have cytoskeletal associations. Mol Cell Proteomics 10(1):M110.003129
149. Williams RR, Azuara V, Perry P, Sauer S, Dvorkina M, Jørgensen H et al (2006) Neural induction promotes large-scale chromatin reorganisation of the Mash1 locus. J Cell Sci 119(1):132–140
150. Winer J (1984) Identification and structure of neurons in the medial geniculate body projecting to primary auditory cortex (AI) in the cat. Neuroscience 13(2):395–413
151. Wong X, Luperchio TR, Reddy KL (2014) NET gains and losses: the role of changing nuclear envelope proteomes in genome regulation. Curr Opin Cell Biol 28:105–120
152. Wood-Trageser MA, Lesniak D, Gambella A, Golnoski K, Feng S, Bucuvalas J et al (2023) Next-generation pathology detection of T cell-antigen-presenting cell immune synapses in human liver allografts. Hepatology 77(2):355–366
153. Worman HJ, Bonne G (2007) "Laminopathies": a wide spectrum of human diseases. Exp Cell Res 313(10):2121–2133
154. Xie W, Burke B (2017) Nuclear networking. Nucleus 8(4):323–330
155. Xie W, Chojnowski A, Boudier T, Lim JS, Ahmed S, Ser Z et al (2016) A-type lamins form distinct filamentous networks with differential nuclear pore complex associations. Curr Biol 26(19):2651–2658
156. Ye Q, Worman HJ (1996) Interaction between an integral protein of the nuclear envelope inner membrane and human chromodomain proteins homologous to drosophila HP1. J Biol Chem 271(25):14653–14656
157. Young JM, Williams DR, Thompson AAR (2019) Thin air, thick vessels: historical and current perspectives on hypoxic pulmonary hypertension. Front Med (Lausanne) 6:93
158. Zheng R, Ghirlando R, Lee MS, Mizuuchi K, Krause M, Craigie R (2000) Barrier-to-autointegration factor (BAF) bridges DNA in a discrete, higher-order nucleoprotein complex. Proc Natl Acad Sci 97(16):8997–9002
159. Zhou X, Groves NR, Meier I (2015) Plant nuclear shape is independently determined by the SUN-WIP-WIT2-myosin XI-i complex and CRWN1. Nucleus 6(2):144–153
160. Zink D, Fischer AH, Nickerson JA (2004) Nuclear structure in cancer cells. Nat Rev Cancer 4(9):677–687
161. Zuleger N, Boyle S, Kelly DA, de Las Heras JI, Lazou V, Korfali N et al (2013) Specific nuclear envelope transmembrane proteins can promote the location of chromosomes to and from the nuclear periphery. Genome Biol 14(2):1–20

# Chapter 4
# Genetics and cancer-related Laminopathies: Involvement of Lamins and Lamin-Chromatin Interactions

**Camille Desgrouas, Diane Frankel, Nathalie Bonello-Palot, Patrice Bourgeois, Annachiara De Sandre-Giovannoli, Patrice Roll, Elise Kaspi, and Catherine Badens**

**Abstract** Lamins are essential for maintaining the mechanical stability of the nucleus and organizing chromatin. B-type lamins are expressed early in embryogenesis, particularly in the central nervous system, where they are crucial for development. In contrast, A-type lamins are predominantly expressed in differentiated cells and are vital for maintaining nuclear stability and chromatin organization. These differences likely account for the distinct clinical characteristics of B-type and A-type laminopathies: B-type laminopathies primarily affect the central nervous system, whereas A-type laminopathies are mainly associated with metabolic dysfunction, cardiopathies, and premature aging.

In this chapter, we explore the broad and heterogeneous group of diseases caused by lamin alterations, detailing the genetic basis of laminopathies, their clinical manifestations, and the current state of research. We also discuss clinical management

C. Desgrouas
Aix Marseille Univ, INSERM, INRAE, C2VN, Marseille, France
e-mail: Camille.desgrouas@univ-amu.fr

D. Frankel · P. Roll · E. Kaspi
Aix Marseille Univ, APHM, INSERM, MMG, Hôpital la Timone, Service de Biologie Cellulaire, Marseille, France
e-mail: Diane.frankel@univ-amu.fr; Patrice.roll@univ-amu.fr; Elise.kaspi@univ-amu.fr

N. Bonello-Palot · P. Bourgeois · A. De Sandre-Giovannoli
Aix Marseille Univ, APHM, INSERM, MMG, Hôpital la Timone, M2GM, Marseille, France
e-mail: Nathalie.bonello@univ-amu.fr; Patrice.bourgeois@ap-hm.fr; Annachiara.desandre@ap-hm.fr

C. Badens (✉)
Aix Marseille Univ, INSERM, INRAE, C2VN, Marseille, France

APHM, Hôpital la Timone, Service de Biochimie, Marseille, France
e-mail: catherine.badens@univ-amu.fr

© The Author(s), under exclusive license to Springer Nature Switzerland AG 2025

K. Sengupta, A. Muchir (eds.), *Lamins in Genome Organization and Diseases*, Subcellular Biochemistry 115, https://doi.org/10.1007/978-3-032-00537-3_4

strategies and the role of chromatin organization in the pathophysiology of these diseases. Finally, we examine the variations in A- and B-type lamin expression in cancer.

**Keywords** Lamins · Laminopathies · Chromatin · Cancer · Myopathies · Cardiomyopathies · Lipodystrophies · Peripheral neuropathies · Progeroid syndromes

## Introduction

Laminopathies constitute a group of rare genetic disorders characterized by abnormalities in the nuclear envelope (NE). Specifically, these pathologies involve mutations in the genes encoding one of the three lamins—lamin A/C, lamin B1, and lamin B2—which are intermediate filament proteins providing structural support to the NE. Clinical manifestations of laminopathies are highly heterogeneous. Firstly, they vary in terms of severity, with some proving lethal in the perinatal period, while others only manifest in adulthood. Secondly, the clinical expression can be tissue-specific, affecting the muscles, heart, and adipose tissue or manifesting as multisystemic disorders. Over 15 pathologies have been classified as laminopathies, and those related to lamin A are the most prevalent, with initial descriptions dating back to 1999 [1, 2]. The complexity of laminopathies arises from the widespread expression of lamins in different tissues and their involvement in various cellular functions, such as maintaining nuclear shape and regulating gene expression. Understanding laminopathies necessitates a multidisciplinary approach, integrating insights from genetics, cell biology, and clinical medicine. Due to their rarity and diverse clinical presentations, the diagnosis and management of laminopathies pose significant challenges. Ongoing research efforts aim to uncover the underlying molecular mechanisms, develop diagnostic tools, and explore potential therapeutic interventions for these intricate and debilitating conditions.

In this chapter, we describe the genetic basis of laminopathies caused by mutations in A-type and B-type lamins, their clinical manifestations, and the current state of research. We also discuss clinical management strategies and the role of chromatin organization in the pathophysiology of these diseases. Finally, we address the variations in A- and B-type lamin expression involved in cancer.

## Laminopathies Caused by Mutations in A-Type Lamins

### *Diseases of Striated Muscle*

In muscle cells, maintaining cellular structure and organelle positioning is particularly crucial. Alterations in lamin A, especially those affecting nuclear positioning within the cell, can have detrimental consequences on myocytes, leading to disruptions in nuclear envelope and changes in gene expression. Pathogenic variants of lamin A are implicated in various muscle dystrophies, including Emery-Dreifuss muscular dystrophy (EDMD2, OMIM #181350), *LMNA*-related congenital muscular dystrophy (L-CMD, OMIM #613205), limb-girdle muscular dystrophy type 1B (LGMD1B, OMIM #150330), and dilated cardiomyopathy with conduction system defects (CMD1A, OMIM #115200). These lamin A-related diseases are among the most frequent and less severe forms of laminopathies. The three dystrophy syndromes are sometimes considered as clinical variants of the same pathology because within certain families, the three types are observed in different family members, due to a common genetic variant [3]. In lamin A-related myopathies, the heart is often affected, with earlier and more severe manifestations observed in EDMD and LGMD1B. Although less frequent, axonal neuropathies can also be observed. The presence of overlapping phenotypes, such as muscle dystrophy with cardiomyopathy or neuropathy, is characteristic of laminopathies. Systemic forms like progeria or acromandibular dysplasia also exhibit muscular and cardiac involvement [4, 5].

## Epidemiology

Dilated cardiomyopathy (DCM) is the second most common cause of heart failure, following coronary disease, constituting approximately one-third of all cases, with a prevalence estimated around 0.4% of the global population worldwide [6]. Among these cases, roughly one quarter to one-third are attributed to familial, most likely genetic origins and predominantly transmitted on a dominant mode [7]. *LMNA* mutations are identified in about 10% of genetic DCM cases, but this prevalence can reach 30% in patients with heart conduction disease. *LMNA*-related myopathies are less frequent with an estimated prevalence ranging between 0.25 and 1 in 100,000.

## Clinical Features

Clinically, *LMNA*-related myopathies commence with contractures, followed by muscle wasting and weakness. Onset occurs either in the neonatal period for the most severe form (L-CMD), in the second decade for EDMD2, or in the fourth decade for milder forms of LGMD1B, potentially resulting in the loss of independent walking ability. The initially affected muscle territories can vary at the disease onset [5].

DCM is characterized by a reduced ejection fraction of the left ventricle (LVEF <45%, normal value >50%), accompanied by a dilated left ventricle at the end of diastole (112–117% of the normal diameter [8]). Typically, patients with *LMNA* variants are relatively young (30–49 years old). A significant concern for these patients revolves around conduction defects, which can lead to severe arrhythmias (brady- or tachycardia) and premature sudden death [6]. The risk of sudden cardiac death is 46% higher in patients with *LMNA* variant compared with controls. Approximately a quarter of patients with *LMNA* variants present with conduction defects only, without other DCM symptoms [6]. While some features of *LMNA*-mutated patients may overlap with striated muscle defects, it has been observed that when neuromuscular defects are present first, cardiac symptoms tend to appear earlier than in patients with cardiac defects only [9].

## Molecular Genetics

Most muscle pathogenic variants of *LMNA* are transmitted in a dominant pattern. *LMNA* missense variants are primarily found in patients with EDMD2 and L-CMD and without cardiac defects, while frameshift variants are more frequent in patients with LGMD1B and cardiac involvement [10, 11]. In patients with EDMD2 and LGMD1B, variants tend to cluster in specific regions of the *LMNA* gene. For EDMD2, these variants primarily cluster in the immunoglobulin-like region spanning exons 7–10, crucial for lamin interactions with inner nuclear membrane proteins. Conversely, in LGMD1B patients, gene variants are predominantly found in the coil 2B region, located in exon 6 of the *LMNA* gene, also essential for lamin interactions with inner nuclear membrane proteins. Variants associated with L-CMD are distributed across the N-terminal and the first part of the rod domains, encompassing exon 1 as well as exons 4 and 5 of the *LMNA* gene. These regions are primarily involved in dimer formation, highlighting their importance in the pathogenesis of L-CMD [5].

There is no mutational hotspot in *LMNA*-mutated DCM patients, with almost 200 variants distributed along the entire gene [12]. However, the majority of the variants occur in the head and in the rod domains, rarely in the tail domain [2, 13]. Pathogenic variants are mainly missense and nonsense, with fewer small deletions/insertions identified [14, 15].

As with other *LMNA*-linked pathologies, the suggested mechanism for missense variants is a dominant negative effect, although no clear common pathophysiological mechanism has been established. Subjects with truncating mutations have been associated with more severe symptoms, including an earlier onset of cardiac conduction defects and atrial arrhythmias, as well as a lower ejection fraction [16]. The precise mechanism by which these variants lead to cardiac defects remains unclear.

## Pathophysiological Mechanisms

Numerous studies utilizing cellular or animal models have provided insights into the mechanisms underlying laminopathies. As anticipated, chromatin disorganization is evident, resulting in abnormality in gene expression pattern [17]. Notably, anomalies in desmin expression, another muscle-specific intermediate filament with pathogenic variants causing muscular dystrophies, have been demonstrated [18].

Similar to other laminopathies, anomalies in the connection between the nuclear envelope and the LINC complex are observed in the muscle, potentially impacting nuclear positioning and envelope integrity by modifying connections with the microtubule network. Research indicates that disorganizing the microtubule network or inhibiting nesprin binding to the nucleoskeleton corrects nuclear envelope anomalies in laminopathies [19]. Clinical trials are underway to overexpress the part of nesprin binding to the nuclear envelope intermediate filaments, known as the KASH domain, aiming to reduce nuclear envelope-microtubule binding through the LINC complex and restore envelope integrity. Another implicated signaling pathway is the PI3-K and MAPK pathway, offering a promising therapeutic avenue [20].

While several mouse models with different *LMNA* variants exist, a definitive model elucidating the pathophysiological mechanism has not been established despite extensive efforts. Notably, LmnaH222P and LmnaN195K transgenic mice represent the most relevant models for studying sarcomere formation and closely mimicking the cardiac phenotypes of *LMNA*-mutated DCM patients [21, 22]. Promising results have emerged from a Cre/Lox model of cardiomyocyte-specific *LMNA* depletion by Stewart and colleagues. They demonstrated that disrupting interactions between nuclear lamina and cytoskeleton by inactivating the LINC complex improved cardiovascular symptoms and lifespan in treated mutant mice [23, 24]. However, this model, while inducing cardiac muscle defects that mimic DCM, may not be directly applicable to human pathology, where DCM signs are typically not caused by a loss-of-function effect of *LMNA* mutations. A recent review by Muchir and colleagues reports approaches exploring the link between both nucleo- and cytoskeleton via the proteins of the LINC complex [18].

Since the advent of iPSC, significant efforts have been invested in producing valuable models, either by dedifferentiating/redifferentiating patients' fibroblasts into cardiac cells [25, 26] or by inducing targeted mutations with CRISPR/Cas9 technology [27, 28]. These cellular models facilitate the relatively easy screening of therapeutic molecules to treat cardiolaminopathies.

Understanding the pathophysiological mechanisms underlying genotype to phenotype correlations is an ongoing challenge. Currently, treatments for cardiolaminopathies are primarily symptomatic, with implantable devices as the first-line option and cardiac transplantation as the last therapeutic resort for patients. *LMNA* is one of the few "actionable" genes in cardiomyopathies. Therefore, for patients with an identified variant, genetic counseling and *LMNA* sequencing of relatives are priorities and can lead to the implantation of a prophylactic cardioverter-defibrillator device or early heart transplantation [29, 30]. Besides this symptomatic treatment, numerous specific approaches are being developed and are based either on gene therapy targeting lamin A/C or lamin-associated proteins, on cell therapy, or on drug therapy (for complete review: Benarroch et al. [31]).

## *Lipodystrophy Syndromes*

### Type 2 Familial Partial Lipodystrophy

Type 2 familial partial lipodystrophy (FPLD2, OMIM #151660), also called Dunnigan syndrome, is a tissue-specific laminopathy affecting the adipose tissue.

#### Epidemiology

FPLD2 is considered as a rare disorder, but it is the most frequent form of FPLD, affecting around 1 in 100,000 people in Europe [32]. The first description was reported by Dunnigan in 1974 [33]. Gonzaga et al. conducted genetic analyses on individuals with a clinical diagnosis of lipodystrophy and available genomic data to identify variants associated with inherited lipodystrophies. They also examined their electronic health records (EHR) for comorbidities linked to lipodystrophy. The study identified 16 individuals carrying the canonical variant p.(Arg482Gln) in *LMNA,* found in 80% of all Dunnigan patients. Among them, only four had a clinical diagnosis of lipodystrophy, while the remaining individuals exhibited metabolic abnormalities but were not formally recognized as lipodystrophy patients. In this cohort of patients with clinical lipodystrophy, the carrier frequency of lipodystrophy-associated variants was estimated to be 1 in 3082 individuals, indicating a substantial burden of metabolic dysregulation. They further estimated a genetic prevalence of the disease at approximately 1 in 7588 in a non-clinical care cohort (ExAc) [34].

#### Clinical and Biological Features

FPLD2 is characterized by the loss of adipose tissue and abnormal fat distribution, typically initiating around puberty, corresponding to the onset of adipocyte degeneration. The reduction in adipose tissue primarily affects the limbs and trunk, while its accumulation is evident in the face, neck, axillae, labia major, and abdominal viscera. Additionally, apparent muscular hypertrophy is often observed [35]. However, phenotypic heterogeneity is noted among carriers of *LMNA* pathogenic variants especially in men for whom the syndrome is probably underdiagnosed. The dysfunction in adipose tissue contributes to the development of metabolic complications associated with insulin resistance, including type 2 diabetes, dyslipidemia, fatty liver disease, cardiovascular disease, and reproductive disorders. The diagnosis of lipodystrophies relies on molecular biology [36].

## Molecular Genetics

Generally, variants in FPLD2 are heterozygous missense variants that impact the globular C-terminal domain of the lamin A/C protein and are transmitted in an autosomal dominant manner. FPLD2 arises from heterozygous or compound heterozygous variants, predominantly affecting exons 8 and 11 of the *LMNA* gene (located at 1q21–22, NC_000001.1) [36], and the recurrent p.(Arg482Trp/Gln) substitutions are responsible for 80% of the cases (Fig. 4.1a).

Typical forms of FPLD2 primarily result from heterozygous missense variants at codon 482 in exon 8, which serves as a mutational hotspot. At this location, arginine is conserved across species but differs from other lamins. In FPLD2, this arginine residue is replaced by a neutral amino acid p.(Arg482Trp/Gln/Leu), with the most frequent variant being p.(Arg482Trp). Unlike *LMNA* variants associated with muscular dystrophies or cardiomyopathies that disrupt the three-dimensional structure of A-type lamins, variants responsible for lipodystrophy alter a positively charged amino acid at the surface of their C-terminal domain [37].

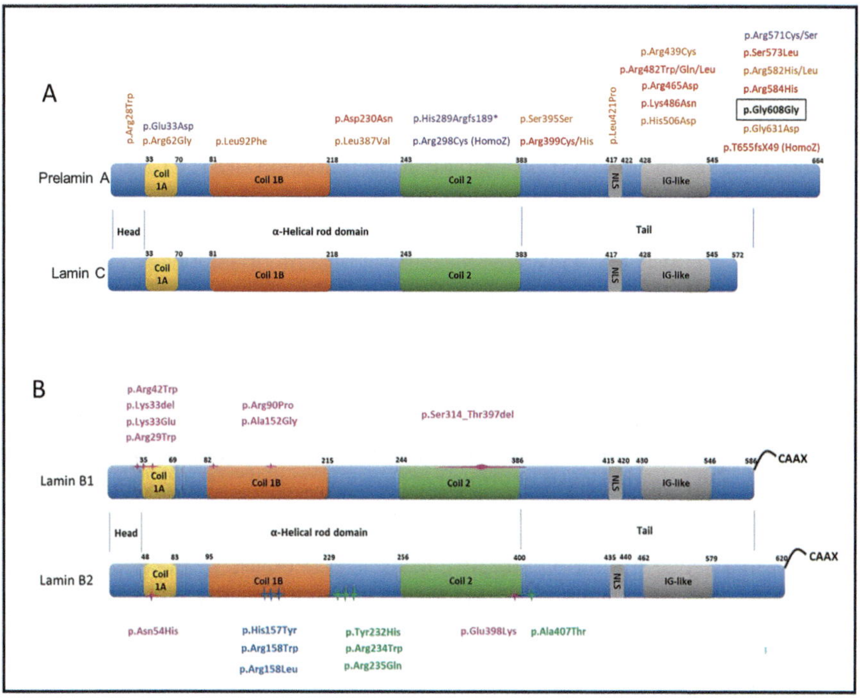


**Fig. 4.1** Laminopathies causing variants (**a**) in lamins A/C—mutations associated with FPLD2 (red), atypical lipodystrophies (orange), CMT (purple), and typical progeria (black box) (for mutations associated with striated muscle and progeroid syndromes, see Storey and Fuller [11] for review)—and (**b**) in lamins B1 and B2: mutations associated with PME with ataxia (blue), partial lipodystrophies (green), and primary microcephaly (pink)

In 2000, Cao and Hegele found the missense variant p.(Arg482Gln) for the first time in 22 Canadian FPLD patients from 5 different families, suggesting a possible founder effect. Carriers of this variant showed a significantly higher incidence of diabetes [38]. That same year, Speckman et al. conducted a study involving 15 families with FPLD and identified various *LMNA* missense variants: p.(Arg482Gln) in 5 families, p.(Arg482Trp) in 7 families, and p.(Arg465Asp) in 1 family [39]. The variant p.(Arg482Trp) was also found in 13 families and 3 isolated cases of partial lipodystrophy [40–42]. The third variant found in position 482 p.(Arg482Leu) was recorded in a family by Shackleton et al. [41]. Still in exon 8, Shackleton identified another variant p.(Lys486Asn), in one family. Several variants located in other exons have been identified in patients with typical forms of FPLD2, including c.688G > A p.(Asp230Asn) in exon 4, c.1195C > T p.(Arg399Cys) in exon 7, and c.1718C > T p.(Ser573Leu) [43] and p.(Arg584His) in exon 11 [42].

All these variants resulting in FPLD2 and metabolic laminopathies are associated with elevated expression of farnesylated prelamin A, increased cellular oxidative stress, and premature senescence [44].

## Pathophysiology

*LMNA*-associated lipodystrophies, characterized by a combination of lipoatrophy and metabolic alterations linked to insulin resistance, may result from defects in adipocyte differentiation and/or fat structure. Notably, a reduction in the expression of genes involved in adipocyte proliferation/differentiation (*RB1*, *CCND1*, *CCND3*, *PPARG2*, *LPL*, *GLUT4*) has been observed in adipose tissue from patients with FPLD2 [45]. The pathogenetic mechanism leading to the loss of subcutaneous white adipose tissue (WAT) involves the early activation of autophagy in adipocyte precursors, followed by impaired autophagic flux. This is associated with altered regulation of PPARG1 and UCP1 (a brown adipose tissue marker). Conversely, the mechanism leading to fat accumulation in the neck area is related to early autophagic activation, PPARG1 downregulation, and aberrant differentiation from brown adipose tissue (BAT) toward the white lineage [46].

In other instances, the accumulation of farnesylated prelamin A at the nuclear envelope is frequently observed in FPLD2 patients' cells, and it has been demonstrated to induce several dysfunctions:

1. Abnormal sequestration of SREBP1 leads to a decrease in the pool of active SREBP1, which normally activates PPARG [47]. This results in the downregulation of Notch signaling pathway, premature senescence of mesenchymal stem cell, and induction of miR-335 expression, both repressing adipocyte differentiation.
2. Sequestration of Sp1 transcription factor alters extracellular matrix (ECM) gene expression and activates TGF-beta signaling that increases MMP9 and endopeptidase activity leading to the degradation of ECM proteins. These phenomena decrease fat storage capacity, inducing an ectopic fat distribution linked to lipotoxicity and metabolic defects.

3. Endothelial dysfunction.
4. Increase of oxidative stress.
5. Cellular senescence. The senescence of vascular smooth muscle cells contributes to atherosclerosis, while osteoblasts' transdifferentiation and calcification promote vascular calcifications. In affected families, metabolic alterations manifest at an earlier age across generations, probably resulting from epigenetic deregulation induced by *LMNA* variants [48, 49].

The variant p.(Arg482Leu) downregulates Notch signaling pathway in mesenchymal stem cells, consequently reducing their adipogenic potential [50], whereas the two variants p.(Gly465Asp) and p.(Lys486Asn) alter the lamin A C-terminal tail SUMOylation, a posttranslational modification known to regulate the localization, interactions, and functions of proteins [51].

The mechanical properties of fibroblasts from patients carrying the p.(Arg482Trp) variant were investigated using microfluidic experiments. These experiments revealed that the mutated fibroblasts exhibited enhanced viscosity that relies to an abnormal link with the microtubule network. Measurement of this parameter could potentially serve as a valuable diagnostic marker for lamin A-related lipodystrophies [52].

The metabolic alterations seen in patients with lipodystrophies result from a reduction in adipose tissue, leading to the decrease of several hormones including leptin. Leptin, a pro-inflammatory adipokine secreted by adipocytes, regulates mediators of appetite and energy consumption. A decrease of this adipokine is responsible of hyperinsulinemia, insulin resistance with possible progression to diabetes, hypertriglyceridemia, and hepatic steatosis [53].

### Therapeutic Approaches

Lipodystrophy syndromes are clinically heterogeneous. It is recommended that patients with lipodystrophy syndrome undergo annual screening for diabetes, dyslipidemia, and conditions affecting the liver, kidney, and heart to manage these comorbidities effectively.

Adopting a healthy lifestyle, including low-calorie diet and regular physical activity, is crucial to reduce insulin resistance and metabolic complications. Notably, a very low-calorie diet followed by a 38-year-old woman allowed an improvement in her metabolic and reproductive dysfunctions [54].

Metreleptin, a recombinant human methionyl leptin, is the only drug approved specifically for lipodystrophy. In the United States, the Food and Drug Administration (FDA) approved metreleptin only for generalized lipodystrophies in 2014. In the European Union, metreleptin was approved in July 2018 by the European Medicines Agency (EMA) as an adjunct to diet as replacement therapy to treat the complications of leptin deficiency in adults and children above the age of 12 years with partial lipodystrophy, when standard treatments have failed. This therapy is effective for metabolic complications in hypoleptinemic patients (leptin <4 ng/mL) with partial lipodystrophy and severe metabolic derangements (HbA1c >8% and/or triglycerides >500 mg/dL). A recent systematic review by Semple et al. showed that the administration of metreleptin in 71 patients with partial lipodystrophy was associated with a decrease in HbA1c, triglyceridemia, and BMI [55]. Currently, there is no treatment available that can reconstitute adipose tissue, but plastic surgery can be proposed when psychological distress and/or physical discomfort is observed [56].

## Atypical Forms of Lipodystrophies

Three additional missense variants in exon 11 of the *LMNA* gene have been documented in atypical forms of lipodystrophies characterized by attenuated lipoatrophy and less precocious type 2 diabetes (Fig. 4.1a). One such variant, c.1745G > A p.(Arg582His), was identified in a heterozygous condition in two sisters with attenuated forms of lipodystrophy [39, 57]. The observation that this variant affects only lamin A suggests a possible explanation for the less severe nature of this form of lipodystrophy. The specific impact on lamin A, sparing a normal lamin C, may contribute to the milder clinical presentation associated with this variant [39]. In another context, the c.1745G > A p.(Arg582His) variant has been identified in a homozygous condition in a patient exhibiting a more severe phenotype characterized by near-total loss of subcutaneous adipose tissue. Interestingly, certain areas such as the retro-orbital region, pal ms and soles, mons pubis, and external genital region retained well-preserved fat. Notably, this patient with the homozygous variant displayed lower leptin levels and experienced an earlier onset of metabolic abnormalities compared to individuals carrying the variant in the heterozygous state and typical cases of FPLD2. This suggests that the homozygous state of this variant may be associated with a more severe and distinct clinical presentation [58].

The *LMNA* variants c.1745G > T p.(Arg582Leu) and c.1892G > A p.(Gly631Asp) have been identified in two patients who suffered from an early onset of diabetes mellitus, hypertriglyceridemia, and non-alcoholic fatty liver disease. In both cases, subcutaneous adipose tissue persisted, particularly on the abdomen, and the serum leptin levels remained within the normal range. These findings suggest a distinctive clinical manifestation associated with these specific *LMNA* variants, characterized by metabolic abnormalities despite the preservation of subcutaneous adipose tissue and normal leptin levels [59]. Functional studies on the fibroblasts of these patients have revealed notable findings. These studies demonstrated a high percentage of nuclear anomalies, an accelerated senescence rate, and a decrease in replication capacity in these fibroblasts. Furthermore, overexpression of mutated prelamin A in control cells induced misshapen nuclei, whereas the partial reduction of lamin A expression in patient cells significantly improved nuclear morphology. These results collectively suggest a link between the expression of the two lamin A variants and phenotypes associated with cellular senescence [60].

Collectively, the three *LMNA* variants, including c.1745G > T p.(Arg582Leu), c.1892G > A p.(Gly631Asp), and c.1745G > A p.(Arg582His), may potentially disrupt interactions between lamin A and chromatin. This hypothesis is supported by the observed nuclear anomalies, accelerated senescence rate, and altered replication capacity in fibroblasts carrying these variants. Notably, it has been demonstrated that C-terminal sequence of lamin A is crucial for the observed lamin-chromatin interaction. Thus, these variants, affecting the C-terminal domain of lamin A, could interfere with these essential interactions, contributing to the cellular and clinical phenotypes observed in patients with these *LMNA* variants [61].

Garg et al. reported two additional missense variants in the first exon of the *LMNA* gene: p.(Arg28Trp) located in the amino-terminal head domain and p.(Arg62Gly) in the alpha-helical rod domain. These variants were identified in two families with FPLD2. Interestingly, affected individuals also presented with cardiac conduction system defects and other manifestations of cardiomyopathy. This suggests the possibility of a multisystemic dystrophy syndrome associated with *LMNA* variants, involving not only adipose tissue but also impacting cardiac and skeletal muscle.

In another study, Decaudain et al. reported a series of nine *LMNA* variants associated with severe metabolic syndrome. These variants include p.(Arg28Trp) and p.(Leu92Phe) in exon 1; p.(Leu387Val), p.(Ser395Ser), p.(Arg399His), p.(Leu421Pro), and p.(Arg439Cys) in exon 7; p.(His506Asp) in exon 9; and the frameshift variant p.(Thr655fsX49) in exon 11. Some patients carrying these variants did not exhibit lipoatrophy. However, the severity of insulin resistance, hypertriglyceridemia, and alterations of cell nuclei were comparable to those observed in FPLD2 patients. Additionally, most patients presented with calf hypertrophy, myalgia, muscle cramps or weakness, and cardiac conduction disturbances. This suggests a broader spectrum of phenotypic manifestations associated with *LMNA* variants, encompassing metabolic dysfunctions resembling the metabolic syndrome, and muscular and cardiac abnormalities [62]. Le Dour et al. described the pathogenic homozygous frameshift variant p.(Thr655fsX49), leading to the synthesis of a prelamin A missing the consensus farnesylation site (CaaX). This variant induces the expression of a non-farnesylated form of prelamin A, without production of mature lamin A, and is responsible for a severe lipodystrophic syndrome [63].

Finally, a severe metabolic syndrome was also reported for a patient carrying a heterozygous missense variant in the *ZMPSTE24* gene, coding for the only metalloprotease that transforms prelamin into mature lamin A. This patient presented a severe metabolic syndrome complicated by cardiomyopathy and no limb lipoatrophy nor limb muscular hypertrophy. He exhibited android obesity, along with fat accumulation in the neck and face, as evidenced by supraclavicular filling and a buffalo neck. This case highlights the diverse and complex phenotypic presentations associated with genetic variants affecting proteins involved in lamin A processing [59, 64].

## *Peripheral Neuropathies*

Charcot-Marie-Tooth (CMT) disease is a heterogeneous group of inherited peripheral neuropathies affecting both sensory and motor nerves, with a prevalence of about 1 in 2500 individuals at birth [65]. Typically, the disease presents in young adults, marked by gait disturbance, wasting, and distal muscular atrophy in the lower limbs. Foot deformities, such as "pes cavus," are common, while children may exhibit flat feet. Neuropathic pain and difficulties to walk are prevalent. Upper limbs are not always affected, but when they are, there is hand atrophy with possible tendon retraction, which can affect writing and gripping. Sensory deficit is mostly present as hypopallesthesia "in gloves" and "in socks." In the most severe form, patients may become wheelchair-bound. CMT is classified into two types—CMT type 1 (demyelinating) and CMT type 2 (axonal)—distinguished by measuring the nerve conduction velocity (NCV) of the median nerve. When the velocity is less than 38 m/s, the CMT is demyelinating, and when the velocity is greater than 42 m/s, the CMT is axonal [66].

Genetically, CMT exhibits high heterogeneity, with over 100 different genes potentially causing the disease [67]. Notable forms include CMT1A due to *PMP22* duplication [68] and the rare CMT2B1 caused by a mutation in the *LMNA* gene (Fig. 4.1a) [69]. The *LMNA* variant p.(Arg298Cys) was first described in 2002 by De Sandre-Giovannoli et al. in Algerian families with autosomal recessive inheritance [70]. Subsequent studies by Tazir et al. expanded the clinical characterization of patients carrying this mutation, revealing marked proximal muscle involvement and variable onset ages, suggesting the influence of modifying genes [71].

In 2004, a case with an autosomal dominant *LMNA* mutation c.99G > T p.(Glu33Asp) presented a CMT "plus" syndrome with features like axonal neuropathy, muscular dystrophy, cardiac disease, and leukonychia. This highlighted the variability within the same genetic abnormality, with symptoms starting in adolescence [72].

In 2005, Benedetti et al. reported two families affected by a combined feature of axonal neuropathy and muscular dystrophy due to one mutation in the *LMNA* gene, so by an autosomal dominant inheritance. The proband of the family 1 originated from Italy was severely affected with an early onset started with frequent falls, weakness, and walk on his heels and tiptoes. At 6 years old, he underwent feet surgery. At 13 years old, he was wheelchair-bound. He had severe muscular atrophy in the distal and moderate in the proximal lower limbs. Muscle biopsy showed myopathic and neuropathic features with atrophic fibers, few necrotic fibers, and increased connective tissue. Tendon reflexes were abolished in all four limbs. He also had fat accumulation in the face and a "buffalo hump" neck. A clinical diagnosis of neuropathy and facio-scapuloperoneal dystrophy was given. At 45 years old, his mother had no complain of motor or sensory deficits. She had a normal neurological examination but some moderate anomalies in the electroneuromyogram (ENMG). She had fat accumulation in the neck and face. His father had no sign of peripheral neuropathy. Molecular analysis identified a missense variant in the *LMNA* gene—c.1711C > T p.(Arg571Cys)—at the heterozygous state. This variant was inherited from his mother who was moderately affected by the disease. These cases illustrated the high variability of the CMT phenotype within the same genetic abnormality in the same family. A variant in the same codon p.(Ar571Ser) has been previously reported in a patient presenting atrial fibrillation and dilated cardiomyopathy without neurological signs.

The proband of the family 2, originated from Filipino, displayed a cardiac defect at 25 years old, which needed a pacemaker implantation because of severe atrioventricular block. At 33 years old, she presented walking difficulties and weakness. Neurologic examination showed proximal weakness of the four limbs, muscle atrophy, hypotonia, and areflexia. A muscle biopsy showed myopathic and neurogenic features, including increased connective tissue and atrophic rounded fibers. At 35 years old, her brother had no complain of neurological deficits. His examination revealed moderate signs of neuropathy as amyotrophy and mild proximal weakness in the lower limbs and hyporeflexia. Their father had walking difficulties and frequent falls since adolescence and died asleep at 40 years old. Molecular analysis identified in the proband and her brother a frameshift deletion in exon 5 (c.864_867del) at heterozygous state, leading to a premature stop codon p.(His289Argfs189*). Clinical diagnosis of limb-girdle myopathy and peripheral neuropathy was given even if no clear neuropathy signs were present. This study described a spectrum of *LMNA*-related disorders and the difficulty in establishing any genotype-phenotype correlation even in the same family with the same mutation [73].

The same year, Walter et al. identified a novel mutation in the 5'UTR of the *LMNA* gene in one family, responsible of a clinical phenotype with features from Emery-Dreifuss muscular dystrophy and Charcot-Marie-Tooth type 2 disease. Molecular analyses were performed in the proband and his brother affected and sharing the same mutation in the *LMNA* gene, c.-3 + 12del, resulting in the loss of 15 nucleotides including the translation ATG initiator codon. The mutation segregated in the patients' family (proband and his brother) with the disease. The proband is a man, and his son, his brother, his father, his uncle, and his aunt are affected by the disease. They presented weakness in proximal and distal muscles. They also had contractures, mild creatine kinase elevation, and marked cardiomyopathy. Muscle biopsy showed a neurogenic atrophy. ENMG showed a neurogenic pattern, but NCV for motor and sensory were in the normal range as described in CMT2 [74].

CMT2B1, initially thought to be restricted to Algerian families, was later identified in Moroccan families with the same variant p.(Arg298Cys) [75]. A study dated the mutation, revealing a common ancestral haplotype and estimating the most recent common ancestor to have lived about 800–900 years ago [76].

Two types of peripheral neuropathies due to *LMNA* mutations are described, the phenotype of CMT2B1 due to the variant p.(Arg298Cys) at homozygous state and CMT "plus" phenotype with some features of muscular dystrophy as described in several families with an autosomal dominant inheritance.

The variability of this laminopathy phenotype is high and seems to be higher in families with only one mutation in the *LMNA* gene, which supposed another factor for understanding the physiopathology of these disorders.

Some studies showed no variant in the *LMNA* gene in patients affected by CMT2 originated from Czech and Beijing [77, 78]. While CMT2B1 is clearly linked to the homozygous p.(Arg298Cys) variant in patients from Algeria and Morocco, autosomal dominant *LMNA* variants appear to contribute to a complex phenotype that may overlap with other laminopathies like muscular dystrophy (LGMB1 or EDMD) or lipodystrophy because the neuropathy seems to be on background.

## *Progeroid Syndromes*

### Hutchinson-Gilford Progeria Syndrome (HGPS)

Hutchinson-Gilford progeria syndrome (HGPS; OMIM #176670), an extremely rare genetic disease (affecting approximately one in four to eight million newborns), was first described in 1886 by Dr. Hutchinson and later by Dr. Gilford in 1897 [79, 80]. The term "progeria" originates from Greek, meaning "prematurely old." This disorder globally affects children, regardless of gender. In 2023, 147 children were diagnosed with HGPS, and the Progeria Research Foundation estimates around 400 children live with HGPS worldwide.

#### Phenotype

While newborns with HGPS typically appear healthy, symptoms become apparent within the first year of life, including postnatal growth retardation. These children gradually develop a characteristic appearance regardless of their country of origin, characterized by hair loss, prominent scalp veins, skin abnormalities (such as dyspigmentation), a pinched nose, micrognathia (potentially causing dental crowding), and midface hypoplasia. Additionally, they also experience loss of subcutaneous fat and lipodystrophy, amyotrophy, joint contractures, and osteolysis of distal phalanges [81]. One-third of patients display extraskeletal calcifications [82]. The average maximum height and weight for these children are 125 cm and 15 kg, respectively.

Over time, these children develop conditions commonly associated with the elderly, exhibiting clinical symptoms reminiscent of aging but occurring prematurely and at an accelerated rate. However, HGPS is considered as a segmental aging disease in that premature aging processes do not exhibit all aspects associated with physiological aging. Patients develop osteoporosis, atherosclerosis, and cardiovascular disease. The most critical issue in HGPS is cardiovascular complications, including loss of vascular smooth muscle cells, increased pulse-wave velocity, and valvular calcification [83, 84]. They may also develop left vascular dysfunction (as left ventricular hypertrophy associated with left ventricular diastolic dysfunction), fibrosis, and generalized atherosclerosis [85–87]. The average lifespan for individuals with HGPS ranges from 8 to 21 years, with most succumbing to myocardial infarction, heart failure, or stroke; over 80% of deaths in HGPS are caused by cardiovascular failure. Notably, cognitive development and immune function appear to be unaffected [81, 88].

### Pathophysiology

HGPS is an autosomal dominant disorder caused by, in the classical form representing 85–90% of cases, a de novo point mutation in exon 11 of the *LMNA* gene (c.1824C > T, p.(Gly608Gly)) (Fig. 4.1a). This mutation, identified in 2003 [89, 90], activates a cryptic donor splice site, resulting in the deletion of 150 nucleotides in the messenger RNA (mRNA) of prelamin A and a corresponding loss of 50 amino acids at the protein level. The mutation disrupts the normal posttranslational processing of prelamin A, by removing the cleavage site of the metalloprotease ZMPSTE24 contained in the deleted region. This abnormal prelamin A, known as progerin [91], retains its farnesylated group at the C-terminus and remains anchored to the inner nuclear membrane rather than soluble in the nuclear compartment. Progerin accumulation disrupts nuclear functions and causes various structural and molecular cellular defects. Progerin exerts a dominant negative effect and progressively accumulates into the nucleus of cells leading to an increase of size and a distortion of shape with the formation of protrusions and invaginations. The nuclear *lamina* is thickened [92], and nuclear pore complexes become disorganized and form clusters [93]. Additionally, the linker of nucleoskeleton and cytoskeleton (LINC) complex, which plays a crucial role in nuclear positioning and interactions with the cytoskeleton, is also altered [94]. These structural changes are accompanied by disruptions in chromatin organization, loss of peripheral heterochromatin, epigenetic changes, and telomere shortening [95–97].

Additionally, progerin interferes with DNA damage response and repair processes by delaying the recruitment of key DNA repair proteins, 53BP1 and Rad51, to sites of DNA damage. This disruption leads to genomic instability and mitotic defects [98–100]. Progerin also inhibits cell proliferation and promotes senescence [93, 101, 102]. HGPS fibroblasts are also more sensitive to mechanical strain [103, 104].

Progerin also contributes to cellular dysregulation by altering the availability and activity of transcription factors and other proteins, such as nuclear factor erythroid 2-related factor 2 (NRF2). NRF2 plays a key role in the regulation of oxidative stress by activating antioxidant enzymes. In HGPS fibroblasts, progerin sequesters NRF2 causing its subnuclear mislocalization and impairing its transcriptional activity. This disruption contributes to chronic oxidative stress. Mitochondrial dysfunctions are also a hallmark of HGPS pathophysiology. HGPS fibroblasts exhibit reduced ATP production, increased levels of reactive oxygen species (ROS), and increased susceptibility to oxidative stress [105, 106].

MicroRNAs (miRNA or miR) play a role in HGPS. A brain-specific miRNA, miR-9, downregulates lamin A and progerin expression in neuronal cells. This selective suppression of progerin in neurons likely contributes to the preservation of cognitive function in HGPS patients [107, 108]. Other overexpressed miRNAs as miR376a-3p, miR-376b-3p, miR-59, and miR-34a are described to contribute to premature senescence [102, 109, 110].

Vascular smooth muscle cells (VSMCs) are one of the principal cellular components of the medial layer of arteries. They regulate contraction and relaxation of blood vessels. Progerin expression in VSMCs of HGPS disrupts cellular processes, resulting in cell death and a severe loss of VSMC that promotes arterial stiffness [111–114]. Progerin expression has also deleterious effects on endothelial cells including shortened telomere, DNA damage, and oxidative stress, triggering premature senescence [115, 116]. Loss of VSMC and alterations of endothelial cells contribute to the development of early atherosclerosis in this disease [117].

All these cellular defects collectively contribute to premature aging and associated pathologies.

## Therapeutic Approaches

Several therapeutic approaches have been developed for patients with HGPS, but currently, lonafarnib, a farnesyltransferase inhibitor, is the only drug approved by Food and Drug Administration (FDA) and European Medicines Agency (EMA). In vitro and in vivo studies have demonstrated the beneficial effects of lonafarnib. This molecule interferes with the posttranslational modification of progerin, reduces nuclear blebbing, enhances bone mineralization and weight, increases longevity, and decreases cardiovascular abnormalities. Lonafarnib extends the HGPS children life expectancy by an average of 1.6 years [118].

Other drugs have been tested, with evidence suggesting that reducing progerin improves nuclear defects [101]. One potential strategy is to activate autophagy. Molecules such as rapamycin, everolimus, temsirolimus, sulforaphane, and MG132 have been shown to reduce progerin-induced alterations in vitro and/or in vivo, including abnormal nuclear shape, cellular senescence, and chromatin abnormalities [119–123]. A phase I/II monocentric trial combining lonafarnib and everolimus (NCT02579044) is ongoing in Boston Children's Hospital.

Additionally, drugs that decrease oxidative stress, such as N-acetyl cysteine, rho-associated protein kinase (ROCK) inhibitor, vitamin D, resveratrol, and MG132, have demonstrated promise in reducing oxidative stress and cellular defects [105, 124]. Vitamin D has been shown to alleviate HGPS symptoms, including premature senescence, nuclear abnormalities, and DNA repair defects [125]. Resveratrol, a SIRT1 activator, restores adult stem cell decline in a progeroid mouse model, by enhancing SIRT1 interaction with lamin A [126].

Morpholino antisense oligonucleotides (AONs), targeting aberrant splicing site in exon 11 and exon 10 (promoting the translation of lamin C rather than lamin A), have been tested in vitro and in vivo by Osorio et al. These AONs effectively reduced progerin levels and improved the progeroid phenotype in *Lmna*$^{G609G/G609G}$ mice [127].

Recent genetic approaches, including in vivo CRISPR/Cas9 gene editing [128, 129] or adenine base editing (ABE) [130, 131], have been shown to reduce progerin accumulation and provide a proof of concept for further therapeutic development.

While lonafarnib has become a treatment option for HGPS children, its ability to fully alleviate the detrimental effects of progerin remains restricted, and HGPS is still a dramatic and deadly disease. However, the emergence of innovative genetic therapies based on AON, CRISPR/Cas9, and ABE techniques gives hope and offers promising avenues for a more effective treatment that directly targets the pathogenic mutation.

### Related Progeroid Syndromes

Closely related syndromes to the classical form of HGPS have been described.

Among them, "progeria-like" or "HGPS-like" syndromes arise from distinct *LMNA* gene mutations leading to the production of progerin or other truncated and farnesylated forms of prelamin A [132]. These conditions also manifest with accelerated and premature aging, but with a less severe phenotype than classical progeria, depending on the type of mutation they carry. Other non-classic HGPS progerin-producing mutations include exon 11 c.1822G > A p.(Gly608Ser) [90], c.1868C > G p.(Thr623Ser), c.1821G > A p.(Val607Val) [133], and heterozygous mutations at the intronic border of exon 11 c.1968 + 5G > C, c.1968 + 2 T > A [134], c.1968 + 2 T > C, c.1968 + 1G > A [133], and c.1968 + 1G > C [135]. The severity of the phenotype is directly proportional to the amount of progerin/farnesylated prelamin A produced.

Compound heterozygous missense mutations in *LMNA* gene p.(Arg527His) and p.(Val440Met) have also been described. The patient presented clinical features commonly associated with mandibuloacral dysplasia type A (MADA) but lacked clavicular dysplasia and exhibited normal metabolic profiles. The patient's cells displayed an accumulation of prelamin A associated with irregular *lamina* thickness [136].

On the other hand, atypical progeroid syndromes (APS) exhibit premature aging symptoms caused by non-truncating mutations in the *LMNA* gene, which do not yield progerin or other forms of truncated and farnesylated prelamin A [55]. These mutations can be either dominant or recessive and induce damages in the nuclear *lamina*.

The clinical features of APS overlap with those of HGPS, including growth retardation, short stature, small mandible, overcrowding teeth, beaked nose, thin lips, loss of hair, joint contractures, and skin changes [137]. Patients could also suffer from lipodystrophy, cardiac anomalies, diabetes mellitus, or glucose intolerance. APS patients exhibit either no acro-osteolysis or only mild changes in the terminal phalanges. They generally have a longer lifespan than HGPS patients.

Numerous variants have been identified, mainly dominant heterozygous missense variations such as p.(Cys588Arg), p.(Pro4Arg), p.(Glu159Lys), p.(Asp136His), p.(Thr10Ile), p.(Glu111Lys), p.(Ala57Pro), p.(Arg133Leu), p.(Glu138Lys), and p.(Glu145Lys). All of the mutations are de novo mutations except p.(Pro4Arg) (autosomal dominant inheritance). For instance, the homozygous p.(Lys542Asp) mutation in *LMNA* gene leads to clinical features of both MADA and HGPS [138]. Skin fibroblasts showed normal localization of lamin A/C, while several nuclei exhibited deformities. The compound heterozygous missense mutations p.(Thr528Met) and p.(Met540Thr) have been described in a 2-year-old boy presented with typical HGPS features. However, neither progerin nor prelamin A was found in the fibroblasts, but major nuclear deformations were observed (honeycomb and blebs) [139].

Given to the production of progerin, HGPS-like patients could potentially benefit from lonafarnib therapy, unlike APS patients.

## Mandibuloacral Dysplasia (MAD)

Mandibuloacral dysplasia (MAD), classified under OMIM #608612 for MADB and OMIM #248370 for MADA, is another very rare laminopathy first described in the 1970s, characterized by a combination of skeletal, skin, and lipodystrophy manifestations [140, 141]. MAD affects both genders equally [142]. To date, around 40 cases have been reported in the literature.

### Phenotype

Patients with MAD exhibit a progeroid appearance, with initial symptoms often appearing in early childhood. Classic facial features include dental abnormalities and micrognathia, prominent eyes, a beaked nose, prominent scalp veins, and atrophic skin. Skeletal abnormalities include mandible and clavicle hypoplasia, acroosteolysis of terminal phalanges, joint contractures, lipodystrophy, cutaneous atrophy, and growth retardation [143]. Approximately half of the patients experience alopecia [142].

MAD is categorized into two main types, primarily distinguished by fat distribution. MADB is characterized by generalized fat loss, while MADA is characterized by restricted fat loss in the extremities and accumulation of fat in the face, neck, and trunk [144]. This abnormal fat distribution is often associated with insulin resistance, impaired glucose tolerance, diabetes mellitus, and hypertriglyceridemia [145].

While MADA patients typically exhibit normal development for the first 5 years, MADB patients develop skeletal abnormalities, skin atrophy, and contractures by the age of 2. Cognitive and social behaviors in MAD patients are usually normal, and the disease does not affect the brain or myocardium, as it does for bone, skin, and adipose tissue.

Molecular Genetics

The mutated genes responsible for MAD, *ZMPSTE24* and *LMNA*, were discovered in 2003 and 2002, respectively [144, 146]. In MADA, the most prevalent *LMNA* variant is the homozygous missense variant c.1580G > A p.(Arg527His). Other homozygous variants include c.1585G > A p.(Ala529Thr) [147], c.1586C > T p.(Ala529Val) [148], c.1623C > T p.(Arg471Cys) [149], c.1579C > T p.(Arg527Cys) [150, 151], c.1580G > T p.(Arg527Leu) [142], and c.1620G > A p.(Met540Ile) [152]. Some cases of compound heterozygous variants have also been reported: c.1318G > A p.(Val440Met) in exon 7 and c.1580G > A p.(Arg527His) in exon 9 [136].

MADB is characterized by loss-of-function compound heterozygous variants in *ZMPSTE24*: c.1018 T > C p.(Trp.340Arg) and c.1085dupT p.(Phe361fsX379) [146], c.1085dupT p.(Phe361fsX379) and c.794A > G p.(Asn265Ser) [153, 154], c.121C > T p.(Gln41X) and c.743C > T p.(Pro248Leu) [155], c.207_208delCT p.(Tyr70fs) and c.794A > G p.(Asp256Ser) [156], c.743C > T p.(Pro248Leu) and c.1349G > A p.(Trp450X), homozygous c.1196A > G p.(Tyr399Cys) [157] [158], c.1085dup p.(Leu362Phefs*19), and c.794A > G p.(Asn265Ser) [159]. Homozygous mutations have also been described: c.1196A > G p.(Tyr399Cys) [160], c.1274 T > C p.(Leu425Pro) [161], and c.281 T > C p.(Leu94Pro) [162].

Pathophysiology

MADA is characterized by the accumulation of farnesylated or full-length prelamin A at the nuclear rim and in intranuclear aggregates, leading to alterations in nuclear architecture (e.g., aberrant honeycomb networks), chromatin defects, and disorganization of nuclear envelope-associated proteins such as SUN2, SUN1, emerin, and LAP2α or decrease of proteins such as SIRT1 [163–165]. MADA cells exhibit DNA damage, altered p53 signaling pathways, genomic instability, and DNA repair defects [166]. Epigenetic alterations are also observed with altered expression of trimethylated H3K9 and a loss of peripheral heterochromatin [165]. Osteoblasts with the p.(Arg527His) *LMNA* variant show increased expression of TGFβ2 and osteoprotegerin. Moreover, the conditioned medium of MADA on human peripheral blood monocytes induces a higher osteoclast differentiation and matrix digestion. The inhibition of TGFβ2 abolished this effect [167]. Mesenchymal stem cells from induced pluripotent stem cells from MADA patients with *LMNA* p.(Arg527Cys) variant exhibit marked senescence and diminished stemness potential. The microRNA miR-311 was found in extracellular vesicle-derived mesenchymal stem cells and was shown to promote premature senescence [168].

Regarding *ZMPSTE24* variant, the disease severity correlates with the residual activity of ZMPSTE24 enzyme. In MADB, unlike in restrictive dermopathy, ZMPSTE24 is partially inactivated, and the residual enzyme activity is sufficient to alleviate the severe symptoms present in restrictive dermopathy, avoiding neonatal death [162].

Therapeutic Approaches

Currently, there are no open clinical trials for MAD patients. Several drug candidates demonstrated efficacy in in vitro studies. Farnesyltransferase inhibitors, for instance, have shown improvement of the cellular phenotype at low passage numbers [163]. Rapamycin has also been tested, and its effects include a reduction in farnesylated prelamin A, increased LAP2α expression, and improved Oct-1 localization [164]. Antisense oligonucleotides (AONs) directed against *LMNA* exon 10 have been tested on MADB patient fibroblasts, leading to decreased lamin A transcript and protein. This reduction is associated with a decrease in nuclear shape abnormalities and senescent cell levels [161]. Additionally, a recent study shows that baricitinib, an FDA-approved JAK/STAT inhibitor, has the potential to improve adipogenesis and lipid droplet formation from skin-derived precursors [169].

Other distinct forms of mandibuloacral dysplasia have been described. Mandibular hypoplasia, deafness, progeroid features, and lipodystrophy syndrome (MDPL) is caused by mutations in *POLD1* gene (c.1812_1814delCTC, p.Ser605del, observed in 80% of the patients) that encodes a DNA polymerase delta catalytic subunit [170, 171]. Symptoms typically appear in early childhood and encompass mandibular hypoplasia, loss of subcutaneous fat, skin abnormalities, insulin resistance, and diabetes mellitus. There are other symptoms that are not specific to progeroid syndromes, such as sensorineural deafness and hypogonadism in males. Mutations in *POLD1* are associated with impaired DNA repair ability [172] and mitochondrial dysfunctions [173].

Recently, a novel type of MAD has been identified implicating another gene, *MTX2*, that encodes metaxin 2, an ubiquitous protein of the outer mitochondrial membrane [174, 175]. This newly recognized MAD is termed MADaM (MAD associated with MTX2) syndrome. Patients affected by MADaM exhibit similar symptoms than other MAD forms including mandibular hypoplasia, growth retardation, lipodystrophy, acro-osteolysis of distal phalanges, and altered skin pigmentation. They also have severe cardiovascular damage, a characteristic that is more similar to HGPS, than to other forms of MAD, and renal dysfunction. MTX2 is a protein located in the outer membrane of the mitochondria and is one of the proteins belonging to the sorting and assembly machinery complex. Several homozygous *MTX2* variants have been found: c.2 T > A p.(Met1Lys), c.544-1G > C (within the splice acceptor site in intron 8) and c.208 + 3_208 + 6del (within the splice donor site in intron 4), c.603del p.(Tyr202Ilefs*26), and c.294_295delTC p.(Leu99*). These variants induce the loss of metaxin 2 expression, leading to the secondary loss of metaxin 1 expression. Patients' fibroblasts displayed mitochondrial network fragmentation and mitochondrial aggregates, resistance to apoptosis induction, enhanced autophagy, increased number of senescent cells, reduced cell proliferation, and a higher percentage of dysmorphic nuclei (blebs and herniations).

To conclude, mandibuloacral dysplasia is another progeroid syndrome presenting with clinical features close to HGPS, albeit in most cases with a milder course and a longer life expectancy for affected individuals. The primary genes implicated in MAD are *LMNA* and *ZMPSTE24*, though other genes such as *POLD1* or *MTX2* can also contribute to this condition.

**Restrictive Dermopathy (RD)**

Restrictive dermopathy (OMIM #275210) is a very rare and lethal neonatal laminopathy. It was named for the first time in 1986 due to distinct epidermal abnormalities in affected children [176, 177]. With fewer than 200 reported cases worldwide to date, this disease affects both genders equally [178]. This condition is classified as an extremely severe progeroid syndrome.

Phenotype

In utero, restrictive dermopathy induces fetal akinesia deformation sequence [179, 180] characterized by reduced fetal movements, joint contractures, polyhydramnios, pulmonary hypoplasia, craniofacial abnormalities, and intrauterine growth retardation. Affected newborns typically undergo preterm cesarean delivery between 30 and 32 weeks of gestation [181]. These newborns exhibit characteristic skin abnormalities: tight, thin, translucent, rigid skin with scleroderma-like lesions and prominent and superficial vasculature. Histopathological examination of the skin reveals hyperkeratosis or hyperorthokeratosis, loss of skin appendages due to abnormal aligned collagen bundles parallel to the epidermal surface, reduced or absence of elastic fibers, a thin dermis, and a flattened dermo-epidermal junction [182, 183]. They display facial dysmorphism including a small, fixed "O"-shaped mouth, a pinched or upturned nose, dysplastic and posteriorly rotated ears, microretrognathia, and large fontanelles. Generalized joint contractures, hypoplastic or dysplastic clavicles, and acro-osteolyses of terminal phalanges are also prevalent. The taut skin impairs normal breathing, leading to respiratory distress and eventual death [178]. No alterations are observed in the central nervous system, nor thoracic and abdominal organs [184].

Molecular Genetics and Pathophysiology

This rare genetic disorder discovered in 2004 and characterized by premature aging and developmental abnormalities is caused by mutations in the *ZMPSTE24* or *LMNA* genes [185]. The *ZMPSTE24* gene encodes a zinc metalloproteinase enzyme responsible for processing prelamin A, a lamin A precursor that is involved in maintaining nuclear structure. A predominant *ZMPSTE24* mutation (c.1085dupT) in exon 9 accounts for nearly 75% of reported cases. This mutation can occur as either a homozygous or compound heterozygous variant: c.1085dupT and c.591dupT in exon 5 [185, 186], c.1085_1086insT and a large deletion r.271_627del, p.(Leu91_Leu209del) on the opposite allele [187], c.1085_1086InsT and c.1249C > T p.(Gln417X) [187], c.1085_1086InsT and c.295delC in exon 3 leading to a shifted reading frame and the creation of a premature termination codon [187], and a 1-bp deletion (c.50delA) in exon 1 and a 2-bp deletion (c.584delAT) in exon 5 predicted to result in premature termination codon [188]. Cases of homozygous variants have also been described: c.50delA [178], 1-bp duplication (c.54dupT) in exon 1 (the parents were both heterozygous for the variant) [186], and 1-bp duplication (c.591dupT) in exon 5 predicted to result in frameshift and premature termination (parents were both heterozygous for the variant) [186, 187].

These variants result in a complete loss of function of *ZMPSTE24* gene and the absence of expression of the zinc metalloproteinase enzyme ZMPSTE24. Consequently, this affects the maturation of lamin A and leads to an accumulation of wild-type, farnesylated prelamin A. The severity of the disease is inversely correlated with the residual activity of ZMPSTE24 [189]. Mandibuloacral dysplasia type B (see paragraph I.4.b) is another progeroid syndrome where biallelic mutations in *ZMPSTE24* lead to residual enzyme activity, allowing patients to survive into adulthood.

Affected cells exhibit characteristic nuclear abnormalities, including blebs, herniations, and honeycomb-like structures, due to the accumulation of prelamin A. Nuclear prelamin A is observed in fibroblasts and the liver, while ZMPSTE24 and lamin A are absent, in contrast to lamin C remaining at normal levels [185, 187].

Heterozygous splicing variants in the *LMNA* gene have also been described. The first one was a heterozygous variant at intron 11 donor splice site (IVS11 + 1G > A) resulting in loss of exon 11 from the transcript and a deletion of 90 amino acids in the protein. This truncated prelamin A cannot undergo posttranslational maturation and accumulates in nuclei. Fibroblasts of this patient express A and C lamins but also a truncated prelamin A (del90 or A$\Delta$90) [132]. Another patient had c.1824C > T variant in *LMNA* exon 11, the same mutation as HGPS children but with a very early and severe presentation [190].

Other *LMNA* mutations give rise to progeroid syndromes with features reminiscent of restrictive dermopathy. For instance, the homozygous *LMNA* p.(Arg435Cys) variant increases DNA double-stranded breaks and delays DNA repair due to decreased recruitment of p53-binding protein 1 (53BP1). Nevertheless, this mutation is not associated with abnormal prelamin A accumulation, but rather with a progressive loss of lamin A [191, 192].

Patients affected with restrictive dermopathy carrying *LMNA* variants tend to live longer (several months) than those with *ZMPSTE24* variants, who typically die within the first weeks of life [178, 185].

#### Therapeutic Approaches

Currently, no treatments are available for restrictive dermopathy. Rapamycin has been tested on affected cells, showing a decrease in prelamin A but without substantial improvements [165].

Restrictive dermopathy is a fatal primary or secondary laminopathy with distinct clinical symptoms. Early genetic testing for *LMNA* and *ZMPSTE24* mutations is critical for diagnosis and family planning for unaffected parents and siblings. To date, no therapeutic options are available, and patients receive compassionate palliative care.

## Laminopathies Caused by Mutations in B-Type Lamins

While much attention has been given to diseases resulting from mutations in A-type lamins, those caused by defects in B-type lamins, namely, lamin B1 and lamin B2, are described more recently and remain to be fully characterized. They represent a significant and distinct subset of laminopathies that are ultra-rare and mostly extremely severe. B-type lamins, comprising lamins B1 and B2 encoded by the *LMNB1* and *LMNB2* genes, respectively, are a fundamental component of the nuclear lamina. Lamins B share common functions with A-type lamins, such as a crucial role in maintaining nuclear structure, regulating gene expression, and ensuring proper cell division. Unlike A-type lamins, lamins B are expressed in all nucleated cells and are essential from the earliest stages of development explaining why diseases associated with lamins B are often severe neurodevelopmental disorders [193, 194].

Mutations in the genes encoding lamins B1 and B2 can lead to a variety of disorders, each characterized by its distinct pathological features. These mutations may cause either a loss of function, an overexpression of the normal protein, or an abnormal gain of function, leading to diverse clinical outcomes (Fig. 4.1b). These defects result in abnormalities in the nuclear envelope, leading to disrupted nuclear integrity and altered nuclear-cytoplasmic interactions. This disruption affects in turn various cellular processes, including DNA replication, RNA transcription, and cell cycle progression, ultimately leading to cell dysfunction and death. The understanding of how specific mutations in *LMNB1* and *LMNB2* lead to disease is a key area of ongoing research.Lamin B-related disorders, like those related to A-type lamin, present a wide spectrum of clinical manifestations, depending on the severity and nature of the genetic mutation. Some of the conditions associated with lamin B defects are specific to a B-type lamin, while others are common to both types [193, 195].

### *A Lamin B1-Specific Laminopathy: The Adult-Onset Autosomal Dominant Leukodystrophy*

Adult-onset autosomal dominant leukodystrophy (ADLD, OMIM #169500) is an extremely rare neurogenetic disorder characterized by symmetrical white matter loss resulting from progressive demyelination and mainly involving the frontoparietal and cerebellar regions of the central nervous system (CNS) [35, 36]. Up to now, less than 100 patients are reported in the literature. It represents a subgroup within the broader category of leukodystrophies, which are a heterogeneous group of inherited disorders primarily affecting the CNS white matter. ADLD typically manifests in adulthood, distinguishing it from other leukodystrophies that often present in childhood.

The clinical manifestation of ADLD exhibits variability, typically manifesting in the fourth to the fifth decade starting with autonomic dysfunction and subsequently progressing to ataxia, tremor, and cognitive decline. The observed clinical heterogeneity in ADLD is ascribed to the progressive nature of demyelination affecting diverse areas of the CNS. Diagnosis relies significantly on magnetic resonance imaging of the brain, which unveils distinctive white matter abnormalities, predominantly localized in the frontal lobes [196, 197].

Molecular basis of ADLD includes duplications of the *LMNB1* gene or gene expression enhancement caused by the deletion of a region localized upstream [198, 199]. The increase in lamin B1 expression has been objectivized in brain tissue, in fibroblasts, and in blood cells [200]. It is the only form of laminopathy, either A-type or B-type related, associated with an overexpression of a normal lamin protein.

Replicating all symptoms of this disease in animal models, especially those indicative of dysautonomia, poses a significant challenge. Nevertheless, transgenic mice, characterized by the overexpression of lamin B1 cDNA in oligodendrocytes – the cell types responsible for CNS myelin production – successfully recapitulate age-dependent motor dysfunction [201, 202]. Additionally, investigations employing ex vivo models derived from human cells have demonstrated distinct morphological and functional abnormalities, notably in astrocytes and fibroblasts [199, 203].

The observation that both humans and animal models are normal at birth and subsequently develop demyelination later in life implies that the pathophysiological mechanism primarily involves pathways responsible for myelin maintenance, rather than exerting a deleterious effect on cell survival [195, 200].

## *Primary Microcephaly Is Caused by Both B-Type Lamins*

Primary microcephaly is defined by a small brain, with normal structure or with cortical malformation, associated with mild to moderate intellectual deficiency. The genetic causes for primary microcephaly are highly heterogeneous.

Lamins B1 and B2 are both responsible for primary microcephaly [204, 205]. Regardless of the specific type of lamin involved, missense mutations or in-frame heterozygous deletions, typically with a dominant effect and mostly occurring de novo, are observed. Notably, certain mutations are recurrent, suggesting the presence of mutational hotspots in these genes. The phenotypes of microcephaly are indistinguishable between those associated with B1 and B2 forms. In both cases, there is evident microcephaly from birth, worsening with age, constant intellectual impairment of variable severity, and sometimes epilepsy (in about half of the cases). The brain maintains a normal structure, although there may be gyral simplification or anomalies in the corpus callosum.

In cells expressing lamin B1 mutations, abnormal lamina with ruptures (depending on the mutations) has been observed, along with nuclei exhibiting abnormal shapes and bi- or tri-lobed nuclei. There is no evidence of mitotic spindle abnormalities [204]. In mice models deficient for B1 or B2, abnormalities in neuronal migration are observed [206]. This phenomenon is observed in a milder aspect in microcephaly related to lamin B, since gyral simplification is observed in certain patients. However, the pathophysiological mechanism is very different since the mutated proteins are expressed, exerting a dominant effect by disrupting the organization of the nuclear lamina.

Despite a few recurrent mutations, various locations for these missense mutations are observed (Fig. 4.1b), occurring more frequently in the rod domain near the head or at the end of this domain, but not exclusively. This preferential localization at the ends of the rod domain could disrupt the binding between lamins B1 and B2, stabilizing lamin dimers.

Some of the mutations are also found pathogenic in the corresponding residues of lamin A (e.g., p.(Glu33Asp)) but result in a muscular or peripheral neuropathy phenotype [207].

## *Two Types of Laminopathy Associated with Lamin B2 Only: A Progressive Myoclonic Epilepsy (PME) and a Partial Lipodystrophy*

### PME with Ataxia

Progressive myoclonic epilepsy (PME) is a rare, mostly recessive disorder presenting with action myoclonus, tonic-clonic seizures and progressive neurological decline, and a high genetic heterogeneity. PME with early ataxia constitutes a subgroup of this disorder with also a genetic heterogeneity. Genes associated with this pathology encode for proteins involved in vesicle trafficking (e.g., GOSR2), for transmembrane channels (KCNC1), or for lamin B2. The form linked to lamin B2 seems to be extremely rare with three independent families described up to now. The first case was described in 2015 in a large family and was due to a recessive missense mutation on residue 157, p.(His157Tyr) [208]. The proband developed normally until the age of 7 when seizures occurred. Disease history was marked by gait deterioration and worsening of epilepsy, while the cognition remains normal. This molecular cause for PME was confirmed in 2021 and 2022 by the description of two other variants in independent families, situated on the adjacent residue 158 p.(Arg158Trp) and p.(Arg158Leu) associated with similar clinical presentation despite an earlier onset of the disease [209, 210]. It was demonstrated that the variant on residue 157 results in abnormal lateral self-assembly of lamin B2 [208]. The expression level of the mutants has not been tested, but the fact that the clinical signs appeared after several years of life suggests a toxic gain of function for the mutant proteins.

### Partial Lipodystrophies

Dysfunctions of lamin B2 have been linked to partial lipodystrophy as soon as 2006, but since the first description, only a very few cases of such disorders have been reported.

Up to now, less than ten patients with lipodystrophy and metabolic disorders have been reported carrying a mutation in *LMNB2*. Among those, three were carriers of the same heterozygous missense variant p.(Arg235Gln) (rs 121912497) and were found after investigation of small cohorts of patients with acquired partial lipodystrophy (APL) or familial partial lipodystrophy (FPLD) [211, 212]. The allelic frequency of this variant is around 1% in the population database gnomAD, and the mutated amino acid is not very conserved across species. This suggests that this variant may not be the only cause for the pathological phenotype but should be considered as a predisposition factor. The patients carrying this mutation may have secondary conditions triggering the occurrence of metabolic disorders. Two other rare heterozygous mutations were described: the p.(Arg234Trp) in a female patient presenting early-onset type 2 diabetes, hypertriglyceridemia, and android fat distribution and the p.(Tyr252His) in a female patient with progressive loss of subcutaneous fat since adolescence [48, 213]. Both mutations modify a highly conserved amino acid.

All variants associated with lipodystrophy are in the same region of lamin B2, which corresponds to the linker 2 of the rod domain of the protein (amino acids 230–256). The function of this region is not well defined, and functional studies are needed to understand the role of this domain on the structure and function of the protein in adipocyte differentiation of maturation process.

In conclusion, advanced genetic testing techniques, such as whole-exome sequencing, have proven essential in identifying mutations in the *LMNB1* and *LMNB2* genes and in the description of novel ultra-rare laminopathies linked specifically to B-type lamins. These findings have resulted in significant insights into the complex interplay between nuclear architecture and cellular function of B-type lamins and underscored the critical role of these proteins in various cellular processes. Further case descriptions are needed to better delineate the clinical presentations and the precise molecular mechanisms underlying these diseases and ultimately to develop targeted treatments.

## The Role of Chromatin Organization in Genetic Laminopathies

One of the major roles of the nuclear lamina is to establish and maintain a chromatin organization that regulates gene expression according to a cell-specific pattern. For this purpose, chromatin is organized into domains: some of them are containing genes expressed at high levels and are located centrally within the nucleus; others, which are gene-poor or containing genes that are expressed at low levels, are located predominantly at the periphery of the nucleus and are stabilized by physical interactions with the lamina, in regions called lamin-associated domains (LADs) [214–216]. Lamins regulate these chromatin domains through interactions with diverse chromatin-binding factors including histones and histone methyltransferases [217]. Most LADs are found in all cell types and are termed constitutive LADs, while some are termed facultative LADs because they are specific to certain cell types. LADs' organization varies between cells and during cell cycles exhibiting complex dynamics at the nuclear lamina [218]. Beyond regulating gene expression patterns, these chromatin-lamina interactions also play a role in the transmission of signals resulting from mechanical stimulation to the nucleus, thereby modulating gene expression in response to mechanical stimuli, a mechanism known as mechanotransduction [219–222].

In cases of lamin defects, alterations in the lamina structure lead to dysfunctions related to chromatin organization. Diseases resulting from lamin abnormalities are due to insufficient lamin production or, more frequently, from mutant lamins caused by heterozygous missense variant in lamin-encoding genes. Consequently, a mutated form of the protein often coexists with the wild-type version within the lamina, leading to structural changes, while the expression of the wild-type protein persists but is insufficient to ensure normal functioning. The number of missense mutations responsible for laminopathies is very high, potentially affecting any residue of the lamins as they are distributed throughout the entire length of the protein and exhibiting considerable variability in nature. For example, more than 700 mutations have been reported in LMNA, of which more than 400 are missense mutations. On some residues, substitutions by different amino acids are observed (e.g., p.Arg482Trp, p.Arg482Leu). Therefore, since the effect of a missense mutation depends on both its location and the nature of the substituted amino acid, each mutation may have a unique effect on lamina functions. This broad diversity is observed in lamin-related pathologies, with some mutations affecting only one tissue and others being multisystemic, without a clear correlation between mutation characteristics and observed clinical effects.

Because of clinical diversity and genetic heterogeneity, it is challenging to identify a single pathophysiological mechanism common to all laminopathies. However, common characteristics of lamin variants' deleterious effects can be identified. Among them, disruption of LADs resulting from abnormal lamina and leading to defects in chromatin organization and dynamics is likely a common mechanism [217, 223]. The nature of the disrupted LADs could explain clinical diversity, and some studies have shown specific alterations of facultative LAD in cardiomyopathies [224], but an exact correlation between domain type and clinical signs has not been established. Moreover, since the majority of LADs are constitutive (i.e., found in all cell types), a homogeneous clinical expression would be expected. However, it is easy to assume that the severity of the lamina alteration can vary significantly from one variant to another. Of note, the contribution of different lamins to the establishment of LADs is not equivalent, and lamin C is particularly involved in this function. Therefore, mutations that preferentially affect lamin C are expected to have a more deleterious effect on this function. However, when the lamina structure is altered, regardless of the variant involved, all lamins are mislocalized, and thus there is no clear correlation between the type of affected lamin and the observed deleterious effect.

Focusing on laminopathies due to heterozygous missense mutations in LMNA, common characteristics of affected cells include an abnormal-shaped nucleus, larger than normal, with decreased or disrupted chromatin attachment [225]. In diseases like progeria, although abnormalities in lamina structure are difficult to demonstrate, anomalies in chromatin organization and LAD binding have been observed [216, 226, 227]. Other studies have shown that lamin A mutants are associated with LAD modifications linked to epigenomic modifications, especially in cardiopathies and myopathies [228–230]. Often, changes in LADs observed in laminopathies are minor, so it is unclear if this is the primary pathophysiological mechanism.

With regard to the role of lamins in mechanotransduction, lamina abnormalities, regardless of the variant involved, prevent the correct transmission of signals resulting from mechanical stimuli. These signals not only fail to induce the expected gene expression adaptation but also worsen the disorders observed in the nuclear envelope and lead eventually to loss of nuclear integrity, nuclear rupture, and DNA damage [223]. This point is supported by the observed overlap of phenotypes produced by pathogenic mutations in lamin A and in nesprins, one of the main proteins of the LINC complex. The more the cell is subjected to mechanical stimuli, the more the nucleus with altered lamina will suffer, explaining why muscular and cardiac clinical expressions are particularly frequently associated with laminopathies [5]. However, the involvement of adipose tissue is not explained yet. Mechanical signal transmission is linked to chromatin dynamics as demonstrated both in normal and affected cells. In normal cells, intense mechanical constraint leads to chromatin reorganization by epigenetic modulation and decrease heterochromatin, to maintain nucleus integrity [231, 232] suggesting that mechanical stimuli could be necessary for a proper chromatin organization. In case of excessive mechanical stress or in laminopathies, the effect on nucleus results in nuclear envelope rupture and major DNA damage [231, 233] as the adaptability of chromatin modification is overtaken.

Another feature common to several lamin A-related laminopathies is that loosening the links between cytoskeleton and the lamina, either by destabilizing the microtubule network or the LINC complex, improves the cellular phenotype of laminopathies and the clinical phenotype in murine models [19, 23, 52, 233–235]. It seems that the normal lamina is stretched on the inner side by its chromatin connections and, on the cytoplasmic side, by its connections to the cytoskeleton. These opposing links maintain the integrity of the nuclear envelope and confer its role in protecting genetic material [236]. If the nuclear side connections are affected, the lamina would be deformed by its intact cytoskeleton connections in the absence of counterbalancing forces. Destabilizing these connections to the microtubule network would release the tension on the lamina and restore a more normal nuclear envelope shape [223]. While the situation would not be fully normalized, there would be an improvement in shape, nuclear envelope integrity, and protection of the lamina from cytoskeletal stimuli. This finding suggests alternative strategies for treating several forms of laminopathies and possibly aging through the modulation of physical interactions between lamina and its protein partners [18, 19, 237] or by acting on mechanical signaling pathways such as the Hippo pathway, an emerging central player in cardiolaminopathies [24, 238, 239].

## Lamins and Cancers

During cytological analysis carried out to diagnose cancer, diagnostic criteria of malignancy are based on morphological changes in nuclear size and shape [240–242]. These nuclear abnormalities, commonly observed in carcinoma cells, can be attributed to changes in lamins' composition and/or quantity. Lamins are the main constituents of the nuclear matrix and are also proteins that interact closely with or are anchored to the nuclear envelope (NE).

Lamins are thought to play a crucial role in tumor development and progression. Most studies reporting the involvement of lamins in cancer concern A-type lamins, while B-type lamins are less frequently studied [243].

Tables 4.1 and 4.2 summarize the variations of A- and B-type lamin expression, at protein level or, at least, at transcriptomic level, depending on the type of cancer, in relation to the prognosis.

### *Lamins and Cell Deformability*

During metastasis process, cancer cells must enhance their mobility and deformability to invade successfully surrounding tissues. Cells have to deform to migrate through very narrow spaces and reach the bloodstream, to finally form a distant metastasis. The nucleus size and stiffness play a crucial role in regulating cellular deformability. Lamin expression directly influences nuclear deformability. Their loss enhances cancer cell deformability [244, 245]. Numerous publications have highlighted the central role of A-type lamins in regulating nuclear deformability and stiffness. Reduced expression of A-type lamins is associated with increased deformability, reduced resistance, and stiffness of the nucleus [104, 246–250], promoting tumor cell invasion and metastasis.

### *Lamins Expression in Carcinoma*

Carcinomas are the best-documented types of cancer, with regard to variations in the localization and expression levels of A-type lamins and their association with disease prognosis. Nevertheless, there is no consistent pattern observed within carcinomas regarding the increase or decrease of this expression and its corresponding clinical prognosis.

**Table 4.1** A-type lamins' expression in patients' tumors, depending on cancer type, and related to prognosis

| A-type lamins' expression | Type of cancer | Prognosis | References |
|---|---|---|---|
| Increase of lamin A and C expression | Prostate carcinoma | Higher-stage tumors<br>Higher lamin A expression: higher-stage tumors | Kong et al. *Carcinogenesis* 2012<br>Zuo et al. *Genes* 2018<br>Skvortsov et al. *J Proteome Res* 2011 |
| | Ovarian carcinoma | Higher-stage tumors | Wang et al. *J Proteome Res* 2009 |
| | Colorectal carcinoma | Reduced overall survival | Willis et al. *PLOs ONE* 2008 |
| | Glioblastoma[a] | Reduced overall survival | Gatti et al. *Biomedicines* 2021 |
| | Glioma (together with B-type lamins) | Disease progression | Pei et al. *J Genet Genomics* 2022 |
| Loss or reduction of both lamin A and C expressions | Breast carcinoma | /<br>Disease recurrence<br>Tumor aggressiveness, reduced overall survival | Capo-Chichi et al. *Chin J Cancer* 2011<br>Matsumoto et al. *Cancer Med* 2015<br>Alhudiri et al. *Breast Cancer Res Treat* 2019 |
| | Gastric carcinoma | Reduced overall survival | Wu et al. *J Exp Clin Cancer Res* 2009 |
| | Small cell lung carcinoma | / | Broers et al. *Histochem Cell Biol* 1997<br>Broers et al. 2014<br>Kaufmann et al. *Cancer Res* 1991<br>Wang et al. *Virchows Arch* 2019 |
| | Cervical carcinoma | Risk factor for developing cervical cancer | Capo-Chichi et al. *Cell Oncol* 2016 |
| | Prostate carcinoma | Increase for lymph node metastasis | Saarinen et al. *PLOs ONE* 2015 |
| | Colon carcinoma | Disease recurrence<br>/ | Belt et al. *Eur J Cancer* 2011<br>Moss et al. *Gut* 1999 |
| | Non-muscle invasive bladder carcinoma | No correlation with clinical prognosis factors, diagnostic marker of malignancy | Setti et al. *Ann Diagn Pathol.* 2021 |
| | Leukemia/ lymphoma[a] | / | Stadelmann et al. *Leuk Res* 1990<br>Agrelo et al. *J Clin Oncol* 2005 |
| | Ewing sarcoma[a] | Reduced overall survival | Chiarini et al. *Cell Death and Disease* 2022 |
| | Glioblastoma | Tumor aggressiveness | Rauschert et al. *PLoS ONE* 2017 |

(continued)

**Table 4.1** (continued)

| A-type lamins' expression | Type of cancer | Prognosis | References |
|---|---|---|---|
| Isolated decrease of lamin A | Lung adenocarcinoma | Increased number of metastases, worsened clinical condition | Kaspi et al. *PLoS ONE* 2017 |
| | Ovarian carcinoma | Increased number of metastases, reduced survival | Gong et al. *Pathol Res Pract* 2015 |
| | Endometrial carcinoma | Tumor aggressiveness | Cicchillitti et al. *Oncotarget* 2017 |
| | Breast carcinoma[a] | / | Aljada et al. *Cell Oncol* 2016 |

Protein expression monitored, except for [a], for which only mRNA levels were measured

## A-type Lamins

Increased expression of both lamin A and lamin C was described in prostate [251–253], ovarian [254], and colorectal [255] carcinoma. In these tumors, this A-type lamins' expression profile was associated with a worse prognosis, including higher tumor stage or reduced overall survival.

Conversely, reduced or absent lamin A and lamin C expression was more commonly reported in breast [241, 256, 257], gastric [258], small cell lung [259–262], cervical [263], prostate [264], colon [265, 266], and bladder [267] carcinoma. In breast, gastric, prostate, and colon carcinoma, this expression pattern was associated with poor prognosis indicators, including disease recurrence, tumor aggressiveness, reduced overall survival, and increased number of metastases. Decrease of A-type lamins' expression was proposed as a predictive risk factor for developing cervical cancer [263]. In non-muscle invasive bladder carcinoma, reduced A-type lamins' expression was not associated with any clinical prognostic factors, but may serve as a diagnostic marker of malignancy [267].

An isolated decrease of lamin A expression was reported in lung adenocarcinoma [268], endometrial carcinoma [269], ovarian carcinoma [270], and breast carcinoma [271]. The high decrease of lamin A, associated with persistent lamin C expression, was linked to a poor prognosis: higher number of metastases, reduced survival, and increased tumor aggressiveness.

Recently, the consequences of high isolated lamin A expression in a high-grade ovarian serous carcinoma cell line were investigated. This cell line exhibits elevated levels of HR (homologous repair) proteins. In this model, elevated lamin A levels may be related to increased chemoresistance and aggressive metastasis [272].

Bell et al. proposed the ratio lamin A to lamin B1 as a prognostic marker for breast carcinoma. A low ratio was indeed associated with a poor outcome, as evidenced by reduced disease-free survival, independently to the ER/PR/HER2 status (i.e., estrogen receptor, progesterone receptor, human epidermal growth factor receptor 2) [273]. Using cell line models, these authors showed that reduced expression of A-type lamins resulted in increased nuclear deformability, increased cell

**Table 4.2** B-type lamins' expression in patients' tumors, depending on cancer type, and related to prognosis

| Type of cancer | B-type lamin expression | Prognosis | References |
|---|---|---|---|
| Lung carcinoma | Decrease of lamin B1 | / | Jia et al. *J Exp Med* 2019 |
| *Small cell lung carcinoma* | Decrease of lamin B1 | Worse prognosis, higher-stage tumors | Jia et al. *J Exp Med* 2019 |
| *Non-small cell lung carcinoma* | Increase of lamin B2 | Lymph node metastasis | Zhang et al. *Exp Cell Res* 2020 |
| Hepatocellular carcinoma | Increase of lamins B1 and B2 | Unfavorable prognostic factors | Yang et al. *J Immunol Res* 2022 |
| | Increase of lamin B1 | / | Sun et al. *J Proteome Res* 2010 |
| Prostate carcinoma | Low expression of lamin B2 | Increase of lymph node metastasis | Saarinen et al. *PLoS ONE* 2015 |
| | Increase of lamin B1[a] | Risk for metastasis, reduced disease-free survival | Luo et al. *Future Oncol* 2021 |
| | Increase of lamin B1 | Higher-stage tumors | Hong et al. *Am J Cancer Res* 2022 |
| Breast carcinoma | Decrease of lamin B1 | *Post-neoadjuvant chemotherapy* | Saleh et al. *Diagnostics* 2022 |
| | | Worse prognosis[a] | Wazir et al. *Cell Mol Biol Lett* 2013 |
| | Increase of lamin B1 | Reduced overall survival, higher-stage tumors | Dong et al. *Exp Cell Res* 2023 |
| | Increase of lamin B2 | Reduced overall survival, lymph node metastasis, higher-stage tumors | Zhao et al. *Biosci Rep* 2021 |
| Gastric carcinoma | Decrease of lamin B1 | Higher clinical stage, nodal stage, and poor prognosis | Yu et al. *Neoplasma* 2020 |
| Colon carcinoma | Low expression of lamin B1 | / | Moss et al. *Gut* 1999 |
| | Increase of lamin B2 | Reduced overall survival, lymph node metastasis | Dong et al. *Cell Death Dis* 2021 |
| Adrenocortical carcinoma | Increase of lamin B1 | / | Chen et al. *Endocr Relat Cancer* 2024 |
| Pancreatic carcinoma | Increase of lamin B1 | Reduced overall survival | Li et al. *Clin Cancer Res* 2013 |
| Bladder carcinoma | Increase of lamin B2 | Reduced overall survival | Ji et al. *Int J Mol Med* 2022 |
| Esophageal carcinoma | Increase of lamin B2 | Worse prognosis | Wang et al. *Ann Clin Lab Sci* 2022 |
| Sarcoma | Increase of lamins B1 and B2 | Reduced overall survival[a] | Wu et al. *Medicine* 2022 |
| Glioma | Increase of lamins B1 and B2 (together with A-type lamins) | Disease progression | Pei et al. *J Genet Genomics* 2022 |

Protein expression monitored, except for [a], for which only mRNA levels were measured

migration, and modulation of gene expression (particularly in genes governing cell-matrix interactions, cell metabolism, and PI3K/Akt signaling) [273].

### B-type Lamins

Increased levels of B-type lamins have been reported in hepatocarcinoma (HCC) [274, 275]. Yang et al. conducted an in silico analysis of transcriptomic and proteomic data already available (the Cancer Genome Atlas and the Human Protein Atlas, respectively) on HCC [275]. Increased expression of both lamins B1 and B2 was identified as an unfavorable prognostic factor, associated with a reduced overall survival. Thus, biological functions of B-type lamins were explored through in vitro functional assay: upregulation of lamins B1 and B2 significantly enhanced cell invasion and migration. In this study, A-type lamins were also monitored, but their variations were not associated with any risk factor [275].

In prostate carcinoma, a low expression of lamin B2 was correlated to an increase of lymph node metastasis. No clinical association with lamin B1 levels was found in this study [264]. In contrast, other teams have demonstrated that a high lamin B1 expression in prostate carcinoma tissues from patients was correlated to a reduced disease-free survival and a higher risk of metastasis. This high lamin B1 expression was thus considered as a marker for high-stage tumors [276, 277]. Ectopic overexpression of lamin B1 in a prostate carcinoma cell line has been shown to promote cell proliferation and invasion [276]. In other prostate cancer cell lines, lamin B1 has been demonstrated to positively regulate FOXD1 (forkhead box D1), a transcription factor that regulates gene expression involved in cell cycle regulation and signal transduction. High FOXD1 expression in tumors is associated with a poor prognosis and stimulates cell viability, migration, and invasion. *LMNB1* silencing has been shown to reduce FOXD1 expression and to inhibit cell migration and invasion [278].

Conflicting literature data on lamin B1 expression were reported in breast carcinoma. One study showed that lamin B1 was highly expressed in tumors, with no difference compared to non-tumoral tissues, and no association with clinical and pathological data. Nevertheless, a strong lamin B1 decrease was described following neoadjuvant chemotherapy [279]. Wazir et al. reported a reduction of lamin B1 expression, which was associated with a worse prognosis [280]. Another study on patient tumor tissues found that increased lamin B1 expression was associated with poorer overall survival and higher-stage tumors. Gene set enrichment analysis showed that *LMNB1* expression interacts with signaling pathways involved in cell cycle regulation, DNA replication, and mismatch repair [281]. Using in vitro cell line model, *LMNB1* silencing inhibited cell proliferation and migration [281]. Similar results were obtained regarding lamin B2 expression in breast tumors from patients. Higher lamin B2 expression was linked to reduced survival, lymph node metastasis, and higher-stage tumors [282]. Again, in vitro functional experiments demonstrated that lamin B2 promoted cell proliferation, tumor growth in a mouse xenograft model, and inhibited apoptosis [282].

In gastric carcinoma, a decrease in lamin B1 expression was linked to a poorer prognosis, including reduced overall survival and higher-stage tumors. Additionally, in vitro experiments demonstrated that *LMNB1* silencing promoted cell proliferation and migration and activated the PI3K/PTEN/Akt and MAPK/ERK signaling pathways. Conversely, mirror experiments with lamin B1 overexpression produced the opposite effects [283].

Few data are available regarding B-type lamins' expression in colon carcinoma. An old study reported a strong reduction of lamin B1 in patients' samples and, in some cases, an aberrant cytoplasmic localization [266]. Increased lamin B2 expression was associated with a poor prognosis and a higher metastatic potential [284].

A recent publication explored lamin B1 expression in adrenocortical carcinoma. The observed increase of lamin B1 was associated with high tumor mutational burden (TMB), a factor known to correlate with a poorer prognosis [285].

In other carcinomas, elevated lamin B1 and lamin B2 expression was associated with a shorter survival in pancreatic cancer [286] and bladder and esophageal carcinoma [287, 288], while reduced lamin B1 expression was linked to worse prognosis in non-small cell lung cancer [289].

### Lamins and Lung Cancer

Oldest studies were purely descriptive, analyzing the expression patterns of lamins in human tumors and cell lines.

The studies of Kaufmann and Broers' teams firstly demonstrated that A-type lamins were strongly expressed in non-small cell lung cancer (NSCLC), while they were decreased or absent in small cell lung cancer (SCLC) cell lines [259, 261, 262].

Machiels et al. described a reduced expression of A-type lamins in certain lung adenocarcinoma cell lines. Nevertheless, no functional tests were performed to investigate consequences of the loss of these proteins [290].

Recent investigations have focused on the expression of A-type lamins in metastatic cells of lung adenocarcinoma from pleural effusions. A strong reduction of lamin A, without lamin C variation, was highlighted in a group of patients, associated with a higher number of metastatic sites and a worsened clinical state [268].

Recent studies have analyzed the consequences of A-type lamins' expression modulation on metastatic potential, epithelial-to-mesenchymal transition (EMT), and response to treatment.

The reduction of A-type lamins in lung adenocarcinoma cell line enhanced their nuclear deformability [248]. Additionally, the isolated loss of lamin A could be associated with the acquisition of an EMT phenotype [268], probably enhancing metastatic potential in lung cancer. Moreover, a study conducted on a lung adenocarcinoma cell line showed that inhibition of *LMNA* gene transcripts (lamin A and lamin C) enhanced the acquisition of EMT and the resistance to the tyrosine kinase inhibitor erlotinib, commonly prescribed in cases with mutation of the *EGFR* gene in malignant cells [35].

Recent studies have shed light on the involvement of B-type lamins in lung cancer development.

As reported for A-type lamins, lamin B1 is not expressed in SCLC, and its reduced expression in NSCLC is linked to a poorer prognosis and more advanced disease stages [289]. In a model of lung epithelial cell line, the knockdown of lamin B1 expression resulted in the acquisition of an epithelial-to-mesenchymal transition (EMT) phenotype, enhanced cell migration, and increased tumor growth and metastasis in a xenograft mouse model [289].

In contrast, an increase of lamin B2 expression is associated with a higher risk of lymph node metastasis [291]. Ectopic overexpression of lamin B2 in a lung adenocarcinoma cell line promoted migration and proliferation by enhancing cell cycle-related proteins (cyclin D1 and CDK4) and inducing EMT. This finding was further supported by the observation that lamin B2 promoted methylation of the *CDH1* (gene encoding E-cadherin) promoter, leading to the silencing of E-cadherin gene, a hallmark of EMT [291].

## *Lamins' Expression in Other Cancers*

### Hemopathies

Unlike lamin B1, A-type lamins are not expressed in acute lymphoblastic leukemia (ALL) or non-Hodgkin's lymphoma (NHL) [292]. This lack of A-type lamins' expression results from the hypermethylation of the *LMNA* gene promoter in leukemias and lymphomas. Moreover, in nodal diffuse large B-cell lymphoma, the degree of *LMNA* promoter hypermethylation is inversely correlated with the overall survival [293].

### Sarcoma

In sarcoma, Wu et al. investigated the mRNA levels of *lamins B1* and *B2* in primitive tumors and cell lines from various sarcoma subtypes. Their findings revealed that both *lamin B1* and *B2* transcripts are mainly overexpressed in sarcoma cells, and this overexpression is associated with a poorer outcome (shorter survival) [294].

Few other studies have analyzed A-type lamins in sarcoma cell lines. For instance, lamin A protein expression was found to be reduced in osteosarcoma cell lines compared to normal osteoblast cell lines, thereby enhancing their migratory capacity [295]. Consistent with these results, *LMNA* mRNA levels were analyzed in primitive Ewing sarcoma tumors and corresponding metastatic sites. Reduced expression of *LMNA* transcripts was linked to a worse overall survival and was more pronounced in metastatic sites relative to the primitive tumors [296]. At the mechanistic level, *LMNA* gene silencing (resulting in the loss of both lamin A and C protein expressions) enhanced Ewing sarcoma cell line migration and led to disorganization of the LINC complex along with nuclear sequestration of the mechanosignaling effector YAP/TAZ [296].

The *LMNA-NTRK1* (neurotrophic tyrosine kinase receptor, type 1) gene fusion, which encodes a functional LMNA-TRK1 fusion oncoprotein, has been frequently reported in sarcoma. This discovery has opened up therapeutic avenues for patients, as targeted therapies that inhibit TRK can be effectively employed [297, 298].

## Nervous Central System (NCS) Tumors

In glioma, the most frequent type of tumor in NCS, elevated levels of both A-type lamins and B-type lamins are associated with disease progression. Functional studies performed on in vitro and in vivo models following the silencing of *LMNA* and *LMNB1* resulted in inhibition of tumor cell migration and growth reduction [299].

Gatti et al. conducted an in silico analysis of transcriptomic data from the Cancer Genome Atlas to investigate glioblastoma subtypes (classical, neural, proneural, and mesenchymal). High *LMNA* gene expression was found to correlate with a reduced overall survival, except in the mesenchymal subtype [300]. Molecular pathways were thus investigated using glioblastoma cell lines in which *LMNA* gene was overexpressed. In this model, overexpression of A-type lamins led to increased tumor aggressiveness (increased cell migration, colony number, and tumor volume in a xenograft mouse model) [300]. A subset of neuroblastoma cell lines has been shown to lose A-type lamins' expression, which was associated with an aggressive tumor phenotype [301, 302].

As described in lymphoma and leukemia (see above), the loss of expression may result from *LMNA* promoter hypermethylation. These observations are consistent with bioinformatic analysis of patients' primary tumor methylation: significant association between the methylation status of *LMNA* promoter and tumor aggressiveness was observed [302].

## Melanoma

To our knowledge, no prior study has reported lamins' expression in human melanoma tumors.

Functional studies on melanoma cell line B16F10 after silencing A-type lamins have demonstrated a reduction of their invasive and metastatic potential [303].

One recent study investigated the silencing of lamin B1, along with its nuclear receptor (LBR), in melanoma cell lines, leading to modifications in chromatin structure (increased heterochromatin foci and histone levels) and induction of senescence [304].

## Conclusion

All these studies pointed out the involvement of both A- and B-type lamins in cancer.

The precise role of lamins and the underlying mechanisms governing their expression likely vary depending on the specific context, cancer type, and other factors including the mutational profile, disease stage, and cell differentiation status, with EMT context.

**Compliance with Ethical Standards** The authors declare no competing interest.

## References

1. Bonne G, Barletta MRD, Varnous S et al (1999) Mutations in the gene encoding lamin A/C cause autosomal dominant Emery-Dreifuss muscular dystrophy. Nat Genet 21:285–288
2. Fatkin D, MacRae C, Sasaki T et al (1999) Missense mutations in the rod domain of the lamin A/C gene as causes of dilated cardiomyopathy and conduction-system disease. N Engl J Med 341:1715–1724
3. Maggi L, Carboni N, Bernasconi P (2016) Skeletal muscle laminopathies: a review of clinical and molecular features. Cells 5:33
4. Bonne G, Leturcq F, Ben Yaou R (1993) Emery-Dreifuss muscular dystrophy. In: Adam MP, Feldman J, Mirzaa GM et al (eds) GeneReviews®. University of Washington, Seattle, WA
5. Maggi L, Mavroidis M, Psarras S et al (2021) Skeletal and cardiac muscle disorders caused by mutations in genes encoding intermediate filament proteins. Int J Mol Sci 22:4256
6. Rosenbaum AN, Agre KE, Pereira NL (2020) Genetics of dilated cardiomyopathy: practical implications for heart failure management. Nat Rev Cardiol 17:286–297
7. Richard P, Fressart V, Charron P, Hainque B (2010) Genetics of inherited cardiomyopathies. Pathol Biol (Paris) 58:343–352
8. Japp AG, Gulati A, Cook SA et al (2016) The diagnosis and evaluation of dilated cardiomyopathy. J Am Coll Cardiol 67:2996–3010
9. Ditaranto R, Boriani G, Biffi M et al (2019) Differences in cardiac phenotype and natural history of laminopathies with and without neuromuscular onset. Orphanet J Rare Dis 14:263
10. Cattin M, Muchir A, Bonne G (2013) 'state-of-the-heart' of cardiac laminopathies. Curr Opin Cardiol 28:297–304
11. Storey EC, Fuller HR (2022) Genotype-phenotype correlations in human diseases caused by mutations of LINC complex-associated genes: a systematic review and meta-summary. Cells 11:4065
12. Tesson F, Saj M, Uvaize MM et al (2014) Lamin A/C mutations in dilated cardiomyopathy. Cardiol J 21:331–342
13. Jakobs PM, Hanson EL, Crispell KA et al (2001) Novel lamin A/C mutations in two families with dilated cardiomyopathy and conduction system disease. J Card Fail 7:249–256
14. Dittmer TA, Sahni N, Kubben N et al (2014) Systematic identification of pathological lamin A interactors. Mol Biol Cell 25:1493–1510
15. Zahr HC, Jaalouk DE (2018) Exploring the crosstalk between LMNA and splicing machinery gene mutations in dilated cardiomyopathy. Front Genet 9:231
16. Nishiuchi S, Makiyama T, Aiba T et al (2017) Gene-based risk stratification for cardiac disorders in LMNA mutation carriers. Circ Cardiovasc Genet 10:e001603
17. Brull A, Morales Rodriguez B, Bonne G et al (2018) The pathogenesis and therapies of striated muscle laminopathies. Front Physiol 9:1533

18. Chatzifrangkeskou M, Le Dour C, Muchir A (2023) Modulation of cytoskeleton in cardiomyopathy caused by mutations in LMNA gene. Am J Physiol Cell Physiol 324:C1223–C1235
19. Leong EL, Khaing NT, Cadot B et al (2023) Nesprin-1 LINC complexes recruit microtubule cytoskeleton proteins and drive pathology in Lmna-mutant striated muscle. Hum Mol Genet 32:177–191
20. Muchir A, Worman HJ (2016) Targeting mitogen-activated protein kinase signaling in mouse models of cardiomyopathy caused by Lamin A/C gene mutations. In: Methods in enzymology. Elsevier, pp 557–580
21. Arimura T, Helbling-Leclerc A, Massart C et al (2005) Mouse model carrying H222P-Lmna mutation develops muscular dystrophy and dilated cardiomyopathy similar to human striated muscle laminopathies. Hum Mol Genet 14:155–169
22. Mounkes LC, Kozlov SV, Rottman JN, Stewart CL (2005) Expression of an LMNA-N195K variant of A-type lamins results in cardiac conduction defects and death in mice. Hum Mol Genet 14:2167–2180
23. Chai RJ, Werner H, Li PY et al (2021) Disrupting the LINC complex by AAV mediated gene transduction prevents progression of Lamin induced cardiomyopathy. Nat Commun 12:4722
24. Tan CY, Chan PS, Tan H et al (2023) Systematic in vivo candidate evaluation uncovers therapeutic targets for LMNA dilated cardiomyopathy and risk of Lamin A toxicity. J Transl Med 21:690
25. Ho JCY, Zhou T, Lai W-H et al (2011) Generation of induced pluripotent stem cell lines from 3 distinct laminopathies bearing heterogeneous mutations in lamin A/C. Aging (Albany NY) 3:380–390
26. Lee Y-K, Lau Y-M, Cai Z-J et al (2017) Modeling treatment response for lamin A/C related dilated cardiomyopathy in human induced pluripotent stem cells. J Am Heart Assoc 6:e005677
27. Bertero A, Fields PA, Smith AST et al (2019) Chromatin compartment dynamics in a haploinsufficient model of cardiac laminopathy. J Cell Biol 218:2919–2944
28. Salvarani N, Crasto S, Miragoli M et al (2019) The K219T-Lamin mutation induces conduction defects through epigenetic inhibition of SCN5A in human cardiac laminopathy. Nat Commun 10:2267
29. Captur G, Arbustini E, Bonne G et al (2018) Lamin and the heart. Heart 104:468–479
30. Kumar S, Baldinger SH, Gandjbakhch E et al (2016) Long-term arrhythmic and nonarrhythmic outcomes of lamin A/C mutation carriers. J Am Coll Cardiol 68:2299–2307
31. Benarroch L, Cohen E, Atalaia A et al (2021) Preclinical advances of therapies for laminopathies. J Clin Med 10:4834
32. Mosbah H, Donadille B, Vatier C et al (2022) Dunnigan lipodystrophy syndrome: French National Diagnosis and Care Protocol (PNDS; Protocole National de Diagnostic et de Soins). Orphanet J Rare Dis 17:170
33. Dunnigan MG, Cochrane MA, Kelly A, Scott JW (1974) Familial lipoatrophic diabetes with dominant transmission. A new syndrome. Q J Med 43:33–48
34. Gonzaga-Jauregui C, Ge W, Staples J et al (2020) Clinical and molecular prevalence of lipodystrophy in an unascertained large clinical care Cohort. Diabetes 69:249–258
35. Fernandez-Pombo A, Diaz-Lopez EJ, Castro AI et al (2023) Clinical spectrum of LMNA-associated type 2 familial partial lipodystrophy: a systematic review. Cells 12
36. Guillín-Amarelle C, Fernández-Pombo A, Sánchez-Iglesias S, Araújo-Vilar D (2018) Lipodystrophic laminopathies: diagnostic clues. Nucleus 9:249–260
37. Krimm I, Östlund C, Gilquin B et al (2002) The Ig-like structure of the C-terminal domain of lamin A/C, mutated in muscular dystrophies, cardiomyopathy, and partial lipodystrophy. Structure 10:811–823
38. Cao H (2000) Nuclear lamin A/C R482Q mutation in Canadian kindreds with Dunnigan-type familial partial lipodystrophy. Hum Mol Genet 9:109–112
39. Speckman RA, Garg A, Du F et al (2000) Mutational and haplotype analyses of families with familial partial lipodystrophy (Dunnigan variety) reveal recurrent missense mutations in the globular C-terminal domain of Lamin A/C. Am J Hum Genet 66:1192–1198

40. Schmidt HH-J, Genschel J, Baier P et al (2001) Dyslipemia in familial partial lipodystrophy caused by an R482W mutation in the LMNA gene. J Clin Endocrinol Metabol 86:2289–2295
41. Shackleton S, Lloyd DJ, Jackson SNJ et al (2000) LMNA, encoding lamin A/C, is mutated in partial lipodystrophy. Nat Genet 24:153–156
42. Vigouroux C, Magré J, Vantyghem MC et al (2000) Lamin A/C gene: sex-determined expression of mutations in Dunnigan-type familial partial lipodystrophy and absence of coding mutations in congenital and acquired generalized lipoatrophy. Diabetes 49:1958–1962
43. Lanktree M, Cao H, Rabkin S et al (2007) Novel *LMNA* mutations seen in patients with familial partial lipodystrophy subtype 2 (FPLD2; MIM 151660). Clin Genet 71:183–186
44. Guénantin AC, Briand N, Bidault G et al (2014) Nuclear envelope-related lipodystrophies. Semin Cell Dev Biol 29:148–157
45. Araujo-Vilar D, Lattanzi G, Gonzalez-Mendez B et al (2008) Site-dependent differences in both prelamin A and adipogenic genes in subcutaneous adipose tissue of patients with type 2 familial partial lipodystrophy. J Med Genet 46:40–48
46. Pellegrini C, Columbaro M, Schena E et al (2019) Altered adipocyte differentiation and unbalanced autophagy in type 2 Familial Partial Lipodystrophy: an in vitro and in vivo study of adipose tissue browning. Exp Mol Med 51:1–17
47. Capanni C, Mattioli E, Columbaro M et al (2005) Altered pre-lamin A processing is a common mechanism leading to lipodystrophy. Hum Mol Genet 14:1489–1502
48. Varlet A-A, Desgrouas C, Jebane C et al (2021) A rare mutation in LMNB2 associated with lipodystrophy drives premature cell senescence. Cells 11:50
49. Vigouroux C, Guénantin A-C, Vatier C et al (2018) Lipodystrophic syndromes due to LMNA mutations: recent developments on biomolecular aspects, pathophysiological hypotheses and therapeutic perspectives. Nucleus 9:235–248
50. Perepelina K, Dmitrieva R, Ignatieva E et al (2018) Lamin A/C mutation associated with lipodystrophy influences adipogenic differentiation of stem cells through interaction with Notch signaling. Biochem Cell Biol 96:342–348
51. Simon DN, Domaradzki T, Hofmann WA, Wilson KL (2013) Lamin A tail modification by SUMO1 is disrupted by familial partial lipodystrophy–causing mutations. MBoC 24:342–350
52. Jebane C, Varlet A-A, Karnat M et al (2023) Enhanced cell viscosity: a new phenotype associated with lamin A/C alterations. iScience 26:107714
53. Quinn K, Chauhan S, Purcell SM (2023) Lipodystrophies. In: StatPearls. StatPearls Publishing, Treasure Island (FL)
54. Foss-Freitas MC, Besci Ö, Meral R et al (2023) A very-low-calorie diet can cause remission of diabetes mellitus and hypertriglyceridemia in familial partial lipodystrophy. Obes Facts 1–6:103
55. Semple RK, Patel KA, Auh S et al (2023) Genotype-stratified treatment for monogenic insulin resistance: a systematic review. Commun Med 3:134
56. Brown RJ, Araujo-Vilar D, Cheung PT et al (2016) The diagnosis and management of lipodystrophy syndromes: a multi-society practice guideline. J Clin Endocrinol Metabol 101:4500–4511
57. Garg A, Vinaitheerthan M, Weatherall PT, Bowcock AM (2001) Phenotypic heterogeneity in patients with familial partial lipodystrophy (dunnigan variety) related to the site of missense mutations in lamin a/c gene. J Clin Endocrinol Metab 86:59–65
58. Soyaltin UE, Simsir IY, Akinci B et al (2020) Homozygous LMNA p.R582H pathogenic variant reveals increasing effect on the severity of fat loss in lipodystrophy. Clin Diabetes Endocrinol 6:13
59. Dutour A, Roll P, Gaborit B et al (2011) High prevalence of laminopathies among patients with metabolic syndrome. Hum Mol Genet 20:3779–3786
60. Desgrouas C, Varlet A-A, Dutour A et al (2020) Unraveling LMNA mutations in metabolic syndrome: cellular phenotype and clinical pitfalls. Cells 9:310

61. Höger TH, Krohne G, Kleinschmidt JA (1991) Interaction of Xenopus lamins A and LII with chromatin in vitro mediated by a sequence element in the carboxyterminal domain. Exp Cell Res 197:280–289
62. Decaudain A, Vantyghem M-C, Guerci B et al (2007) New metabolic phenotypes in laminopathies: *LMNA* mutations in patients with severe metabolic syndrome. J Clin Endocrinol Metabol 92:4835–4844
63. Le Dour C, Schneebeli S, Bakiri F et al (2011) A homozygous mutation of prelamin-A preventing its farnesylation and maturation leads to a severe lipodystrophic phenotype: new insights into the pathogenicity of nonfarnesylated prelamin-A. J Clin Endocrinol Metabol 96:E856–E862
64. Galant D, Gaborit B, Desgrouas C et al (2016) A heterozygous ZMPSTE24 mutation associated with severe metabolic syndrome, ectopic fat accumulation, and dilated cardiomyopathy. Cells 5:21
65. Skre H (1974) Genetic and clinical aspects of Charcot-Marie-Tooth's disease. Clin Genet 6:98–118
66. Harding AE, Thomas PK (1980) The clinical features of hereditary motor and sensory neuropathy types I and II. Brain 103:259–280
67. Bacquet J, Stojkovic T, Boyer A et al (2018) Molecular diagnosis of inherited peripheral neuropathies by targeted next-generation sequencing: molecular spectrum delineation. BMJ Open 8:e021632
68. Roa BB, Garcia CA, Lupski JR (1991) Charcot-Marie-Tooth disease type 1A: molecular mechanisms of gene dosage and point mutation underlying a common inherited peripheral neuropathy. Int J Neurol 25–26:97–107
69. De Sandre-Giovannoli A, Chaouch M, Kozlov S et al (2002) Homozygous defects in LMNA, encoding lamin A/C nuclear-envelope proteins, cause autosomal recessive axonal neuropathy in human (Charcot-Marie-Tooth disorder type 2) and mouse. Am J Hum Genet 70:726–736
70. Chaouch M, Allal Y, De Sandre-Giovannoli A et al (2003) The phenotypic manifestations of autosomal recessive axonal Charcot–Marie–Tooth due to a mutation in Lamin A/C gene. Neuromuscul Disord 13:60–67
71. Tazir M (2004) Phenotypic variability in autosomal recessive axonal Charcot-Marie-Tooth disease due to the R298C mutation in lamin A/C. Brain 127:154–163
72. Goizet C, Yaou RB, Demay L et al (2004) A new mutation of the lamin A/C gene leading to autosomal dominant axonal neuropathy, muscular dystrophy, cardiac disease, and leuconychia. J Med Genet 41:e29
73. Benedetti S, Bertini E, Iannaccone S et al (2005) Dominant LMNA mutations can cause combined muscular dystrophy and peripheral neuropathy. J Neurol Neurosurg Psychiatry 76:1019–1021
74. Walter MC, Witt TN, Weigel BS et al (2005) Deletion of the LMNA initiator codon leading to a neurogenic variant of autosomal dominant Emery-Dreifuss muscular dystrophy. Neuromuscul Disord 15:40–44
75. Bouhouche A, Birouk N, Azzedine H et al (2007) Autosomal recessive axonal Charcot-Marie-Tooth disease (ARCMT2): phenotype-genotype correlations in 13 Moroccan families. Brain 130:1062–1075
76. Hamadouche T, Poitelon Y, Genin E et al (2008) Founder effect and estimation of the age of the c.892C>T (p.Arg298Cys) mutation in *LMNA* associated to charcot-marie-tooth subtype CMT2B1 in families from North Western Africa. Ann Hum Genet 72:590–597
77. Lassuthová P, Baránková L, Haberlová J et al (2009) Mutations in the LMNA gene do not cause axonal CMT in Czech patients. J Hum Genet 54:365–368
78. Song S, Zhang Y, Chen B et al (2006) No mutation was detected in the LMNA gene among sporadic Charcot-Marie-Tooth patients. Beijing Da Xue Xue Bao 38:78–79
79. Gilford H (1897) On a condition of mixed premature and immature development. Med Chir Trans 80(17–46):25

80. Hutchinson J (1886) Congenital absence of hair and mammary glands with atrophic condition of the skin and its appendages, in a boy whose mother had been almost wholly bald from alopecia areata from the age of six. Med Chir Trans 69:473–477
81. Ullrich NJ, Gordon LB (2015) Hutchinson-Gilford progeria syndrome. Handb Clin Neurol 132:249–264
82. Gordon CM, Cleveland RH, Baltrusaitis K et al (2019) Extraskeletal calcifications in Hutchinson-Gilford progeria syndrome. Bone 125:103–111
83. Murtada S-I, Kawamura Y, Cavinato C et al (2023) Biomechanical and transcriptional evidence that smooth muscle cell death drives an osteochondrogenic phenotype and severe proximal vascular disease in progeria. Biomech Model Mechanobiol 22:1333–1347
84. Stehbens WE, Wakefield SJ, Gilbert-Barness E et al (1999) Histological and ultrastructural features of atherosclerosis in progeria. Cardiovasc Pathol 8:29–39
85. Benedicto I, Dorado B, Andrés V (2021) Molecular and cellular mechanisms driving cardiovascular disease in Hutchinson-Gilford progeria syndrome: lessons learned from animal models. Cells 10:1157
86. Olsen FJ, Gordon LB, Smoot L et al (2023) Progression of cardiac abnormalities in Hutchinson-Gilford progeria syndrome: a prospective longitudinal study. Circulation 147:1782–1784
87. Prakash A, Gordon LB, Kleinman ME et al (2018) Cardiac abnormalities in patients with Hutchinson-Gilford progeria syndrome. JAMA Cardiol 3:326–334
88. Hennekam RCM (2006) Hutchinson-Gilford progeria syndrome: review of the phenotype. Am J Med Genet A 140:2603–2624
89. De Sandre-Giovannoli A, Bernard R, Cau P et al (2003) Lamin a truncation in Hutchinson-Gilford progeria. Science 300:2055
90. Eriksson M, Brown WT, Gordon LB et al (2003) Recurrent de novo point mutations in lamin A cause Hutchinson–Gilford progeria syndrome. Nature 423:293–298
91. Cau P, Navarro C, Harhouri K et al (2014) Nuclear matrix, nuclear envelope and premature aging syndromes in a translational research perspective. Semin Cell Dev Biol 29:125–147
92. Goldman RD, Shumaker DK, Erdos MR et al (2004) Accumulation of mutant lamin A causes progressive changes in nuclear architecture in Hutchinson-Gilford progeria syndrome. Proc Natl Acad Sci USA 101:8963–8968
93. Röhrl JM, Arnold R, Djabali K (2021) Nuclear pore complexes cluster in dysmorphic nuclei of normal and progeria cells during replicative senescence. Cells 10:153
94. Chen Z-J, Wang W-P, Chen Y-C et al (2014) Dysregulated interactions between lamin A and SUN1 induce abnormalities in the nuclear envelope and endoplasmic reticulum in progeric laminopathies. J Cell Sci 127:1792–1804
95. Chojnowski A, Ong PF, Foo MXR et al (2020) Heterochromatin loss as a determinant of progerin-induced DNA damage in Hutchinson-Gilford Progeria. Aging Cell 19:e13108
96. Scaffidi P, Misteli T (2006) Lamin A-dependent nuclear defects in human aging. Science 312:1059–1063
97. Shumaker DK, Dechat T, Kohlmaier A et al (2006) Mutant nuclear lamin A leads to progressive alterations of epigenetic control in premature aging. Proc Natl Acad Sci USA 103:8703–8708
98. Cao K, Capell BC, Erdos MR et al (2007) A lamin A protein isoform overexpressed in Hutchinson-Gilford progeria syndrome interferes with mitosis in progeria and normal cells. Proc Natl Acad Sci USA 104:4949–4954
99. Cobb AM, Murray TV, Warren DT et al (2016) Disruption of PCNA-lamins A/C interactions by prelamin A induces DNA replication fork stalling. Nucleus 7:498–511
100. Liu B, Wang J, Chan KM et al (2005) Genomic instability in laminopathy-based premature aging. Nat Med 11:780–785
101. Chojnowski A, Ong PF, Wong ESM et al (2015) Progerin reduces LAP2α-telomere association in Hutchinson-Gilford progeria. elife 4:e07759

102. Frankel D, Delecourt V, Novoa-Del-Toro E-M et al (2022) miR-376a-3p and miR-376b-3p overexpression in Hutchinson-Gilford progeria fibroblasts inhibits cell proliferation and induces premature senescence. iScience 25:103757
103. Dahl KN, Scaffidi P, Islam MF et al (2006) Distinct structural and mechanical properties of the nuclear lamina in Hutchinson-Gilford progeria syndrome. Proc Natl Acad Sci USA 103:10271–10276
104. Lammerding J, Schulze PC, Takahashi T et al (2004) Lamin A/C deficiency causes defective nuclear mechanics and mechanotransduction. J Clin Invest 113:370–378
105. Richards SA, Muter J, Ritchie P et al (2011) The accumulation of un-repairable DNA damage in laminopathy progeria fibroblasts is caused by ROS generation and is prevented by treatment with N-acetyl cysteine. Hum Mol Genet 20:3997–4004
106. Viteri G, Chung YW, Stadtman ER (2010) Effect of progerin on the accumulation of oxidized proteins in fibroblasts from Hutchinson Gilford progeria patients. Mech Ageing Dev 131:2–8
107. Jung H-J, Coffinier C, Choe Y et al (2012) Regulation of prelamin A but not lamin C by miR-9, a brain-specific microRNA. Proc Natl Acad Sci USA 109:E423–E431
108. Nissan X, Blondel S, Navarro C et al (2012) Unique preservation of neural cells in Hutchinson- Gilford progeria syndrome is due to the expression of the neural-specific miR-9 microRNA. Cell Rep 2:1–9
109. Hu Q, Zhang N, Sui T et al (2023) Anti-hsa-miR-59 alleviates premature senescence associated with Hutchinson-Gilford progeria syndrome in mice. EMBO J 42:e110937
110. Manakanatas C, Ghadge SK, Agic A et al (2022) Endothelial and systemic upregulation of miR-34a-5p fine-tunes senescence in progeria. Aging (Albany NY) 14:195–224
111. Coll-Bonfill N, Mahajan U, Shashkova EV et al (2023) Progerin induces a phenotypic switch in vascular smooth muscle cells and triggers replication stress and an aging-associated secretory signature. Geroscience 45:965–982
112. Olive M, Harten I, Mitchell R et al (2010) Cardiovascular pathology in Hutchinson-Gilford progeria: correlation with the vascular pathology of aging. Arterioscler Thromb Vasc Biol 30:2301–2309
113. Stehbens WE, Delahunt B, Shozawa T, Gilbert-Barness E (2001) Smooth muscle cell depletion and collagen types in progeric arteries. Cardiovasc Pathol 10:133–136
114. Varga R, Eriksson M, Erdos MR et al (2006) Progressive vascular smooth muscle cell defects in a mouse model of Hutchinson-Gilford progeria syndrome. Proc Natl Acad Sci USA 103:3250–3255
115. Bidault G, Garcia M, Capeau J et al (2020) Progerin expression induces inflammation, oxidative stress and senescence in human coronary endothelial cells. Cells 9:1201
116. Xu Q, Mojiri A, Boulahouache L et al (2022) Vascular senescence in progeria: role of endothelial dysfunction. Eur Heart J Open 2:oeac047
117. Hamczyk MR, Andrés V (2019) Vascular smooth muscle cell loss underpins the accelerated atherosclerosis in Hutchinson-Gilford progeria syndrome. Nucleus 10:28–34
118. Gordon LB, Kleinman ME, Miller DT et al (2012) Clinical trial of a farnesyltransferase inhibitor in children with Hutchinson-Gilford progeria syndrome. Proc Natl Acad Sci USA 109:16666–16671
119. Cao K, Graziotto JJ, Blair CD et al (2011) Rapamycin reverses cellular phenotypes and enhances mutant protein clearance in Hutchinson-Gilford progeria syndrome cells. Sci Transl Med 3:89ra58
120. Cenni V, Capanni C, Columbaro M et al (2011) Autophagic degradation of farnesylated prelamin A as a therapeutic approach to lamin-linked progeria. Eur J Histochem 55:e36
121. Gabriel D, Gordon LB, Djabali K (2016) Temsirolimus partially rescues the Hutchinson-Gilford progeria cellular phenotype. PLoS One 11:e0168988
122. Gabriel D, Roedl D, Gordon LB, Djabali K (2015) Sulforaphane enhances progerin clearance in Hutchinson-Gilford progeria fibroblasts. Aging Cell 14:78–91
123. Harhouri K, Navarro C, Depetris D et al (2017) MG132-induced progerin clearance is mediated by autophagy activation and splicing regulation. EMBO Mol Med 9:1294

124. Kang HT, Park JT, Choi K et al (2017) Chemical screening identifies ROCK as a target for recovering mitochondrial function in Hutchinson-Gilford progeria syndrome. Aging Cell 16:541–550
125. Kreienkamp R, Croke M, Neumann MA et al (2016) Vitamin D receptor signaling improves Hutchinson-Gilford progeria syndrome cellular phenotypes. Oncotarget 7:30018–30031
126. Liu B, Ghosh S, Yang X et al (2012) Resveratrol rescues SIRT1-dependent adult stem cell decline and alleviates progeroid features in laminopathy-based progeria. Cell Metab 16:738–750
127. Osorio FG, Navarro CL, Cadiñanos J et al (2011) Splicing-directed therapy in a new mouse model of human accelerated aging. Sci Transl Med 3:106ra107
128. Beyret E, Liao H-K, Yamamoto M et al (2019) Single-dose CRISPR-Cas9 therapy extends lifespan of mice with Hutchinson-Gilford progeria syndrome. Nat Med 25:419–422
129. Santiago-Fernández O, Osorio FG, Quesada V et al (2019) Development of a CRISPR/Cas9-based therapy for Hutchinson-Gilford progeria syndrome. Nat Med 25:423–426
130. Koblan LW, Erdos MR, Wilson C et al (2021) In vivo base editing rescues Hutchinson-Gilford progeria syndrome in mice. Nature 589:608–614
131. Whisenant D, Lim K, Revêchon G et al (2022) Transient expression of an adenine base editor corrects the Hutchinson-Gilford progeria syndrome mutation and improves the skin phenotype in mice. Nat Commun 13:3068
132. Barthélémy F, Navarro C, Fayek R et al (2015) Truncated prelamin A expression in HGPS-like patients: a transcriptional study. Eur J Hum Genet 23:1051–1061
133. Moulson CL, Fong LG, Gardner JM et al (2007) Increased progerin expression associated with unusual LMNA mutations causes severe progeroid syndromes. Hum Mutat 28:882–889
134. Bar DZ, Arlt MF, Brazier JF et al (2017) A novel somatic mutation achieves partial rescue in a child with Hutchinson-Gilford progeria syndrome. J Med Genet 54:212–216
135. Gordon LB, Brown WT, Collins FS (2023) Hutchinson-Gilford progeria syndrome. In: GeneReviews® [Internet]. University of Washington, Seattle
136. Lombardi F, Gullotta F, Columbaro M et al (2007) Compound heterozygosity for mutations in LMNA in a patient with a myopathic and lipodystrophic mandibuloacral dysplasia type A phenotype. J Clin Endocrinol Metab 92:4467–4471
137. Garg A, Subramanyam L, Agarwal AK et al (2009) Atypical progeroid syndrome due to heterozygous missense LMNA mutations. J Clin Endocrinol Metabol 94:4971–4983
138. Plasilova M, Chattopadhyay C, Pal P et al (2004) Homozygous missense mutation in the lamin A/C gene causes autosomal recessive Hutchinson-Gilford progeria syndrome. J Med Genet 41:609–614
139. Verstraeten VLRM, Broers JLV, van Steensel MAM et al (2006) Compound heterozygosity for mutations in LMNA causes a progeria syndrome without prelamin A accumulation. Hum Mol Genet 15:2509–2522
140. Danks DM, Mayne V, Wettenhall NB, Hall RK (1974) Craniomandibular dermatodysostosis. Birth Defects Orig Artic Ser 10:99–105
141. Young LW, Radebaugh JF, Rubin P et al (1971) New syndrome manifested by mandibular hypoplasia, acroosteolysis, stiff joints and cutaneous atrophy (mandibuloacral dysplasia) in two unrelated boys. Birth Defects Orig Artic Ser 7:291–297
142. Jéru I, Nabil A, El-Makkawy G et al (2021) Two decades after mandibuloacral dysplasia discovery: additional cases and comprehensive view of disease characteristics. Genes (Basel) 12:1508
143. Cenni V, D'Apice MR, Garagnani P et al (2018) Mandibuloacral dysplasia: a premature ageing disease with aspects of physiological ageing. Ageing Res Rev 42:1–13
144. Novelli G, Muchir A, Sangiuolo F et al (2002) Mandibuloacral dysplasia is caused by a mutation in LMNA-encoding lamin A/C. Am J Hum Genet 71:426–431
145. Simha V, Garg A (2002) Body fat distribution and metabolic derangements in patients with familial partial lipodystrophy associated with mandibuloacral dysplasia. J Clin Endocrinol Metab 87:776–785

146. Agarwal AK, Fryns J-P, Auchus RJ, Garg A (2003) Zinc metalloproteinase, ZMPSTE24, is mutated in mandibuloacral dysplasia. Hum Mol Genet 12:1995–2001
147. Kosho T, Takahashi J, Momose T et al (2007) Mandibuloacral dysplasia and a novel LMNA mutation in a woman with severe progressive skeletal changes. Am J Med Genet A 143A:2598–2603
148. Ozer L, Unsal E, Aktuna S et al (2016) Mandibuloacral dysplasia and LMNA A529V mutation in Turkish patients with severe skeletal changes and absent breast development. Clin Dysmorphol 25:91–97
149. Zirn B, Kress W, Grimm T et al (2008) Association of homozygous LMNA mutation R471C with new phenotype: mandibuloacral dysplasia, progeria, and rigid spine muscular dystrophy. Am J Med Genet A 146A:1049–1054
150. Agarwal AK, Kazachkova I, Ten S, Garg A (2008) Severe mandibuloacral dysplasia-associated lipodystrophy and progeria in a young girl with a novel homozygous Arg527Cys LMNA mutation. J Clin Endocrinol Metab 93:4617–4623
151. Luo D-Q, Wang X-Z, Meng Y et al (2014) Mandibuloacral dysplasia type A-associated progeria caused by homozygous LMNA mutation in a family from Southern China. BMC Pediatr 14:256
152. Yassaee VR, Khojaste A, Hashemi-Gorji F et al (2016) A novel homozygous LMNA mutation (p.Met540Ile) causes mandibuloacral dysplasia type A. Gene 577:8–13
153. Agarwal AK, Zhou XJ, Hall RK et al (2006) Focal segmental glomerulosclerosis in patients with mandibuloacral dysplasia owing to ZMPSTE24 deficiency. J Investig Med 54:208–213
154. Shackleton S, Smallwood DT, Clayton P et al (2005) Compound heterozygous ZMPSTE24 mutations reduce prelamin A processing and result in a severe progeroid phenotype. J Med Genet 42:e36
155. Miyoshi Y, Akagi M, Agarwal AK et al (2008) Severe mandibuloacral dysplasia caused by novel compound heterozygous ZMPSTE24 mutations in two Japanese siblings. Clin Genet 73:535–544
156. Cunningham VJ, D'Apice MR, Licata N et al (2010) Skeletal phenotype of mandibuloacral dysplasia associated with mutations in ZMPSTE24. Bone 47:591–597
157. Haye D, Dridi H, Levy J et al (2016) Failure of ossification of the occipital bone in mandibuloacral dysplasia type B. Am J Med Genet A 170:2750–2755
158. Ahmad Z, Zackai E, Medne L, Garg A (2010) Early onset mandibuloacral dysplasia due to compound heterozygous mutations in ZMPSTE24. Am J Med Genet A 152A:2703–2710
159. Alarcón PI, Mujica I, Sanz P et al (2019) Mandibuloacral dysplasia with type B lipodystrophy in a patient from Chile. Am J Med Genet A 179:893–895
160. Hitzert MM, van der Crabben SN, Baldewsingh G et al (2019) Mandibuloacral dysplasia type B (MADB): a cohort of eight patients from Suriname with a homozygous founder mutation in ZMPSTE24 (FACE1), clinical diagnostic criteria and management guidelines. Orphanet J Rare Dis 14:294
161. Harhouri K, Navarro C, Baquerre C et al (2016) Antisense-based Progerin downregulation in HGPS-like patients' cells. Cells 5
162. Ben Yaou R, Navarro C, Quijano-Roy S et al (2011) Type B mandibuloacral dysplasia with congenital myopathy due to homozygous ZMPSTE24 missense mutation. Eur J Hum Genet 19:647–654
163. Camozzi D, D'Apice MR, Schena E et al (2012) Altered chromatin organization and SUN2 localization in mandibuloacral dysplasia are rescued by drug treatment. Histochem Cell Biol 138:643
164. Cenni V, Capanni C, Mattioli E et al (2014) Rapamycin treatment of Mandibuloacral dysplasia cells rescues localization of chromatin-associated proteins and cell cycle dynamics. Aging (Albany NY) 6:755–770
165. Filesi I, Gullotta F, Lattanzi G et al (2005) Alterations of nuclear envelope and chromatin organization in mandibuloacral dysplasia, a rare form of laminopathy. Physiol Genomics 23:150–158

166. di Masi A, D'Apice MR, Ricordy R et al (2008) The R527H mutation in LMNA gene causes an increased sensitivity to ionizing radiation. Cell Cycle 7:2030–2037
167. Avnet S, Pallotta R, Perut F et al (2011) Osteoblasts from a mandibuloacral dysplasia patient induce human blood precursors to differentiate into active osteoclasts. Biochim Biophys Acta 1812:711–718
168. He L, Liao J, Liu Z et al (2023) Multi-omic analysis of mandibuloacral dysplasia type A patient iPSC-derived MSC senescence reveals miR-311 as a novel biomarker for MSC senescence. Hum Mol Genet 32:2872–2886
169. Hartinger R, Lederer E-M, Schena E et al (2023) Impact of combined Baricitinib and FTI treatment on adipogenesis in Hutchinson-Gilford progeria syndrome and other lipodystrophic laminopathies. Cells 12:1350
170. Elouej S, Beleza-Meireles A, Caswell R et al (2017) Exome sequencing reveals a de novo POLD1 mutation causing phenotypic variability in mandibular hypoplasia, deafness, progeroid features, and lipodystrophy syndrome (MDPL). Metabolism 71:213–225
171. Weedon MN, Ellard S, Prindle MJ et al (2013) An in-frame deletion at the polymerase active site of POLD1 causes a multisystem disorder with lipodystrophy. Nat Genet 45:947–950
172. Murdocca M, Spitalieri P, De Masi C et al (2021) Functional analysis of POLD1 p.ser605del variant: the aging phenotype of MDPL syndrome is associated with an impaired DNA repair capacity. Aging (Albany NY) 13:4926–4945
173. Murdocca M, Spitalieri P, Cappello A et al (2022) Mitochondrial dysfunction in mandibular hypoplasia, deafness and progeroid features with concomitant lipodystrophy (MDPL) patients. Aging (Albany NY) 14:1651–1664
174. Elouej S, Harhouri K, Le Mao M et al (2020) Loss of MTX2 causes mandibuloacral dysplasia and links mitochondrial dysfunction to altered nuclear morphology. Nat Commun 11:4589
175. Yeter Doğan B, Günay N, Ada Y, Doğan ME (2023) A novel MTX2 gene splice site variant resulting in exon skipping, causing the recently described mandibuloacral dysplasia progeroid syndrome. Am J Med Genet A 191:173–182
176. Toriello HV (1986) Restrictive dermopathy and report of another case. Am J Med Genet 24:625–629
177. Witt DR, Hayden MR, Holbrook KA et al (1986) Restrictive dermopathy: a newly recognized autosomal recessive skin dysplasia. Am J Med Genet 24:631–648
178. Scott JB, Yanes AF, Vivar KL et al (2021) Restrictive dermopathy: three new patients with ZMPSTE24 mutations and a review of the literature. Pediatr Dermatol 38:1535–1540
179. Hall JG, Kiefer J (2016) Arthrogryposis as a syndrome: gene ontology analysis. Mol Syndromol 7:101–109
180. Niles KM, Blaser S, Shannon P, Chitayat D (2019) Fetal arthrogryposis multiplex congenita/fetal akinesia deformation sequence (FADS)-Aetiology, diagnosis, and management. Prenat Diagn 39:720–731
181. Mau U, Kendziorra H, Kaiser P, Enders H (1997) Restrictive dermopathy: report and review. Am J Med Genet 71:179–185
182. Morais P, Magina S, Ribeiro M d C et al (2009) Restrictive dermopathy–a lethal congenital laminopathy. Case report and review of the literature. Eur J Pediatr 168:1007–1012
183. Pradeep I, Gowrishankar K, Shanmugasundaram L (2022) Lethal restrictive dermopathy with ZMPSTE24 mutation. Pediatr Dev Pathol 25:327–329
184. Nijsten TEC, De Moor A, Colpaert CG et al (2002) Restrictive dermopathy: a case report and a critical review of all hypotheses of its origin. Pediatr Dermatol 19:67–72
185. Navarro CL, De Sandre-Giovannoli A, Bernard R et al (2004) Lamin A and ZMPSTE24 (FACE-1) defects cause nuclear disorganization and identify restrictive dermopathy as a lethal neonatal laminopathy. Hum Mol Genet 13:2493–2503
186. Moulson CL, Go G, Gardner JM et al (2005) Homozygous and compound heterozygous mutations in ZMPSTE24 cause the laminopathy restrictive dermopathy. J Invest Dermatol 125:913–919

187. Navarro CL, Cadiñanos J, De Sandre-Giovannoli A et al (2005) Loss of ZMPSTE24 (FACE-1) causes autosomal recessive restrictive dermopathy and accumulation of Lamin A precursors. Hum Mol Genet 14:1503–1513
188. Smigiel R, Jakubiak A, Esteves-Vieira V et al (2010) Novel frameshifting mutations of the ZMPSTE24 gene in two siblings affected with restrictive dermopathy and review of the mutations described in the literature. Am J Med Genet A 152A:447–452
189. Barrowman J, Wiley PA, Hudon-Miller SE et al (2012) Human ZMPSTE24 disease mutations: residual proteolytic activity correlates with disease severity. Hum Mol Genet 21:4084–4093
190. Sevenants L, Wouters C, De Sandre-Giovannoli A et al (2005) Tight skin and limited joint movements as early presentation of Hutchinson-Gilford progeria in a 7-week-old infant. Eur J Pediatr 164:283–286
191. Starke S, Meinke P, Camozzi D et al (2013) Progeroid laminopathy with restrictive dermopathy-like features caused by an isodisomic LMNA mutation p.R435C. Aging (Albany NY) 5:445–459
192. Youn GJ, Uzunyan M, Vachon L et al (2010) Autosomal recessive LMNA mutation causing restrictive dermopathy. Clin Genet 78:199–200
193. Evangelisti C, Rusciano I, Mongiorgi S et al (2022) The wide and growing range of lamin B-related diseases: from laminopathies to cancer. Cell Mol Life Sci 79:126
194. Odell J, Lammerding J (2023) Lamins as structural nuclear elements through evolution. Curr Opin Cell Biol 85:102267
195. Neri I, Ramazzotti G, Mongiorgi S et al (2023) Understanding the ultra-rare disease autosomal dominant leukodystrophy: an updated review on morpho-functional alterations found in experimental models. Mol Neurobiol 60:6362–6372
196. Kim Y (2023) The impact of altered lamin B1 levels on nuclear lamina structure and function in aging and human diseases. Curr Opin Cell Biol 85:102257
197. Raininko R, Gosky M, Padiath QS (1993) LMNB1-related autosomal dominant leukodystrophy. In: Adam MP, Feldman J, Mirzaa GM et al (eds) GeneReviews®. University of Washington, Seattle (WA)
198. Giorgio E, Robyr D, Spielmann M et al (2015) A large genomic deletion leads to enhancer adoption by the lamin B1 gene: a second path to autosomal dominant adult-onset demyelinating leukodystrophy (ADLD). Hum Mol Genet 24:3143–3154
199. Padiath QS, Saigoh K, Schiffmann R et al (2006) Lamin B1 duplications cause autosomal dominant leukodystrophy. Nat Genet 38:1114–1123
200. Padiath QS (2019) Autosomal dominant leukodystrophy: a disease of the nuclear lamina. Front Cell Dev Biol 7:41
201. Heng MY, Lin S-T, Verret L et al (2013) Lamin B1 mediates cell-autonomous neuropathology in a leukodystrophy mouse model. J Clin Invest 123:2719–2729
202. Rolyan H, Tyurina YY, Hernandez M et al (2015) Defects of lipid synthesis are linked to the age-dependent demyelination caused by Lamin B1 overexpression. J Neurosci 35:12002–12017
203. Ferrera D, Canale C, Marotta R et al (2014) Lamin B1 overexpression increases nuclear rigidity in autosomal dominant leukodystrophy fibroblasts. FASEB J 28:3906–3918
204. Cristofoli F, Moss T, Moore HW et al (2020) De Novo variants in LMNB1 cause pronounced syndromic microcephaly and disruption of nuclear envelope integrity. Am J Hum Genet 107:753–762
205. Parry DA, Martin C-A, Greene P et al (2021) Heterozygous lamin B1 and lamin B2 variants cause primary microcephaly and define a novel laminopathy. Genet Med 23:408–414
206. Coffinier C, Jung H-J, Nobumori C et al (2011) Deficiencies in lamin B1 and lamin B2 cause neurodevelopmental defects and distinct nuclear shape abnormalities in neurons. Mol Biol Cell 22:4683–4693
207. Dittmer TA, Misteli T (2011) The lamin protein family. Genome Biol 12:222
208. Damiano JA, Afawi Z, Bahlo M et al (2015) Mutation of the nuclear lamin gene LMNB2 in progressive myoclonus epilepsy with early ataxia. Hum Mol Genet 24:4483–4490

209. Farajzadeh Valilou S, Karimzad Hagh J, Salimi Asl M et al (2021) A novel biallelic LMNB2 variant in a patient with progressive myoclonus epilepsy and ataxia: a case of laminopathy. Clin Case Rep 9:e04520
210. Soleimanipour F, Razmara E, Rahbarizadeh F et al (2022) A novel missense variant in the LMNB2 gene causes progressive myoclonus epilepsy. Acta Neurol Belg 122:659–667
211. Akinci B, Onay H, Demir T et al (2017) Clinical presentations, metabolic abnormalities and end-organ complications in patients with familial partial lipodystrophy. Metab Clin Exp 72:109–119
212. Hegele RA, Cao H, Liu DM et al (2006) Sequencing of the reannotated LMNB2 gene reveals novel mutations in patients with acquired partial lipodystrophy. Am J Hum Genet 79:383–389
213. Gao J, Li Y, Fu X, Luo X (2012) A Chinese patient with acquired partial lipodystrophy caused by a novel mutation with LMNB2 gene. J Pediatr Endocrinol Metab 25:375–377
214. Guelen L, Pagie L, Brasset E et al (2008) Domain organization of human chromosomes revealed by mapping of nuclear lamina interactions. Nature 453:948–951
215. Meuleman W, Peric-Hupkes D, Kind J et al (2013) Constitutive nuclear lamina-genome interactions are highly conserved and associated with A/T-rich sequence. Genome Res 23:270–280
216. van Steensel B, Belmont AS (2017) Lamina-associated domains: links with chromosome architecture, heterochromatin, and gene repression. Cell 169:780–791
217. Camozzi D, Capanni C, Cenni V et al (2014) Diverse lamin-dependent mechanisms interact to control chromatin dynamics. Nucleus 5:427–440
218. Alagna NS, Thomas TI, Wilson KL, Reddy KL (2023) Choreography of lamina-associated domains: structure meets dynamics. FEBS Lett 597:2806–2822
219. Cho S, Irianto J, Discher DE (2017) Mechanosensing by the nucleus: from pathways to scaling relationships. J Cell Biol 216:305–315
220. Davidson PM, Lammerding J (2014) Broken nuclei—lamins, nuclear mechanics, and disease. Trends Cell Biol 24:247–256
221. Kalukula Y, Stephens AD, Lammerding J, Gabriele S (2022) Mechanics and functional consequences of nuclear deformations. Nat Rev Mol Cell Biol 23:583–602
222. Tajik A, Zhang Y, Wei F et al (2016) Transcription upregulation via force-induced direct stretching of chromatin. Nat Mater 15:1287–1296
223. Shah PP, Santini GT, Shen KM, Jain R (2023) InterLINCing chromatin organization and mechanobiology in laminopathies. Curr Cardiol Rep 25:307–314
224. Shah PP, Lv W, Rhoades JH et al (2021) Pathogenic LMNA variants disrupt cardiac lamina-chromatin interactions and de-repress alternative fate genes. Cell Stem Cell 28:938–954.e9
225. Buxboim A, Kronenberg-Tenga R, Salajkova S et al (2023) Scaffold, mechanics and functions of nuclear lamins. FEBS Lett 597:2791–2805
226. Handoko L, Xu H, Li G et al (2011) CTCF-mediated functional chromatin interactome in pluripotent cells. Nat Genet 43:630–638
227. Pickersgill H, Kalverda B, de Wit E et al (2006) Characterization of the Drosophila melanogaster genome at the nuclear lamina. Nat Genet 38:1005–1014
228. Cheedipudi SM, Matkovich SJ, Coarfa C et al (2019) Genomic reorganization of Lamin-associated domains in cardiac myocytes is associated with differential gene expression and DNA methylation in human dilated cardiomyopathy. Circ Res 124:1198–1213
229. Harr JC, Schmid CD, Muñoz-Jiménez C et al (2020) Loss of an H3K9me anchor rescues laminopathy-linked changes in nuclear organization and muscle function in an Emery-Dreifuss muscular dystrophy model. Genes Dev 34:560–579
230. Perovanovic J, Dell'Orso S, Gnochi VF et al (2016) Laminopathies disrupt epigenomic developmental programs and cell fate. Sci Transl Med 8:335ra58
231. Gonzalo S, Kreienkamp R (2015) DNA repair defects and genome instability in Hutchinson-Gilford Progeria Syndrome. Curr Opin Cell Biol 34:75–83
232. Nava MM, Miroshnikova YA, Biggs LC et al (2020) Heterochromatin-driven nuclear softening protects the genome against mechanical stress-induced damage. Cell 181:800–817.e22

233. Earle AJ, Kirby TJ, Fedorchak GR et al (2020) Mutant lamins cause nuclear envelope rupture and DNA damage in skeletal muscle cells. Nat Mater 19:464–473
234. Chang W, Wang Y, Luxton GWG et al (2019) Imbalanced nucleocytoskeletal connections create common polarity defects in progeria and physiological aging. Proc Natl Acad Sci USA 116:3578–3583
235. Larrieu D, Britton S, Demir M et al (2014) Chemical inhibition of NAT10 corrects defects of laminopathic cells. Science 344:527–532
236. Lee G-E, Byun J, Lee C-J, Cho Y-Y (2023) Molecular mechanisms for the regulation of nuclear membrane integrity. IJMS 24:15497
237. Balmus G, Larrieu D, Barros AC et al (2018) Targeting of NAT10 enhances healthspan in a mouse model of human accelerated aging syndrome. Nat Commun 9:1700
238. Guénantin A-C, Jebeniani I, Leschik J et al (2021) Targeting the histone demethylase LSD1 prevents cardiomyopathy in a mouse model of laminopathy. J Clin Invest 131(e136488):136488
239. Veltrop RJA, Kukk MM, Topouzidou K et al (2024) From gene to mechanics: a comprehensive insight into the mechanobiology of LMNA mutations in cardiomyopathy. Cell Commun Signal 22:197
240. Butin-Israeli V, Adam SA, Goldman AE, Goldman RD (2012) Nuclear lamin functions and disease. Trends Genet 28:464–471
241. Capo-chichi CD, Cai KQ, Smedberg J et al (2011) Loss of A-type lamin expression compromises nuclear envelope integrity in breast cancer. Chin J Cancer 30:415–425
242. Zink D, Fischer AH, Nickerson JA (2004) Nuclear structure in cancer cells. Nat Rev Cancer 4:677–687
243. Sakthivel KM, Sehgal P (2016) A novel role of lamins from genetic disease to cancer biomarkers. Oncol Rev 10
244. Denais C, Lammerding J (2014) Nuclear mechanics in cancer. Adv Exp Med Biol 773:435–470
245. Tsai J-W, Chen Y, Kriegstein AR, Vallee RB (2005) LIS1 RNA interference blocks neural stem cell division, morphogenesis, and motility at multiple stages. J Cell Biol 170:935–945
246. Davidson PM, Sliz J, Isermann P et al (2015) Design of a microfluidic device to quantify dynamic intra-nuclear deformation during cell migration through confining environments. Integr Biol 7:1534–1546
247. Lammerding J, Fong LG, Ji JY et al (2006) Lamins A and C but not lamin B1 regulate nuclear mechanics. J Biol Chem 281:25768–25780
248. Pajerowski JD, Dahl KN, Zhong FL et al (2007) Physical plasticity of the nucleus in stem cell differentiation. Proc Natl Acad Sci USA 104:15619–15624
249. Ungricht R, Kutay U (2017) Mechanisms and functions of nuclear envelope remodelling. Nat Rev Mol Cell Biol 18:229–245
250. Yabuki M, Miyake T, Doi Y et al (1999) Role of nuclear lamins in nuclear segmentation of human neutrophils. Physiol Chem Phys Med NMR 31:77–84
251. Kong L, Schäfer G, Bu H et al (2012) Lamin A/C protein is overexpressed in tissue-invading prostate cancer and promotes prostate cancer cell growth, migration and invasion through the PI3K/AKT/PTEN pathway. Carcinogenesis 33:751–759
252. Skvortsov S, Schäfer G, Stasyk T et al (2011) Proteomics profiling of microdissected low- and high-grade prostate tumors identifies Lamin A as a discriminatory biomarker. J Proteome Res 10:259–268
253. Zuo L, Zhao H, Yang R et al (2018) Lamin A/C might be involved in the EMT signalling pathway. Gene 663:51–64
254. Wang Y, Wu R, Cho KR et al (2009) Differential protein mapping of ovarian serous adenocarcinomas: identification of potential markers for distinct tumor stage. J Proteome Res 8:1452–1463
255. Willis ND, Cox TR, Rahman-Casañs SF et al (2008) Lamin A/C is a risk biomarker in colorectal cancer. PLoS One 3:e2988

256. Alhudiri IM, Nolan CC, Ellis IO et al (2019) Expression of Lamin A/C in early-stage breast cancer and its prognostic value. Breast Cancer Res Treat 174:661–668
257. Matsumoto A, Hieda M, Yokoyama Y et al (2015) Global loss of a nuclear lamina component, lamin A/C, and LINC complex components SUN 1, SUN 2, and nesprin-2 in breast cancer. Cancer Med 4:1547–1557
258. Wu Z, Wu L, Weng D et al (2009) Reduced expression of lamin A/C correlates with poor histological differentiation and prognosis in primary gastric carcinoma. J Exp Clin Cancer Res 28:8
259. Broers JL, Raymond Y, Rot MK et al (1993) Nuclear A-type lamins are differentially expressed in human lung cancer subtypes. Am J Pathol 143:211–220
260. Broers JLV, Ramaekers FCS (2014) The role of the nuclear lamina in cancer and apoptosis. In: Schirmer EC, de las Heras JI (eds) Cancer biology and the nuclear envelope. Springer, New York, NY, pp 27–48
261. Kaufmann SH, Mabry M, Jasti R, Shaper JH (1991) Differential expression of nuclear envelope lamins A and C in human lung cancer cell lines. Cancer Res 51:581–586
262. Wang J, Kondo T, Nakazawa T et al (2019) Constitutional abnormality of nuclear membrane proteins in small cell lung carcinoma. Virchows Arch 475:407–414
263. Capo-chichi CD, Aguida B, Chabi NW et al (2016) Lamin A/C deficiency is an independent risk factor for cervical cancer. Cell Oncol (Dordr) 39:59–68
264. Saarinen I, Mirtti T, Seikkula H et al (2015) Differential predictive roles of A- and B-type nuclear Lamins in prostate cancer progression. PLoS One 10:e0140671
265. Belt EJT, Fijneman RJA, van den Berg EG et al (2011) Loss of lamin A/C expression in stage II and III colon cancer is associated with disease recurrence. Eur J Cancer 47:1837–1845
266. Moss SF, Krivosheyev V, de Souza A et al (1999) Decreased and aberrant nuclear lamin expression in gastrointestinal tract neoplasms. Gut 45:723–729
267. Setti Boubaker N, Gurtner A, Trabelsi N et al (2021) The diagnostic applicability of A-type Lamin in non-muscle invasive bladder cancer. Ann Diagn Pathol 54:151808
268. Kaspi E, Frankel D, Guinde J et al (2017) Low lamin A expression in lung adenocarcinoma cells from pleural effusions is a pejorative factor associated with high number of metastatic sites and poor Performance status. PLoS One 12:e0183136
269. Cicchillitti L, Corrado G, Carosi M et al (2017) Prognostic role of NF-YA splicing isoforms and Lamin A status in low grade endometrial cancer. Oncotarget 8:7935–7945
270. Gong G, Chen P, Li L et al (2015) Loss of lamin A but not lamin C expression in epithelial ovarian cancer cells is associated with metastasis and poor prognosis. Pathol Res Pract 211:175–182
271. Aljada A, Doria J, Saleh AM et al (2016) Altered Lamin A/C splice variant expression as a possible diagnostic marker in breast cancer. Cell Oncol 39:161–174
272. Sengupta D, Mukhopadhyay A, Sengupta K (2023) Elevated levels of Lamin A promote HR and NHEJ-mediated repair mechanisms in high-grade ovarian serous carcinoma cell line. Cells 12:757
273. Bell ES, Shah P, Zuela-Sopilniak N et al (2022) Low lamin A levels enhance confined cell migration and metastatic capacity in breast cancer. Oncogene 41:4211–4230
274. Sun S, Xu MZ, Poon RT et al (2010) Circulating Lamin B1 (LMNB1) biomarker detects early stages of liver cancer in patients. J Proteome Res 9:70–78
275. Yang Y, Xiao W, Liu R et al (2022) A Lamin family-based signature predicts prognosis and immunotherapy response in hepatocellular carcinoma. J Immunol Res 2022:1–23
276. Hong J-H, Liang S-T, Wang AS-S et al (2022) LMNB1, a potential marker for early prostate cancer progression. Am J Cancer Res 12:3390–3404
277. Luo F, Han J, Chen Y et al (2021) Lamin B1 promotes tumor progression and metastasis in primary prostate cancer patients. Future Oncol 17:663–673
278. Huang Y, Zhang L, Liu T, Liang E (2023) LMNB1 targets FOXD1 to promote progression of prostate cancer. Exp Ther Med 26:513

279. Saleh T, Alhesa A, El-Sadoni M et al (2022) The expression of the senescence-associated biomarker Lamin B1 in human breast cancer. Diagnostics 12:609
280. Wazir U, Ahmed M, Bridger J et al (2013) The clinicopathological significance of lamin A/C, lamin B1 and lamin B receptor mRNA expression in human breast cancer. Cell Mol Biol Lett 18:595
281. Dong J, Ru Y, Zhai L et al (2023) LMNB1 deletion in ovarian cancer inhibits the proliferation and metastasis of tumor cells through PI3K/Akt pathway. Exp Cell Res 426:113573
282. Zhao C-C, Chen J, Zhang L-Y et al (2021) Lamin B2 promotes the progression of triple negative breast cancer via mediating cell proliferation and apoptosis. Biosci Rep 41:BSR20203874
283. Yu ZY, Jiang XY, Zhao RR et al (2021) Lamin B1 deficiency promotes malignancy and predicts poor prognosis in gastric cancer. Neoplasma 67:1303–1313
284. Dong C-H, Jiang T, Yin H et al (2021) LMNB2 promotes the progression of colorectal cancer by silencing p21 expression. Cell Death Dis 12:331
285. Chen Y, Chen J, Shi Y et al (2024) Lamin B1: a novel biomarker in adult and pediatric adrenocortical carcinoma. Endocr Relat Cancer 31(3):e230178
286. Li L, Du Y, Kong X et al (2013) Lamin B1 is a novel therapeutic target of betulinic acid in pancreatic cancer. Clin Cancer Res 19:4651–4661
287. Ji J, Li H, Chen J, Wang W (2022) Lamin B2 contributes to the proliferation of bladder cancer cells via activating the expression of cell division cycle-associated protein 3. Int J Mol Med 50:111
288. Wang J, Ma M, Liu X (2022) Lamins B2 promotes esophageal cancer by stimulating proliferation and inhibiting apoptosis. Ann Clin Lab Sci 52:202–212
289. Jia Y, Vong JS-L, Asafova A et al (2019) Lamin B1 loss promotes lung cancer development and metastasis by epigenetic derepression of RET. J Exp Med 216:1377–1395
290. Machiels BM, Broers JL, Raymond Y et al (1995) Abnormal A-type lamin organization in a human lung carcinoma cell line. Eur J Cell Biol 67:328–335
291. Zhang M-Y, Han Y-C, Han Q et al (2020) Lamin B2 promotes the malignant phenotype of non-small cell lung cancer cells by upregulating dimethylation of histone 3 lysine 9. Exp Cell Res 393:112090
292. Stadelmann B, Khandjian E, Hirt A et al (1990) Repression of nuclear lamin A and C gene expression in human acute lymphoblastic leukemia and non-Hodgkin's lymphoma cells. Leuk Res 14:815–821
293. Agrelo R, Setien F, Espada J et al (2005) Inactivation of the *Lamin A/C* gene by CpG Island promoter hypermethylation in hematologic malignancies, and its association with poor survival in nodal diffuse large B-cell lymphoma. JCO 23:3940–3947
294. Wu G, Tian Q, Liu J et al (2022) Comprehensive analysis of expression and prognosis for LMNB family genes in human sarcoma. Medicine 101
295. Evangelisti C, Paganelli F, Giuntini G et al (2020) Lamin A and prelamin A counteract migration of osteosarcoma cells. Cells 9:774
296. Chiarini F, Paganelli F, Balestra T et al (2022) Lamin A and the LINC complex act as potential tumor suppressors in Ewing Sarcoma. Cell Death Dis 13:346
297. Doebele RC, Davis LE, Vaishnavi A et al (2015) An oncogenic *NTRK* fusion in a patient with soft-tissue sarcoma with response to the tropomyosin-related kinase inhibitor LOXO-101. Cancer Discov 5:1049–1057
298. Haller F, Knopf J, Ackermann A et al (2016) Paediatric and adult soft tissue sarcomas with *NTRK1* gene fusions: a subset of spindle cell sarcomas unified by a prominent myopericytic/haemangiopericytic pattern. J Pathol 238:700–710
299. Pei S, Wang X, Wang X et al (2022) Aberrant nuclear lamina contributes to the malignancy of human gliomas. J Genet Genomics 49:132–144
300. Gatti G, Vilardo L, Musa C et al (2021) Role of Lamin A/C as candidate biomarker of aggressiveness and tumorigenicity in glioblastoma multiforme. Biomedicine 9:1343
301. Maresca G, Natoli M, Nardella M et al (2012) LMNA knock-down affects differentiation and progression of human neuroblastoma cells. PLoS One 7:e45513

302. Rauschert I, Aldunate F, Preussner J et al (2017) Promoter hypermethylation as a mechanism for Lamin A/C silencing in a subset of neuroblastoma cells. PLoS One 12:e0175953
303. Roncato F, Regev O, Feigelson SW et al (2021) Reduced Lamin A/C does not facilitate cancer cell transendothelial migration but compromises lung metastasis. Cancers (Basel) 13:2383
304. Lämmerhirt L, Kappelmann-Fenzl M, Fischer S et al (2022) Knockdown of lamin B1 and the corresponding lamin B receptor leads to changes in heterochromatin state and senescence induction in malignant melanoma. Cells 11:2154

# Chapter 5
# Differential Expression of Lamins and Modulation of Chromatin Architecture in Carcinogenesis: *tous pour un et un pour tous*

**Shreyasi Dey Sarkar, Duhita Sengupta, and Kaushik Sengupta**

**Abstract** Lamins are intermediate filament proteins of the nucleus that are present in nuclear lamina as well as nucleoplasm. They play diverse roles in maintaining the structure and rigidity of the nucleus as well as nuclear homeostasis. Lamins are of two main types—A and B. B-type lamins are expressed from the embryonic stage, whereas A-type lamins are expressed during cell differentiation. Both A- and B-type lamins form distinct but interacting networks that contribute to differential chromosome tethering and distribution within the nucleus. A- and B-type lamins maintain the euchromatin-to-heterochromatin ratio in health and disease. Interestingly, lamin A/B itself varies largely and distinctly in different types of cancer. Likewise, the lamina-associated domains of the chromatin network get significantly altered in the process of carcinogenesis. We have discussed here the differential expression of lamin proteins in different cancers, contributing to distinct genome organization, ultimately precipitating into diverse neoplastic transformation.

**Keywords** Lamin · Chromosome territory · LIPs · LADs · Cancer · Epigenome

## Genesis of "Chromosome Territory"

The human genome exists as a complex hierarchical folding of the very basic unit molecule—DNA. DNA gets wrapped around the octamer of core histone proteins to make the nucleosome. Multiple nucleosomes connected in a stack with DNA form a beads-on-a-string fiber of ~10 nanometers, which supercoils on itself to form

---

S. D. Sarkar · D. Sengupta · K. Sengupta (✉)
Biophysics and Structural Genomics Division, Saha Institute of Nuclear Physics, Kolkata, West Bengal, India

Homi Bhabha National Institute, Mumbai, Maharashtra, India
e-mail: kaushik.sengupta@saha.ac.in

© The Author(s), under exclusive license to Springer Nature Switzerland AG 2025 
K. Sengupta, A. Muchir (eds.), *Lamins in Genome Organization and Diseases*, Subcellular Biochemistry 115, https://doi.org/10.1007/978-3-032-00537-3_5

fibers of increasing diameters [1–3]. At the final level of organization, chromatin fiber folds into a distinguishable sub-chromosomal domain of ~1 Mb size, which folds and undergoes a compaction to give rise to interphase chromosomes [4, 5]. Interphase chromosomes generally occupy specific 3D nuclear volumes within the nucleus, termed chromosome territories (CTs) [6]. The hypothesis of chromosome territory was put forward by Carl Rabl (1885) and Theodor Boveri (1909) who speculated a nonrandom and preserved localization of chromosomes during their experiments with different cell cycle stages, respectively, on plant cells of *Salamandra maculata* and animal cells of horse's roundworm [7, 8]. Boveri who also coined the term "chromosome territories" came up with three strikingly novel observations of that time, which had similar conclusions to that made by Rabl who observed the orientation of centromere-telomere of chromosomes during anaphase to remain conserved throughout the cell cycle in salamander cell division [8]. Boveri noticed that the nuclear material conserves a similar position and neighborhood at the end of one mitosis to another beginning of prophase, and these territories occupied by chromosomes are surrounded by interchromatin spaces [7]. In 1905, Eduard Strasburger published his findings of the similar territorial distribution of chromosomes in plant cell nuclei [9]. In 1973, the idea of chromosome territory was questioned by the experiments using electron microscopy, which showed that interphase chromatin randomly intermingles within the nucleus like a "bowl of spaghetti." [10] The first experimental evidence to support the chromosome territory theory was published by Stephen M. Stack (1977) [11], who found a patch-like territorial distribution of chromosomes in fixed Chinese hamster ovary cells (CHO) and cells of *Allium cepa*. His observations of chromosomal dynamics during cell division fell in line with the previous observations of Boveri [7, 11]. During the 1980s, concrete shreds of evidence in favor of CTs came from Thomas Cremer group who, in their laser UV microbeam experiments, showed that pulse-labeled recovery of micro laser beam ablated nuclear damage, was restricted to specific chromosomes in nuclear space, and was conserved, not intermingling randomly [12, 13]. The first visualization of a chromosome territorial organization became possible in 1988, with the advent of the fluorescence in situ hybridization (FISH) technique. In vivo, live tracing of chromosomes for a short period was made possible using sequence-specific DNA probes composed of biotin-labeled nucleotides and an anti-biotin antibody-mediated detection system, which was essentially non-radioactive in nature [14, 15]. The interchromosome domain (ICD) model of nuclear architecture was put forward in 1993 by Zirbel et al. [16], and the term interchromatin domain compartment was coined by the T. Cremer group in 1995. This gradually established the presence of specific territories of the nonrandom chromosome intermingling and interchromatin compartment (IC) running in between CTs. The finest branches emanating from the CTs, detectable at the ultrastructural level between ~1 Mb and ~ 100 kb chromatin domains, harbor nuclear bodies or speckles enriched in splicing factors [17–19]. For the first time in 2005, after many advances in technologies, with the association of chromosome painting probes, 3D FISH protocols for fixed and live cell imaging, confocal microscopy, and 3D and 4D image analysis, all chromosome territories could be studied with 3D nuclear positioning of all the chromosomes [20–25].

## From Loops to Knots: Evolving Models of Chromosome Territories (CTs)

During compaction, the tightly coiled 100–130 nm "chromonema" fibers of early G1 condense to form 60–80 nm chromonema fibers, which can be traced as discrete fibers up to 2–3 microns in length [5]. The fibers undergo further compaction from 10 to 30 nm fibers. According to different schools of thought, the loop formation and formation of interphase chromatin domains may involve small-scale chromatin loops (SLs) [26], which are 50–200 kbp long, or giant loops (GLs) [27, 28] comprising of one to several Mbp DNA. Initial models that supported SLs as the building blocks for chromosome territories (CTs) include the ICD model [16]. The ICD model depicts the CTs to have a smooth surface with considerable infoldings where the active genes are peripherally located thereby allowing the genes to interact with functional components such as nuclear speckles and nuclear bodies [6]. The infoldings of the loops extend into neighboring territories expanding the interchromosome domain area.

A modified model backed by several in silico experiments [29, 30], proposed by the Thomas Cremer group, is the CT-IC model (chromosome territory-interchromatin compartment), which argues for the distribution of the territories to be in a sponge-like architecture with condensed chromatin fibers, accommodating DNA-free channels or lacunae that starts at the nuclear pores, extending in between the chromatin structures forming IC (interchromatin compartment) [19]. This model outgrew the concept of the active domains only existing at the periphery of CTs, saying that CTs are formed from 100 kbp SLs, whose configurations are subjected to the active or silent status of their genes. According to this model, a series of SLs representing non-contiguous DNA segments from a given CT or even from different CTs form rosette-like ~1 Mbp chromatin domains that are connected by linker chromatin domains, and several of such connected 1 Mbp domains constitute chromatin fibers. The poised or actively transcribed genes remain directly exposed at the IC linings via partial or complete decondensation of ~100 *kbp chromatin domains*. This is also known as the multi-loop sub-compartment (MLS) model [26]. While IC is synonymous with ICD, the only thing is IC continues as DNA-free space between neighboring CTs and also between compact chromatin domains within CTs [31].

The first model favoring the chromosome territory formation from the giant chromatin loops (CT-GL) was the random walk/giant loop (RW/GL) model proposed by Sachs et al. [28] According to the GL model, continuous chromatin fiber comprising several Mb DNA forms GLs, and one such giant loop remains connected to other such loops with 200 kb chromatin linker within the same CT. Chromatin dispersion or compaction due to functional requirements is taken care of by the modulation of the loop size or linker size [26]. Expansion of GLs from corresponding CTs has been reported by several groups in numerous experiments where the gene-dense regions or actively transcribed gene clusters [32] of compact CT cores looped out with a higher frequency [33–36]. Open chromatin fibers that show looping out are essentially found to be enriched with the highest gene density but do not always represent highly transcribed genes [37]. The GL model provides provision for the intermingling of neighboring chromosome territories where actively transcribed regions from different chromosome territories loop out to come to proximity and can condense back onto CT cores when transcription ceases [33]. When required, the decondensed intermingling GLs can come across the nuclear speckles, RNP clusters, and other regulatory proteins and elements such as enhancers and promoters.

To accommodate the observations that, at a given point in time, most of the nuclear material remains in a higher-order structure above the level of 10 to 30 nm fibers, we will have to take into account a dynamic, context-based transition of both the models, which, when necessary, allows chromatin to get detangled to allow transcription. Although the intermingling of chromosome territories and formation of transcription hub are better explained by the CT-GL model, the dynamic nature of the interchromatin compartments is more favored by the CT-IC model, which accommodates extensive changes in the IC lacunae to allow spatial interactions between two distantly located genes.

Various models of chromosome territory formation rely on inter and intra-chromosome looping. Such looping depends on various other regulators. As reviewed previously [38], intra-chromosomal looping can be of four types correlating with their functional consequences relating to transcription. The first is the looping of the 5′ end of the transcribed gene to come to the vicinity of the termination site, which ensures efficient recycling of the RNA polymerase II from the termination site back onto the promoter making it possible to rapidly reactivate gene transcription of such functionally active genes. Such gene loops preserve a short-term memory of the transcriptionally active genes [39] and establish transcription directionality [40] for the coding genes. Such loops were reported in yeast [41] for the first time and subsequently in some mammalian promoters [42]. The second type of functionally relevant looping brings the distant enhancer in contact with promoters, and this might involve gathering different distant enhancers to a single promoter [43, 44]. Targeting of transcription cofactors, noncoding regulatory RNAs [45, 46] may be involved in such loop formation and transcriptional activation. Not only do enhancers loop in to interact with promoters but also distal promoters may be involved in loop formation to access and regulate the proximal promoter [47]. The third type of looping essentially involves transcription silencing and is mediated by the looping of the Polycomb response element sequence in the proximity of the promoter along with the juxtaposition of the Polycomb proteins and promoter-associated factors [48–50]. This type of looping is reported in mammalian cells [51, 52] as well as in fruit fly [48, 53]. The fourth type of looping is mediated by CTCF (CCCTC-binding factor) and cohesins [54]. Genome-wide CTCF-binding sites (CBSs) are discovered in mouse and human genomes [55]. This part of the genome is highly influenced by methylation profile depending on the required transcriptional state [56]. CBSs act as insulator sequences that prevent enhancers from accessing promoters located in between them [56]. Insulators, along with the bound proteins and other factors, promote loop formation that acts as a structural scaffold to isolate specific gene sequences regulating their activation or repression.

Interchromosomal contacts are influenced by functional interactions between different chromosomes mediated by transcription factors, regulatory RNAs, and epigenetic modifications that bring together different chromosomes in proximity. Some examples of interchromosomal interactions involve X-chromosome inactivation, individual monoallelically expressed genes such as at imprinted loci, arrays of ribosomal RNA genes clustered in transcription factories in the nucleolus, tethering of telomeric and centromeric regions in the nuclear periphery, etc. The long-range movements of chromosomes in mammals and *Drosophila* provide experimental evidence for long-range movements of chromosomes and adjacent territories across the nucleus [24]. Genome topology is not exclusive to its functional role, and epigenetics plays a deterministic role in shaping the spatiotemporal topological variations. Hence, the way forward to uncover the principles of genome organization is to revise the key role players who hold the picture together in the scenario of genome distribution.

Protein complexes such as structural maintenance of chromosomes (SMCs) [57, 58] proteins, which actively play a role in modulating different aspects of looping [59] and organization of chromatin, intricately work alongside cohesin [60, 61] and condensin complexes [62]. Self-interacting regions of chromatin, which are mainly segregated by CTCF and cohesin complexes, comprise the topologically associated domains (TADs) [63–65] where the functionally active or inactive regions of chromatin maintain a regulated proximity. Multiple TADs and loops organize themselves into transcriptionally active or inactive compartments, referred to as A and B compartments [66], respectively, which are not mutually exclusive as per recent findings [67]. This compartmentalization of chromatin depends on several factors [68] such as properties of chromatin or DNA sequence [69–72], which makes way for the methylation profile as well as other epigenetic modifications [73–76], which in turn control the differential recruitment of transcription factors [74, 75, 77] and finally the physical location of chromatin within the nucleus and the relation of chromatin with nuclear lamina [78].

## Nuclear Lamina and Regulation of Chromosomal Landscape

The hierarchy of higher-order chromatin architecture begins with the double helix DNA being wrapped around the nucleosomes and finally culminates into whole chromosomes, which, in interphase nuclei, occur as specific 3D entities, i.e., CTs [6]. As the functional domains of such territories remain within an ever-interacting network of the nuclear matrix, a huge role of the nuclear matrix as a tool for the organization and functional regulation and dynamics of CTs is inevitable.

Among the three principal components of a typical metazoan nucleus are the nuclear pore complex (NPC), nucleoplasm, and nuclear envelope consisting of an outer nuclear membrane (ONM) and an inner nuclear membrane (INM). INM facing the nucleoplasm consists of the lamina, which is contributed majorly by a family of type V intermediate filament (IF) proteins called lamins, which form a meshwork on the nucleoplasmic face of the inner nuclear membrane [79]. Vertebrates evolved to come up with four lamin genes, LMNA, LMNB1, LMNB2, and LIII [80], which encode four major and three minor types of lamin proteins. The major mammalian lamins are classified as lamins B1, B2, A, and C, among which the LMNA gene encodes the A-type lamins, which include two major isoforms lamins A and C. The B-type lamins include the major isoform B1, which is encoded by the LMNB1 gene, and the major isoform B2, encoded by the LMNB2 gene [81, 82]. A- and B-type lamins form distinct interacting networks [83] providing microdomains for chromatin organization and transcription regulation.

Chromosomes, depending on their transcriptional status and regulatory roles, can be classified into euchromatin and heterochromatin [66]. Euchromatin includes transcriptionally active loosely folded genome regions that interact with NPCs at the nuclear periphery as well as remains situated toward the center of the nuclei comprising the A compartment. On the other hand, the heterochromatin part consists of the transcriptionally silent genes, displaying a tightly packed arrangement in individual cell types. This heterochromatin remains tethered to the INM and interacts with nuclear lamins (lamins A/C, B1, and B2). More than one-third of the human genome shows an association with the nuclear lamina and constitutes a major part of the epigenome, termed lamina-associated domains (LADs), and LADs generally form the B compartments. All chromosomes contribute to the LADs, making the count ~1000–1500 LADs, typically 10 kb–10 Mb in size. LADs can be facultative or constitutive (fLADs and cLADs, respectively), of which fLADs can loop out to interact with A compartments and get expressed [84]. Nuclear lamins regulate the distribution of chromatin in 3D space alongside many other proteins that contribute directly or indirectly to the regulation of chromosomal distribution and neighborhood. Nuclear lamina intricately associates with the maintenance of chromatin organization within the nucleus [83, 85–87]. Lamin meshwork plays an important role in chromatin packaging and its accessibility to epigenetic and transcription factors [75]. Several studies by tethering integrated reporter genes to the nuclear periphery and checking the expression profile of the reporter gene showed that translocation of a reporter gene to the nuclear lamina can contribute to gene repression, which to some extent is also dependent on the strength of the promoter and the type of cell [88]. Nuclear lamins can directly [89–91] mediate this interaction with chromatin via its tail domain or by binding either to histones [92, 93] or other lamin-interacting proteins (LIPs) that interact with chromatin such as the ones having LAP2-emerin-MAN1-domain (LEM-D) as emerin, MAN1 [94], lamin B receptor (LBR) [95–97], and LAP2$\beta$ [98] and by interacting with chromatin-associated proteins as BAF [97, 99], histone deacetylase 3 (HDAC3) [100], and heterochromatin protein 1 (HP1) [101]. The lamin A/C within the nucleus is associated with LAP2$\boldsymbol{\alpha}$, [102] pRb, and cyclin D3 [103] complexes. Studies showed direct or indirect binding of lamins to promoter regions or lamin A artificially tethered to promoters, which contributes to gene silencing [104]. The LADs [78, 105] are often part of the heterochromatin pool of the genome [106], and LADs harbor heterochromatin-like epigenetic modifications such as histone H3 lysine 9 di- or trimethylation (H3K9me2/3) or histone H3 lysine 27 trimethylation (H3K27me3), alongside lacking the active transcriptional modifications such as histone H3 or H4 lysine acetylation, histone H3 lysine 4 trimethylation (H3K4me3), or histone H3 lysine 36 trimethylation (H3K36me3) [107, 108]. Both A- and B-type lamin depletions were found to cause mislocalization of chromosome territories [109–114], and after the advent of Hi-C and similar methods, the role of lamins in territorial distribution of chromatin became more prominent [114]. In the case of laminopathies, severe disorganization of chromatin in the 3D nuclear space was reported. [115, 116] In Hutchinson-Gilford progeria syndrome (HGPS), a premature aging disease caused by a mutation in lamin A, SAMMY-seq, a high-throughput sequencing, showed an altered epigenetic landscape and associated heterochromatin regulation [117–119]. Alteration of radial positioning of LADs was found in lamin A mutants linked to a progeroid syndrome that aligned with the prediction from 3D genome modeling data [115, 120]. Pathogenesis in muscular dystrophy was found to be caused by dysfunctional Polycomb transcriptional repression and altered heterochromatin distribution due to mutations in lamin A [121]. Cardiomyopathy conditions are also accompanied by altered distribution of chromatin and transcriptional landscape [122, 123]. Distinct distribution of lamin A- and B-type filaments within the nucleus leads to differential contribution by them in chromatin distribution [124, 125]. Also, it is reported that the post-mitotic organization of the globular LADs in association with nuclear envelopes is regulated by lamin C distinctively and not lamin A [126]. Experiments on triple knockout mouse embryonic stem cells, which are devoid of all three major lamins, showed the contribution of lamins in the organization and delineation of LADs, TADs, and active and repressed state of chromatin [114].

## Involvement of Nuclear Lamina in Regulating Chromosomal Organizations in Malignancy

Cancer is one of the leading causes of human morbidity [127], and the 3D chromatin architecture has a significant impact on the hallmarks of cancer, as it is involved in the induction and repression of genes through multiple levels [128]. This dynamicity in chromatin arrangement and its commands on gene expression actively regulate the hallmarks of cancers, such as (i) sustenance of proliferative signaling, (ii) evasion from growth suppressors, (iii) resistance to cell death, (iv) activation of invasion and metastasis, (v) replicative immortality, (vi) induction of angiogenesis, (vii) reprogramming of energy metabolism, (viii) formation of the tumor microenvironment, (ix) inflammation, (x) evading immune system, and (xi) aberrant genome stability due to oncogenic mutations [129]. In regular situations, the relocation of human promoters from their natural LAD area to a neutral environment within the same cell activates them [130] causing deregulations of the involved genes. According to Ji et al. [131], the tumor suppressor TOPORS (TOP1 binding arginine/serine rich protein) depletion lowers chromatin-lamina contacts and the coverage of

LADs. Changes in chromatin organization can hasten the development of cancer and impact genomic organization at all levels. Hi-C measurements and microscopic observations confirm heterochromatin redistribution and decompaction in the scenario of carcinogenesis [132]. Genomic rearrangements in the carcinogenesis as duplications, deletions, inversions, and translocations can also cause a reshuffling of TADs in the nucleus, which might lead to activation of oncogenes by "enhancer hijacking" [133] leading to malignant phenotypes in several cancers [134–136]. However, it remains unclear if the global alteration in gene expressions in tumor cells is the cause of the reorganization of spatial chromatin architecture or if it precedes these alterations. It also is a matter of debate as to how nuclear lamins play their role in this context in various types of cancer as a biomarker [137]. Gaining a comprehensive understanding of how malignancy impacts chromatin architecture through lamin is probably crucial to understanding how cancer develops, progresses, and responds to treatment [138]. The molecules that bridge the lamin-chromatin crosstalk, which reportedly regulates the territorial distribution of chromatin, are versatile, and in different pathological scenarios including cancer, the expression of these molecules also gets dysregulated. Some of the discussed interactors are found to be direct mediators influencing the lamin-chromatin relationship, whereas others, showing dysregulated expression in pathological scenarios, take part in chromatin organization and are found to be the physical interactors of nuclear lamins in different high-throughput experiments. These molecules might be the potential links between lamin and altered chromatin landscape in cancer. CTCF as mentioned earlier is a multifunctional zinc finger protein with a highly conserved DNA-binding domain. It is implicated in multivalent roles related to gene expression as it acts with cohesin and other SMCs in the formation of loops and compartments and modulates higher-order chromatin organization. It also acts as a tumor suppressor as well as is associated with the regulation of several other tumor suppressor genes, such as p16, BRCA1, RB, RASSF1A, and CDH1(E-cadherin) in a methylation-sensitive fashion [139, 140].

CTCF shows insulator function by preventing interactions between enhancers and promoters and hence plays a major role in preventing the spread of heterochromatin by forming homodimers and multimers on specific chromatin sequences. It also helps in the maintenance of proper histone modifications. Epigenetic silencing of the tumor suppressor genes has been reported to correlate with the abolished interaction of CTCF and its binding partners at the upstream of the gene promoters in several different types of human cancer [139]. CTCF was also found to play a profound role in maintaining telomere homeostasis [141]. A relatively unchartered territory is the possible interaction between lamin A and CTCF. The initial hypothesis of a functional lamin/CTCF interaction came from the discovery that CTCF interacts with lamin A to mediate the peripheral tethering of chromosome 4q telomere within the nucleus [142, 143]. Other than this, CTCF and lamin A were previously copurified as interacting complexes from CTCF overexpressing HeLa cells [144], and recently via co-fractionation/mass spectrometry (CF/MS), these two proteins were found to interact in different breast cancer cells [145].

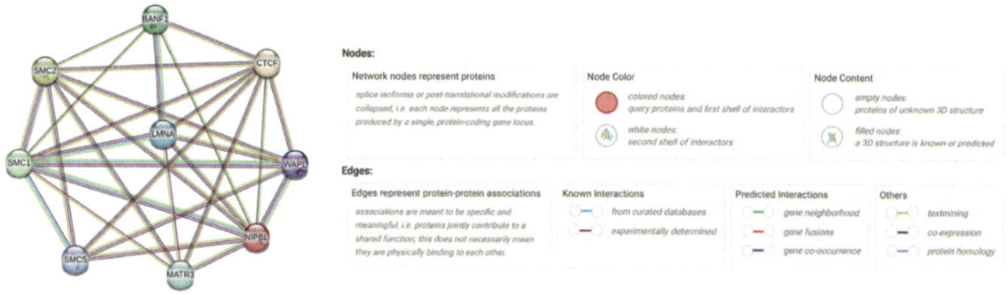


**Fig. 5.1** Known and predicted interactions of lamin A/C with other potential chromatin organization modulators derived from STRING (version: 12.0)

Other than CTCF, SATB1 (special AT-rich sequence binding protein 1) also determines various cellular responses by recruiting several histone modifiers and chromatin remodeling complexes, which promote loop formation and hence dictate the higher-order chromatin structures. SATB1 might act as the structural scaffold that provides the space for chromatin looping, establishing the proximal presence of the genes that share a similar regulatory environment. Different SMCs were found to be either colocalizing or cross-linked together with lamin A [146–148]. A recently highlighted protein Matrin 3 (MATR3), which interacts with chromatin remodeling factors like CTCF and cohesin, was also found in the lamin A interactome [149, 150]. Lamin A/C thus could be a potential candidate bridging different chromatin organization modulators as is indicated from the interactome analysis from STRING (Fig. 5.1).

Bringing together the available cues regarding different types of cancers showcasing differential expression of nuclear lamins and the molecules along with the associated pathways might lead us to understand lamin better as the modulator of genome organization onset of malignancy.

## Lamins: Modulators of Genome Organization in Different Cancers

### *Breast Cancer*

Breast cancer comprises 32% of the estimated newly diagnosed cancer cases in females and stands second in causing cancer morbidity in females globally [127]. Nuclear size and ploidy alterations in breast cancer are reported previously [151], which often is associated with atypical nuclear morphology and deregulation of expression levels of nuclear lamin proteins. Reportedly in breast cancer, both A-type lamins and B-type lamins were found to be depleted [152–154], making the expression level of nuclear lamin a prognostic marker for breast cancer. The significant depletion of lamin A in breast cancer, in line with CpG island promoter hypermethylation of other tumor suppressor genes during breast tumorigenesis, is indicative of lamin A to serve as a tumor suppressor in breast cancer scenario [152, 155]. A significant decrease in lamin A expression in breast cancer can lead to malignancy in different pathways. Several pathways are altered in association with LMNA downregulations in breast cancer, such as the Hippo signaling pathway involving YAP/TAZ, which responds to mechanical stresses at the extracellular matrix-cytoskeleton-lamin interconnectivity [156], PI3K/Akt signaling [157] pathway, etc. Along with nuclear atypia, chromosomal territorial distribution and inter-chromosomal associations also alter significantly in breast carcinoma [128, 158], showing relative peripheral positioning of chromosomes 4, 12, 15, 16, and 21 in breast cancer cells, harboring mostly the downregulated genes. Hi-C analysis on MCF-10A mammary epithelial and MCF-7 breast cancer cell lines revealed compartment switching of chromatin [159]. Along with that, severe disruption in local and global 3D chromatin architectures, including TADs, contact domains, and chromatin loops were also reported in breast cancer [160]. From another perspective, CTCF, which co-fractionates with lamin A/C, plays a key role in the modulation of breast cancer [161]. Other than its canonical role of maintaining the loops in TADs, CTCF also acts as a transcription factor to control the metastasis in breast cancer [162]. Intriguingly, the $p16^{INK4a}$ tumor suppressor gene, which remains aberrantly silenced in breast cancer and strongly co-expresses with lamin A/C [163], was found to be associated with loss of CTCF binding [140, 163].

## *Ovarian Cancer*

Ovarian cancer is one of the leading causes of cancer morbidity in females claiming 4% of total estimated deaths worldwide [127]. Expression of lamin A/C as well as lamin B1 was found to be increased in many reports [155, 164], while some observations reported a decrease [165] in lamin A/C depending on the cohort of patient populations and subtype of ovarian cancer. Deregulated expression of lamin A/C in ovarian carcinoma might influence the severe nuclear deformities as well as the emphatically altered epigenetic landscape [166, 167]. Epithelial ovarian cancer showed global hypomethylation of the genome, which essentially overlapped with late-replicating regions, LADs, and the H3K27me3 histone marks. Other than the involvement of nuclear lamin proteins in delineating LADs, a few genes among the plethora of downregulated genes were even found to correlate with lamin A/C upon interactome analysis from primary databases [163, 168]. Along with vast epigenetic alterations, serous ovarian carcinoma also harbors severe physical chromosome instability [169].

## *Cervical Cancer*

Cervical cancer is the fourth most common cancer in women globally [170]. Cervical cancer cells show extensive deformation in nuclear shape with large nuclei and specked chromatin distribution [171]. Expression of nuclear lamin in cervical cancer is not much explored, but an increased risk of human papillomavirus (HPV) infection in lamin A/C-deficient patients was reported [172]. Hi-C data of cervical cancer reported structural variations in chromatin showing rearrangement of the 3D genome architecture in comparison with the normal tissue, where almost 24% of the total genome was found to switch their A/B compartments [173]. However, the direct contribution of nuclear lamin in delineating the genome reorganization is yet unexplored in the context of cervical cancer.

## *Prostate Cancer*

Nuclear lamins act as potential biomarkers for prostate cancer [174, 175]. Lamin A/C expression in prostate cancer correlates with the Gleason score of the tumors and is indicative of the risk of the disease [176]. The deregulation of genomic organization in prostate cancer and associated pathways such as PI3k-PTEN-Akt leading to more aggressive cancer were found to be directly associated with lamin A/C expression [176]. Lamin A/C expression level also helps to delineate the EMT or MET transitions in prostate cancer. [177] On the other hand, lamin B1 was found to promote prostate cancer progression and metastasis [178]. Also, lamin B1-deficient microdomains were found to host chromosomal regions associated with prostate cancer risk, exhibiting reduced gene transcription [179]. The deregulated and differential expression of nuclear lamin proteins might be the answer to TAD mislocalization and partitioning of the existing TADs into smaller ones in the case of prostate cancer [180]. Hence, clearly nuclear lamin plays a major role in oncogenesis and related reorganization of the genome of prostate cancer, although the related mechanisms are yet to be uncovered.

## *Lung Cancer*

Non-small cell lung cancers (NSCLC) and small cell lung cancers (SCLC) are two common subtypes of lung cancer that makeup approximately 85% and 15% of all lung cancers, respectively. Varying amounts of lamin A/C were found in lung tumor cells, depending on the subtype-specific differences. Compared to NSCLC tumor tissues, SCLC tumor tissues express LMNA at lower levels [181, 182]. There are contradictory findings on lamin levels in the context of lung cancers. While Stefanello et al. [183] observed reduced lamin A/C levels in the lung adenocarcinoma (LUAD) cell line A549 in preclinical settings, Rubporn et al. [184] showed overexpression of LMNA in the same cell line. There have also been conflicting reports about the effects of lamin B1 on the biology of lung cancer. Jia et al. [182] found that reduced lamin B1 expression was a significant factor that promoted lung cancer, while a report by Li et al. [185] suggested a nice correlation between lamin B1 overexpression and advanced-stage non-small cell lung cancer. On the other hand, it is interesting to note that lamin B2 expression levels were consistently linked to a higher clinical stage [186–188]. Jia et al. proposed that transcriptional control and chromatin accessibility mediated by H3K27me3 depend significantly on lamin B1 levels [182]. The finding highlights the possibility that dysregulation of cellular senescence and epigenetic modifiers involved in cell division may be caused by altered lamin levels in lung cancer. Along with that, a parallel study in lung epithelial cells found low levels of lamin B1 to be associated with the upregulation of RET proto-oncogene and its coreceptor Gfra1 (glial cell line-derived neurotrophic factor family receptor α-1). This phenomenon was further confirmed by investigating the role of lamins in regulating the rearrangement of chromosome territories. The spatial gene positioning of the RET (Chr 6) and its coreceptor Gfra1 gene loci (Chr 19) in control and lamin B1-depleted MLE12 cells were analyzed. In the control cells, RET and Gfra1 are found at the nuclear periphery as seen in the FISH studies, but their contact with the lamina got lost in the lamin B1 knockdown background. Interestingly, Chr 6 contained most of the genes that are upregulated in the lamin B1 knockdown condition, and Chr 10 contained the least number of differentially expressed genes. Thus, the territorial reorganization of these two chromosomes was studied by chromosome painting in lamin B1 knockdown condition. In normal conditions, chr 6 was found to be located near the nuclear periphery, which shifted more toward the center upon lamin B1 knockdown. No significant change in the localization of Chr 10 was observed, which was a significant finding to demonstrate the role of lamin B1 in chromosome positioning in lung epithelial cells [182]. Another recent study on lung adenocarcinoma showed that chromosome accessibility in the telomere region increases significantly in LMNB1 knockdown cells, which corroborates the cause of telomere shortening and accumulation of irreversible DNA damage in the telomeres. The study found that the loss of LMNB1 resulted in a decrease in the expression of H3K9me3 level, thereby increasing the accessibility of chromosomes in those regions [185].

## *Pancreatic Cancer*

Research on nuclear lamina proteins in pancreatic cancer is quite limited. According to reports, lamin B1 is overexpressed in pancreatic cancer as compared to the normal pancreas, and high expression of lamin B1 is linked to less differentiated tumors and a shorter life span of patients [189]. Additionally, this study found that depleting lamin B1 inhibited the development of pancreatic cancer cells in mouse xenografts.

[189] Another study examined pancreatic specimens for nuclear morphology and lamin A expression in intraductal papillary mucinous adenoma (IPMA) and intraductal papillary mucinous carcinoma (IPMC), which recorded a decreased lamin A positive ratio facilitating nuclear enlargement, thereby promoting carcinogenesis. It proposes lamin A expression to be a reliable biomarker for distinguishing between IPMC and IPMA [190]. Although there are currently few studies examining the mechanisms behind the altered nuclear morphology and chromatin organization in PDAC, these studies are already offering insights into the pathology of cancer cells, revealing significant epigenetic differences and chromatin organization between PDAC subtypes, presenting new therapeutic avenues for the treatment of pancreatic cancer and other malignancies [191].

## *Hepatocellular Carcinoma*

In conjunction with several clinical observations, several findings suggested that the LMNA gene may function as an oncogene in hepatocellular carcinoma (HCC) [192]. Research has shown that LMNA reduces the breakdown of the extracellular matrix and prevents cell invasion by suppressing the production of MMP2/9 (matrix metalloproteases). Conversely, LMNA causes the tumor suppressor gene P16 to be expressed at a lower level, which increases the expression of CDK1 (cyclin-dependent kinase 1). It was also confirmed in a parallel study that core histone macroH2A1 (which is linked to restrictive chromatin) and lamin B1 are connected. In addition, it was discovered that the LADs of mouse liver cells are enriched for macroH2A. More intriguingly, knocking down macroH2A1 results in the decondensation of global chromatin structure and the release of heterochromatin foci identified by histone H3 lysine 9 trimethylation (H3K9me3). However, macroH2A1 was redistributed as a result of lamin B1 downregulation. When considered collectively, these findings suggested that macroH2A1 is spatially connected to the lamina and serves as an important factor for the preservation of chromatin architecture in mouse liver cells [193]. In a FISH study in hepatocellular carcinoma model, it was confirmed that lamin B association influences the radial distribution of lamina-associated loci more prominently than lamin A [124].

## *Melanoma*

Studies have found that the downregulation of LMNB1 and the LBR can control mitotic processes in malignant melanoma, which can result in alterations in chromatin state and cellular senescence due to replication errors. By affecting gene regulation, LMNB1 and the LBR proved to be functionally significant for the prevention of cell cycle arrest in melanoma cells [194]. It was also observed by FISH studies that altered lamin A localization in the nucleus and/or altered nuclear size actively influences the nuclear positioning of several genes like KIT, PADI2, RAMP1, etc. through another important factor called NTF2 (nuclear transport factor 2) in melanoma cell line model [195].

## Sarcoma

The impact of lamin A modulation on nucleo-cytoskeleton dynamics in OS (osteosarcoma) and EWS (Ewing sarcoma) cancer models was studied, which showed that low levels of lamin A are associated with a markedly more pronounced distortion in nuclear shapes in both the cancer types. Reduced lamin A levels in these tumors enhanced their ability to migrate, according to in vitro investigations. Furthermore, gene expression studies conducted on EWS patient samples demonstrated that patients with low LMNA had considerably shorter overall survival and that LMNA levels were significantly lower in metastatic lesions compared to initial tumors. Interestingly, studies have shown in pediatric osteosarcoma cell lines that nuclear lamin A/C expression is inversely correlated with tumor aggressiveness, which validates its role as an onco-suppressor [196]. Furthermore, reintroducing lamin A/C expression in the high metastatic osteosarcoma cell line 143B decreases tumor aggressiveness and affects adhesion properties on particular matrices [138]. On the other hand, Chiarini et al. have shown a strong negative link between tumor aggressiveness and *LMNA* gene expression, as well as a bad prognosis for patients with Ewing sarcoma. Lamin A's function as a tumor suppressor was highlighted by *LMNA* overexpression in EWS cell lines, which supported these findings [197]. In U2OS osteosarcoma cells, the slow anomalous diffusion of telomeres was noticed in all Lmna+/+ cells, whereas the diffusion of telomeres was normal in Lmna−/− U2OS cells in which lamin A was knocked down [198]. Some cases of congenital infantile fibrosarcoma with LMNA-NTRK1(neurotrophic tyrosine kinase receptor 1) gene fusion were observed in infants that showed significant effects on tyrosine kinase inhibitors [199]. Break-apart FISH probes for the NTRK1 gene in a case of spindle cell sarcoma with LMNA-NTRK1 gene fusion showed subtly split signals, reflecting an intra-chromosomal rearrangement [200].

## Colorectal Cancer

Lamin A/C is a poor prognosis biomarker for patients with colorectal cancer as it is found to be positively correlated with increased cell migration in a pre-metastatic colon adenocarcinoma cell line. This is because the expression of lamin A/C facilitates cell invasiveness and a more stem cell-like phenotype [138, 201]. LMNB2 was discovered to be highly elevated in primary colorectal cancer tissues and cell lines when non-cancerous tissues and normal colorectal epithelium were evaluated. There exists a substantial correlation between the elevated expression of LMNB2 in the tissues of colorectal cancer and the clinicopathological features of the patients, as well as the reduced overall and disease-free cumulative survival. Recent observations propose that LMNB2 stimulates colorectal cancer growth by controlling the course of the p21-mediated cell cycle, but they also point to the role of LMNB2 as a molecular therapeutic target and clinical prognostic indicator [202]. In a study in DLD-1 colorectal cells, co-depletion of lamin A/C and emerin was found to perturb territories of chromosomes 18 and 19 since CT18 (gene density ~ 12.35 genes/Mbp, size 80 Mbp) and CT19 (gene density ~ 42.05 genes/Mbp, size 59 Mbp) have comparable DNA content but contrasting gene densities. The spatial positions of both chromosomes remain unaltered in conditions of lamin A/C or emerin knockdown. Interestingly, co-depletion of lamin A/C and emerin significantly repositioned gene-rich CT19, but the peripheral nuclear location of CT18 was unaltered. Positions of gene-rich CT17 and gene-poor CT19 were also determined. CT17 shifted more toward the nuclear interior upon lamin A/C knockdown and in lamin A/C-emerin co-depleted cells. Similar to previous findings, lamin A/C knockdown was found to affect the topology of chromosome territories. So, it was evident that lamin A/C and emerin play a combined role in maintaining the conserved positions of chromosome territories in colorectal carcinoma [110]. 3D FISH studies confirmed that aneuploid chromosome territories (CT11, CT17, CT18, and CT19) were found to be mislocalized upon lamin B2 depletion [203].

## *Gastric Cancer*

Lamin A/C expression was not inhibited in gastrointestinal malignancies, nor was it controlled by DNA methylation. A negative correlation was found between the expression of lamin B1, the depth of invasion, and the clinical stages of gastric cancer. Lamin B1 expression was downregulated in gastric cancer tissues, and there was a strong positive correlation between low lamin B1 expression and higher clinical stage, depth of invasion, nodal stage, and unfavorable prognosis. While lamin B1 overexpression decreased the proliferation and migration of gastric cancer cells, lamin B1 knockdown enhanced these processes. Lamin B1 overexpression contributed to the opposite outcomes, whereas lamin B1 knockdown may increase the activity of the PI3K/PTEN/Akt and MAPK/ERK pathway with a reduction in the p53/p21WAF1/CIP1 expression [204]. Distinct deregulation of lamin expression might correlate with the aberrant structural variations in gastric cancer. Distinct structural variations in gastric cancer cause rewiring of cis-regulatory elements. Hi-C analysis along with ChIP-sequencing data [205] revealed that CCNE1, a key modulator of therapy resistance, was found to be frequently juxtaposed with diverse distal enhancers. Gastric cancers associated with high CCNE1 expression were found to harbor disrupted CCNE1 topologically associating domain (TAD) boundaries, along with novel TAD interactions.

## *Brain Cancer*

According to Pei et al., nuclear lamins may be used as oncogene targets in therapies for brain tumors in children. According to the research, glioma tissues have much higher levels of expression for all three human genes that encode lamin isoforms (LMNA, LMNB1, and LMNB2) than normal brain cells. Additionally, the absence of these genes has been shown to greatly limit the growth of glioma both in in vitro and in vivo animal models [206]. Simultaneously, Gatti et al. proposed lamin A/C as a biomarker of the aggressiveness and tumorigenicity in glioblastoma multiforme, since the expression of this nuclear protein was found to be primarily involved in cell adhesion and migration processes and correlated with a lower overall survival rate in glioma patients [207]. The most prevalent extracranial solid tumor seen in children is neuroblastoma, which completely reverses the previously indicated function that lamin A/C plays as a cancer-favoring factor. According to another group, decreasing lamin A/C expression of the SH-SYS5 cell line (high lamin A/C-expressing line) prevents retinoic acid-induced differentiation and promotes the growth of tumor-initiating cells, leading to a more aggressive phenotype [208].

Considering this, lamin A/C acts as an onco-suppressor factor in this kind of malignancy. In a parallel study, it was shown that CpG promoter hypermethylation is the mechanism behind the downregulation of lamin A/C in a subset of neuroblastoma cells [209]. Another study shed light on the interaction of nuclear lamina with mid-to-late replicating chromatin in neuroblastoma cell lines. Immunofluorescence of lamin B on two-color replication labeled neuroblastoma cell nuclei revealed specific attachment of the mid-to-late replicating chromatin not just along the periphery but also in the nuclear interior along the invaginations of the lamina [210].

# Conclusion

Three noteworthy roles of lamins have been discovered, which may be dysregulated in cancer cells and appear to rely on the differential expression of various lamin subtypes in various cancers: (1) cellular stiffness and mobility can be linked to lamin interaction with proteins regulating EMT [211], (2) cell cycle regulation is induced by protein kinase signaling that affects lamin polymerization by phosphorylation, and most importantly, (3) chromatin organization and compactness can impact transcriptional activity and genome maintenance [212, 213]. Over the past few decades, fundamental concepts of chromatin organization and their significance for genome function have been largely reported. The differential expression of lamin proteins and concomitant alterations in the chromatin landscapes in cancers open a new vista of investigation. Lamins in tandem with its numerous interacting partners play decisive roles in neoplastic transformation much of which is hitherto in oblivion. Future research into these conduits might reveal an avenue for therapeutic intervention in the coming years.

## References

1. Berezney R (2002) Regulating the mammalian genome: the role of nuclear architecture. Adv Enzym Regul 42:39–52
2. Oudet P, Gross-Bellard M, Chambon P (1975) Electron microscopic and biochemical evidence that chromatin structure is a repeating unit. Cell 4:281–300
3. Zinchenko A, Berezhnoy NV, Chen Q, Nordenskiöld L (2018) Compaction of Single-Molecule Megabase-Long Chromatin under the Influence of Macromolecular Crowding. Biophys J 114:2326–2335
4. Adkins NL, Watts M, Georgel PT (2004) To the 30-nm chromatin fiber and beyond. Biochim Biophys Acta (BBA) 1677:12–23
5. Belmont AS, Bruce K (1994) Visualization of G1 chromosomes: a folded, twisted, supercoiled chromonema model of interphase chromatid structure. J Cell Biol 127:287–302
6. Cremer T, Cremer C (2001) Chromosome territories, nuclear architecture and gene regulation in mammalian cells. Nat Rev Genet 2:292
7. Boveri T. Die Blastomerenkerne von Ascaris Megalocephala und die Theorie der Chromosomenindividualität
8. Uber Zelltheilung | WorldCat.org. https://search.worldcat.org/title/uber-zelltheilung/oclc/71236921
9. Die stofflichen Grundlagen der Vererbung im organischen Reich. Versuch einer gemeinverständlichen Darstellung. by Strasburger, Eduard:: Gr.-8°, Original-Broschur. (1905) Erste Ausgabe. | Antiquariat Dennis R. Plummer. https://www.abebooks.com/first-edition/stofflichen-Grundlagen-Vererbung-organischen-Reich-Versuch/31407837695/bd
10. Wischnitzer S (1973) The Submicroscopic Morphology of the Interphase Nucleus. Int Rev Cytol 34:1–48
11. Stack SM, Brown DB, Dewey WC (1977) Visualization of interphase chromosomes. J Cell Sci 26:281–299
12. Zorn C, Cremer C, Cremer T, Zimmer J (1979) Unscheduled DNA synthesis after partial UV irradiation of the cell nucleus: Distribution in interphase and metaphase. Exp Cell Res 124:111–119
13. Cremer T, Baumann H, Nakanishi K, Cremen C. Correlation between interphase and metaphase chromosome arrangements as studied by laser-uv-microbeam experiments
14. Manuelidis L (1985) Individual interphase chromosome domains revealed by in situ hybridization. Hum Genet 71:288–293
15. Lichter P, Cremer T, Borden J, Manuelidis L, Ward DC (1988) Delineation of individual human chromosomes in metaphase and interphase cells by in situ suppression hybridization using recombinant DNA libraries. Hum Genet 80:224–234
16. Zirbel RM, Mathieu UR, Kurz A, Cremer T, Lichter P (1993) Evidence for a nuclear compartment of transcription and splicing located at chromosome domain boundaries. Chromosom Res 1:93–106
17. Cremer C et al (1996) Nuclear architecture and the induction of chromosomal aberrations. Mutat Res/Rev Genet Toxicol 366:97–116
18. Fakan S (2004) The functional architecture of the nucleus as analysed by ultrastructural cytochemistry. Histochem Cell Biol 122:83–93
19. Cremer T et al (2006) Chromosome territories—a functional nuclear landscape. Curr Opin Cell Biol 18:307–316
20. plbi-03-05-14 826..842 _cremer2005
21. Cremer M et al (2008) Multicolor 3D fluorescence in situ hybridization for imaging interphase chromosomes. Methods Mol Biol 463:205–239
22. Pliss A, Malyavantham KS, Bhattacharya S, Berezney R (2013) Chromatin dynamics in living cells: identification of oscillatory motion. J Cell Physiol 228:609–616
23. Pliss A, Malyavantham K, Bhattacharya S, Zeitz M, Berezney R (2009) Chromatin dynamics is correlated with replication timing. Chromosoma 118:459–470

24. Gasser SM (2002) Visualizing chromatin dynamics in interphase nuclei. Science 296:1412–1416
25. Zink D et al (1998) Structure and dynamics of human interphase chromosome territories in vivo. Hum Genet 102:241–251
26. Münkel C et al (1999) Compartmentalization of interphase chromosomes observed in simulation and experiment. J Mol Biol 285:1053–1065
27. Yokota H, Van Den Engh G, Hearst JE, Sachs RK, Trask BJ Evidence for the organization of chromatin in megabase pair-sized loops arranged along a random walk path in the human G0/G1 interphase nucleus
28. Sachs RK, Van Den Engh G, Trask B, Yokota H, Hearst JE (1995) A random-walk/giant-loop model for interphase chromosomes. Proc Natl Acad Sci 92:2710–2714
29. Kreth G, Finsterle J, Von Hase J, Cremer M, Cremer C (2004) Radial arrangement of chromosome territories in human cell nuclei: a computer model approach based on gene density indicates a probabilistic global positioning code. Biophys J 86:2803–2812
30. Walter J, Schermelleh L, Cremer M, Tashiro S, Cremer T (2003) Chromosome order in HeLa cells changes during mitosis and early G1, but is stably maintained during subsequent interphase stages. J Cell Biol 160:685–697
31. Cremer T et al (2000) Chromosome territories, interchromatin domain compartment, and nuclear matrix: an integrated view of the functional nuclear architecture. Crit Rev Gene Expr 10:179–212
32. Caron H et al (2001) The human transcriptome map: clustering of highly expressed genes in chromosomal domains. Science 291:1289–1293
33. Mahy NL, Perry PE, Bickmore WA (2002) Gene density and transcription influence the localization of chromatin outside of chromosome territories detectable by FISH. J Cell Biol 159:753
34. Williams RRE, Broad S, Sheer D, Ragoussis J (2002) Subchromosomal positioning of the epidermal differentiation complex (EDC) in keratinocyte and lymphoblast interphase nuclei. Exp Cell Res 272:163–175
35. Volpi EV et al (2000) Large-scale chromatin organization of the major histocompatibility complex and other regions of human chromosome 6 and its response to interferon in interphase nuclei. J Cell Sci 113:1565–1576
36. Chubb JR, Bickmore WA (2003) Considering nuclear compartmentalization in the light of nuclear dynamics. Cell 112:403–406
37. Gilbert N et al (2004) Chromatin architecture of the human genome: gene-rich domains are enriched in open chromatin fibers. Cell 118:555–566
38. Cavalli G, Misteli T (2013) Functional implications of genome topology. Nat Struct Mol Biol 20:290
39. Tan-Wong SM, Wijayatilake HD, Proudfoot NJ (2009) Gene loops function to maintain transcriptional memory through interaction with the nuclear pore complex. Genes Dev 23:2610–2624
40. Tan-Wong SM et al (2012) Gene loops enhance transcriptional directionality. Science 338:671–675
41. O'Sullivan JM et al (2004) Gene loops juxtapose promoters and terminators in yeast. Nat Genet 36:1014–1018
42. Tan-Wong SM, French JD, Proudfoot NJ, Brown MA (2008) Dynamic interactions between the promoter and terminator regions of the mammalian BRCA1 gene. Proc Natl Acad Sci USA 105:5160–5165
43. Deng W et al (2012) Controlling long-range genomic interactions at a native locus by targeted tethering of a looping factor. Cell 149:1233–1244
44. Hou C, Corces VG (2012) Throwing transcription for a loop: expression of the genome in the 3D nucleus. Chromosoma 121:107–116
45. Ørom UA, Shiekhattar R (2011) Noncoding RNAs and enhancers: complications of a long-distance relationship. Trends Genet 27:433–439

46. Wang KC et al (2011) A long noncoding RNA maintains active chromatin to coordinate homeotic gene expression. Nature 472:120–126
47. Li G et al (2012) Extensive promoter-centered chromatin interactions provide a topological basis for transcription regulation. Cell 148:84–98
48. Saurin AJ, Shao Z, Erdjument-Bromage H, Tempst P, Kingston RE (2001) A Drosophila Polycomb group complex includes Zeste and dTAFII proteins. Nature 412:655–660
49. Lanzuolo C, Orlando V (2007) The function of the epigenome in cell reprogramming. Cell Mol Life Sci 64:1043–1062
50. Breiling A, Turner BM, Bianchi ME, Orlando V (2001) General transcription factors bind promoters repressed by Polycomb group proteins. Nature 412:651–655
51. Tiwari VK, Cope L, McGarvey KM, Ohm JE, Baylin SB (2008) A novel 6C assay uncovers Polycomb-mediated higher order chromatin conformations. Genome Res 18:1171–1179
52. Tiwari VK et al (2008) PcG proteins, DNA methylation, and gene repression by chromatin looping. PLoS Biol 6:2911
53. Lanzuolo C, Roure V, Dekker J, Bantignies F, Orlando V (2007) Polycomb response elements mediate the formation of chromosome higher-order structures in the bithorax complex. Nat Cell Biol 9:1167–1174
54. Yang J, Corces VG (2012) Insulators, long-range interactions, and genome function. Curr Opin Genet Dev 22:86–92
55. Bao L, Zhou M, Cui Y (2008) CTCFBSDB: a CTCF-binding site database for characterization of vertebrate genomic insulators. Nucleic Acids Res 36:D83
56. Phillips JE, Corces VG (2009) CTCF: Master weaver of the genome. Cell 137:1194–1211
57. Harvey SH, Krien MJ, O'Connell MJ (2002) Structural maintenance of chromosomes (SMC) proteins, a family of conserved ATPases. Genome Biol 3, reviews3003.1:reviews3003.1
58. Yoshinaga M, Inagaki Y (2021) Ubiquity and origins of Structural Maintenance of Chromosomes (SMC) proteins in eukaryotes. Genome Biol Evol 13
59. Patta I et al (2024) Nuclear morphology is shaped by loop-extrusion programs. Nature 627(8002):196–203
60. Agarwal A, Korsak S, Choudhury A, Plewczynski D (2023) The dynamic role of cohesin in maintaining human genome architecture. BioEssays 45:2200240
61. Davidson IF et al (2019) DNA loop extrusion by human cohesin. Science 1979(366):1338–1345
62. Skibbens RV (2019) Condensins and cohesins—one of these things is not like the other! J Cell Sci 132
63. da Costa-Nunes JA, Noordermeer D (2023) TADs: Dynamic structures to create stable regulatory functions. Curr Opin Struct Biol 81:102622
64. Rowley MJ, Corces VG (2018) Organizational principles of 3D genome architecture. Nat Rev Genet 19(12):789–800
65. Fudenberg G et al (2016) Formation of chromosomal domains by loop extrusion
66. Lieberman-Aiden E et al (2009) Comprehensive mapping of long-range interactions reveals folding principles of the human genome. Science (1979) 326:289–293
67. Harris HL et al (2023) Chromatin alternates between A and B compartments at kilobase scale for subgenic organization. Nat Commun 14(1):1–17
68. Harris HL, Rowley MJ (2024) Mechanistic drivers of chromatin organization into compartments. Curr Opin Genet Dev 86:102193
69. Imakaev M et al (2012) Iterative correction of Hi-C data reveals hallmarks of chromosome organization. Nat Methods 9(10):999–1003
70. Bouwman BA, Crosetto N, Bienko M (2023) A GC-centered view of 3D genome organization. Curr Opin Genet Dev 78:102020
71. Biswas A, Basu A (2023) The impact of the sequence-dependent physical properties of DNA on chromatin dynamics. Curr Opin Struct Biol 83:102698
72. Lucini F et al (2024) Biochemical properties of chromatin domains define genome compartmentalization. Nucleic Acids Res 52:e54–e54

73. Akilli N, Cheutin T, Cavalli G (2024) Phase separation and inheritance of repressive chromatin domains. Curr Opin Genet Dev 86:102201
74. Chen B et al (2024) Stratifying TAD boundaries pinpoints focal genomic regions of regulation, damage, and repair. Brief Bioinform 25
75. Kant A et al (2024) Active transcription and epigenetic reactions synergistically regulate meso-scale genomic organization. Nat Commun 15(1):1–19
76. Grewal SIS (2023) The molecular basis of heterochromatin assembly and epigenetic inheritance. Mol Cell 83:1767–1785
77. Senapati S, Irshad IU, Sharma AK, Kumar H (2023) Fundamental insights into the correlation between chromosome configuration and transcription. Phys Biol 20:051002
78. van Steensel B, Belmont AS (2017) Lamina-associated domains: links with chromosome architecture, heterochromatin, and gene repression. Cell 169:780–791
79. Goldman RD, Goldman AE, Shumaker DK (2005) Nuclear lamins: building blocks of nuclear structure and function. Novartis Found Symp 264(3–21):227–230
80. Dechat T, Adam SA, Taimen P, Shimi T, Goldman RD (2015) Nuclear lamins. 1–23
81. Shumaker DK, Kuczmarski ER, Goldman RD (2003) The nucleoskeleton: lamins and actin are major players in essential nuclear functions. Curr Opin Cell Biol:358–366
82. Turgay Y et al (2017) The molecular architecture of lamins in somatic cells. Nature 543:261–264
83. Shimi T et al (2008) The A- and B-type nuclear lamin networks: Microdomains involved in chromatin organization and transcription. Genes Dev 22:3409–3421
84. Alagna NS, Thomas TI, Wilson KL, Reddy KL (2023) Choreography of lamina-associated domains: structure meets dynamics. FEBS Lett 597:2806–2822
85. Lamins organize the global three-dimensional genome from the nuclear periphery _ Enhanced Reader
86. Bera M, Sengupta K (2020) Nuclear filaments: role in chromosomal positioning and gene expression. Nucleus 11 99–110
87. Gaillard MC, Reddy KL (2017) The nuclear lamina and genome organization. In: Nuclear architecture and dynamics. Elsevier, pp 321–343
88. Gruenbaum Y, Foisner R (2015) Lamins: nuclear intermediate filament proteins with fundamental functions in nuclear mechanics and genome regulation
89. Ludérus MEE et al (1992) Binding of matrix attachment regions to lamin B1. Cell 70:949–959
90. Lund EG, Duband-Goulet I, Oldenburg A, Buendia B, Collas P (2015) Distinct features of lamin A-interacting chromatin domains mapped by ChIP-sequencing from sonicated or micrococcal nuclease-digested chromatin. Nucleus 6:30–39
91. Luderus MEE et al (1994) Binding of matrix attachment regions to lamin polymers involves single-stranded regions and the minor groove. Mol Cell Biol 14:6297
92. Taniura H, Glass C, Gerace L (1995) A chromatin binding site in the tail domain of nuclear lamins that interacts with core histones. J Cell Biol 131:33–44
93. Goldberg M et al (1999) The tail domain of lamin Dm0 binds histones H2A and H2B. Proc Natl Acad Sci USA 96:2852–2857
94. Barton LJ, Soshnev AA, Geyer PK (2015) Networking in the nucleus: a spotlight on LEM-domain proteins. Curr Opin Cell Biol 34:1–8
95. Ye Q, Worman HJ (1996) Interaction between an integral protein of the nuclear envelope inner membrane and human chromodomain proteins homologous to Drosophila HP1. J Biol Chem 271:14653–14656
96. Solovei I et al (2013) LBR and lamin A/C sequentially tether peripheral heterochromatin and inversely regulate differentiation. Cell 152:584–598
97. Kind J, van Steensel B (2014) Stochastic genome-nuclear lamina interactions. Nucleus 5:124–130
98. Foisner R, Gerace L (1993) Integral membrane proteins of the nuclear envelope interact with lamins and chromosomes, and binding is modulated by mitotic phosphorylation. Cell 73:1267–1279

99. Furukawa K (1999) LAP2 binding protein 1 (L2BP1/BAF) is a candidate mediator of LAP2-chromatin interaction. J Cell Sci 112:2485–2492
100. Nili E et al (2001) Nuclear membrane protein LAP2beta mediates transcriptional repression alone and together with its binding partner GCL (germ-cell-less). J Cell Sci 114:3297–3307
101. Ye Q, Callebaut I, Pezhman A, Courvalin JC, Worman HJ (1997) Domain-specific interactions of human HP1-type chromodomain proteins and inner nuclear membrane protein LBR. J Biol Chem 272:14983–14989
102. Dechat T, Vlcek S, Foisner R (2000) Review: lamina-associated polypeptide 2 isoforms and related proteins in cell cycle-dependent nuclear structure dynamics. J Struct Biol 129:335–345
103. Mariappan I, Gurung R, Thanumalayan S, Parnaik VK (2007) Identification of cyclin D3 as a new interaction partner of lamin A/C. Biochem Biophys Res Commun 355:981–985
104. Lee DC, Welton KL, Smith ED, Kennedy BK (2009) A-type nuclear lamins act as transcriptional repressors when targeted to promoters. Exp Cell Res 315:996–1007
105. Madsen-Østerbye J, Abdelhalim M, Pickering SH, Collas P (2023) Gene regulatory interactions at lamina-associated domains. Genes (Basel) 14:334
106. Siegenfeld AP et al (2022) Polycomb-lamina antagonism partitions heterochromatin at the nuclear periphery. Nat Commun 13(1):1–16
107. Wen B, Wu H, Shinkai Y, Irizarry RA, Feinberg AP (2009) Large histone H3 lysine 9 dimethylated chromatin blocks distinguish differentiated from embryonic stem cells. Nat Genet 41(2):246–250
108. Guelen L et al (2008) Domain organization of human chromosomes revealed by mapping of nuclear lamina interactions. *Nature* 453(7197):948–951
109. Camps J et al (2014) Loss of lamin B1 results in prolongation of S phase and decondensation of chromosome territories. FASEB J 28:3423–3434
110. Ranade D, Pradhan R, Jayakrishnan M, Hegde S, Sengupta K (2019) Lamin A/C and Emerin depletion impacts chromatin organization and dynamics in the interphase nucleus. BMC Mol Cell Biol 20:1–20
111. Malhas A, Lee CF, Sanders R, Saunders NJ, Vaux DJ (2007) Defects in lamin B1 expression or processing affect interphase chromosome position and gene expression. J Cell Biol 176:593–603
112. Chang L et al (2022) Nuclear peripheral chromatin-lamin B1 interaction is required for global integrity of chromatin architecture and dynamics in human cells. Protein Cell 13:258–280
113. Pujadas Liwag EM et al (2024) Depletion of lamins B1 and B2 promotes chromatin mobility and induces differential gene expression by a mesoscale-motion-dependent mechanism. Genome Biol 25:1–42
114. Zheng X et al (2018) Lamins organize the global three-dimensional genome from the nuclear periphery. Mol Cell 71:802–815.e7
115. Briand N, Collas P (2018) Laminopathy-causing lamin A mutations reconfigure lamina-associated domains and local spatial chromatin conformation. Nucleus 9 216–226
116. Puckelwartz MJ, Depreux FFS, McNally EM (2011) Gene expression, chromosome position and lamin A/C mutations. Nucleus 2:162–167
117. Sebestyén E et al (2020) SAMMY-seq reveals early alteration of heterochromatin and deregulation of bivalent genes in Hutchinson-Gilford Progeria Syndrome. Nat Commun 11:1–16
118. McCord RP et al (2013) Correlated alterations in genome organization, histone methylation, and DNA–lamin A/C interactions in Hutchinson-Gilford progeria syndrome. Genome Res 23:260–269
119. Parnaik VK (2008) Role of nuclear lamins in nuclear organization, cellular signaling, and inherited diseases. Int Rev Cell Mol Biol 266:157–206
120. Paulsen J et al (2017) Chrom3D: Three-dimensional genome modeling from Hi-C and nuclear lamin-genome contacts. Genome Biol 18:1–15
121. Bianchi A et al (2020) Dysfunctional polycomb transcriptional repression contributes to lamin A/C–dependent muscular dystrophy. J Clin Invest 130:2408–2421

122. Tiwari V, Alam MJ, Bhatia M, Navya M, Banerjee SK (2024) The structure and function of lamin A/C: Special focus on cardiomyopathy and therapeutic interventions. Life Sci 341:122489
123. Bertero A, Rosa-Garrido M (2021) Three-dimensional chromatin organization in cardiac development and disease. J Mol Cell Cardiol 151:89
124. Forsberg F, Brunet A, Ali TML, Collas P (2019) Interplay of lamin A and lamin B LADs on the radial positioning of chromatin. Nucleus 10:7–20
125. Wang B et al (2024) The molecular basis of lamin-specific chromatin interactions. bioRxiv:2024.08.05.604734
126. Wong X et al (2021) Lamin C is required to establish genome organization after mitosis. Genome Biol 22:1–27
127. Siegel RL, Giaquinto AN, Jemal A (2024) Cancer statistics, 2024. CA Cancer J Clin 74:12–49
128. Fritz AJ, Sehgal N, Pliss A, Xu J, Berezney R (2019) Chromosome territories and the global regulation of the genome. Genes Chromosomes Cancer 58:407
129. Hanahan D (2022) Hallmarks of cancer: new dimensions. Cancer Discov 12
130. Leemans C et al (2019) Promoter-intrinsic and local chromatin features determine gene repression in LADs. Cell 177:852–864.e14
131. Ji L et al (2020) TOPORS, a tumor suppressor protein, contributes to the maintenance of higher-order chromatin architecture. Biochim Biophys Acta 1863:194518
132. Yoon I, Kim U, Song Y, Park T, Lee DS (2024) 3C methods in cancer research: recent advances and future prospects. Exp Mol Med 56(4):788–798
133. Spielmann M, Lupiáñez DG, Mundlos S (2018) Structural variation in the 3D genome. Nat Rev Genet 19(7):453–467
134. Valentijn LJ et al (2015) TERT rearrangements are frequent in neuroblastoma and identify aggressive tumors. Nat Genet 47:1411–1414
135. Krijger PHL, De Laat W (2016) Regulation of disease-associated gene expression in the 3D genome. Nat Rev Mol Cell Biol 17:771–782
136. Northcott PA et al (2014) Enhancer hijacking activates GFI1 family oncogenes in medulloblastoma. Nature 511:428–434
137. Foster CR, Przyborski SA, Wilson RG, Hutchison CJ (2010) Lamins as cancer biomarkers. Biochem Soc Trans 38:297–300
138. Urciuoli E, Peruzzi B (2022) The paradox of nuclear lamins in pathologies: apparently controversial roles explained by tissue-specific mechanobiology. Cells 11
139. Misteli, T. Higher-order genome organization in human disease. *Cold Spring Harbor perspectives in biology* vol. 2 Preprint at 10.1101/cshperspect.a000794 (2010)
140. Witcher M, Emerson BM (2009) Epigenetic silencing of the p16(INK4a) tumor suppressor is associated with loss of CTCF binding and a chromatin boundary. Mol Cell 34:271–284
141. Deng Z et al (2012) A role for CTCF and cohesin in subtelomere chromatin organization, TERRA transcription, and telomere end protection. EMBO J 31:4165–4178
142. Ottaviani A et al (2009) Identification of a perinuclear positioning element in human subtelomeres that requires A-type lamins and CTCF. EMBO J 28:2428–2436
143. Ottaviani A et al (2009) The D4Z4 Macrosatellite Repeat Acts as a CTCF and A-Type Lamins-Dependent Insulator in Facio-Scapulo-Humeral Dystrophy. PLoS Genet 5:e1000394
144. Yusufzai TM, Tagami H, Nakatani Y, Felsenfeld G (2004) CTCF tethers an insulator to subnuclear sites, suggesting shared insulator mechanisms across species. Mol Cell 13:291–298
145. Havugimana PC et al (2022) Scalable multiplex co-fractionation/mass spectrometry platform for accelerated protein interactome discovery. Nat Commun 13:4043
146. Serebryannyy LA, Misteli T (2019) HiPLA: High-throughput imaging Proximity Ligation Assay. Methods 157:80
147. Gao H et al (2022) In-Depth In Vivo Crosslinking in Minutes by a Compact, Membrane-Permeable, and Alkynyl-Enrichable Crosslinker. Anal Chem 94:7551–7558
148. Oravcová M et al (2022) The Nse5/6-like SIMC1-SLF2 complex localizes SMC5/6 to viral replication centers. elife 11

149. Iradi MCG et al (2018) Characterization of gene regulation and protein interaction networks for Matrin 3 encoding mutations linked to amyotrophic lateral sclerosis and myopathy. Sci Rep 8:4049
150. Depreux FF et al (2015) Disruption of the lamin A and matrin-3 interaction by myopathic LMNA mutations. Hum Mol Genet 24:4284–4295
151. Jevtić P, Levy DL (2014) Mechanisms of Nuclear Size Regulation in Model Systems and Cancer. Adv Exp Med Biol 773:537–569
152. Alhudiri IM et al (2019) Expression of Lamin A/C in early-stage breast cancer and its prognostic value. Breast Cancer Res Treat 174:661–668
153. Wazir U et al (2013) The clinicopathological significance of lamin A/C, lamin B1 and lamin B receptor mRNA expression in human breast cancer. Cell Mol Biol Lett 18:595–611
154. Saleh T et al (2022) The expression of the senescence-associated biomarker Llamin B1 in human breast cancer. Diagnostics 12:609
155. Irianto J, Pfeifer CR, Ivanovska IL, Swift J, Discher DE (2016) Nuclear lamins in cancer. Cell Mol Bioeng 9:258–267
156. Dupont S et al (2011) Role of YAP/TAZ in mechanotransduction. Nature 474:179–183
157. Bell, E. S. *et al.* Low lamin A levels enhance confined cell migration and metastatic capacity in breast cancer. *Oncogene 2022 41:36* 41, 4211–4230 (2022)
158. Marella NV, Bhattacharya S, Mukherjee L, Xu J, Berezney R (2009) Cell type specific chromosome territory organization in the interphase nucleus of normal and cancer cells. J Cell Physiol 221:130–138
159. Barutcu AR et al (2015) Chromatin interaction analysis reveals changes in small chromosome and telomere clustering between epithelial and breast cancer cells. Genome Biol 16:1–14
160. Kim T et al (2022) Comparative characterization of 3D chromatin organization in triple-negative breast cancers. Exp Mol Med 54(5):585–600
161. Oh S, Oh C, Yoo KH (2017) Functional roles of CTCF in breast cancer. BMB Rep 50:445
162. Wong KM, Song J, Wong YH (2021) CTCF and EGR1 suppress breast cancer cell migration through transcriptional control of Nm23-H1. Sci Reports 11:1–18
163. Liu X et al (2018) An AP-MS- and BioID-compatible MAC-tag enables comprehensive mapping of protein interactions and subcellular localizations. Nat Commun 9
164. Sengupta D et al (2022) A deep hybrid learning pipeline for accurate diagnosis of ovarian cancer based on nuclear morphology. PLoS One 17:e0261181
165. Wang Y, Jiang J, He L, Gong G, Wu X (2019) Effect of lamin-A expression on migration and nuclear stability of ovarian cancer cells. Gynecol Oncol 152:166–176
166. Asadollahi R, Hyde CAC, Zhong XY (2010) Epigenetics of ovarian cancer: From the lab to the clinic. Gynecol Oncol 118:81–87
167. Yang Q et al (2018) Epigenetics in ovarian cancer: Premise, properties, and perspectives. Mol Cancer 17:1–21
168. Balch C, Fang F, Matei DE, Huang THM, Nephew KP, Minireview: Epigenetic changes in ovarian cancer (2009) Endocrinology 150:4003–4011
169. Morden CR et al (2021) Chromosome instability is prevalent and dynamic in high-grade serous ovarian cancer patient samples. Gynecol Oncol 161:769–778
170. Bray F et al (2018) Global cancer statistics 2018: GLOBOCAN estimates of incidence and mortality worldwide for 36 cancers in 185 countries. CA Cancer J Clin 68
171. Papanicolaou GN, Traut HF (1997) The diagnostic value of vaginal smears in carcinoma of the uterus. 1941. Arch Pathol Lab Med 121
172. Capo-chichi CD et al (2016) Lamin A/C deficiency is an independent risk factor for cervical cancer. Cell Oncol 39:59
173. Adeel MM et al (2021) Structural variations of the 3D genome architecture in cervical cancer development. Front Cell Dev Biol 9
174. Martin RS, Das P, Hossain S, McCord RP (2023) Abstract A061: Nuclear lamin content as a biomarker for prostate cancer progression. Cancer Res 83:A061–A061

175. Hong J-H et al (2022) LMNB1, a potential marker for early prostate cancer progression. Am J Cancer Res 12:3390
176. Dubik N, Mai S (2020) Lamin A/C: Function in normal and tumor cells. Cancers 12:1–21
177. Zuo L et al (2018) Lamin A/C might be involved in the EMT signalling pathway. Gene 663:51
178. Luo F et al (2021) Lamin B1 promotes tumor progression and metastasis in primary prostate cancer patients. Future Oncol 17:663
179. Helfand BT et al (2012) Chromosomal regions associated with prostate cancer risk localize to lamin B-deficient microdomains and exhibit reduced gene transcription. J Pathol 226:735
180. Rhie SK et al (2019) A high-resolution 3D epigenomic map reveals insights into the creation of the prostate cancer transcriptome. Nat Commun 10:4154
181. Wang J et al (2019) Constitutional abnormality of nuclear membrane proteins in small cell lung carcinoma. Virchows Arch 475
182. Jia Y et al (2019) Lamin B1 loss promotes lung cancer development and metastasis by epigenetic derepression of RET. J Exp Med 216:1377
183. Stefanello ST et al (2021) Impact of the Nuclear Envelope on Malignant Transformation, Motility, and Survival of Lung Cancer Cells. Adv Sci 8:e2102757
184. Rubporn A et al (2009) Comparative proteomic analysis of lung cancer cell line and lung fibroblast cell line. Cancer Genomics Proteomics 6:229
185. Li J et al (2022) Knockdown of LMNB1 inhibits the proliferation of lung adenocarcinoma cells by inducing DNA damage and cell senescence. Front Oncol 12
186. Jiao X et al (2022) Kruppel like factor 16 promotes lung adenocarcinoma progression by upregulating lamin B2. Bioengineered 13:9483
187. Ma Y et al (2017) Lamin B2 binding to minichromosome maintenance complex component 7 promotes non-small cell lung carcinogenesis. Oncotarget 8
188. Zhang MY et al (2020) Lamin B2 promotes the malignant phenotype of non-small cell lung cancer cells by upregulating dimethylation of histone 3 lysine 9. Exp Cell Res 393:112090
189. Li L et al (2013) Lamin B1 is a novel therapeutic target of betulinic acid in pancreatic cancer. Clin Cancer Res 19:4651
190. Hiroe T et al (2022) Negative correlation between the nuclear size and nuclear Lamina component Lamin A in intraductal papillary mucinous neoplasms of the pancreas. Pathol Oncol Res 28
191. Flores LF et al (2021) Nuclear dynamics and chromatin structure: Implications for pancreatic cancer. Cells 10
192. Liu H et al (2020) LMNA functions as an oncogene in hepatocellular carcinoma by regulating the proliferation and migration ability. J Cell Mol Med 24:12008–12019
193. Fu Y et al (2015) MacroH2A1 associates with nuclear lamina and maintains chromatin architecture in mouse liver cells. Sci Rep 5
194. Lämmerhirt L et al (2022) Knockdown of lamin B1 and the corresponding lamin B receptor leads to changes in heterochromatin state and senescence induction in malignant melanoma. Cells 11
195. Vuković LD et al (2021) Nuclear Transport Factor 2 (NTF2) suppresses WM983B metastatic melanoma by modifying cell migration, metastasis, and gene expression. Sci Rep 11:23586
196. Urciuoli E et al (2020) Nuclear lamins and emerin are differentially expressed in osteosarcoma cells and scale with tumor aggressiveness. Cancers (Basel) 12
197. Chiarini F et al (2022) Lamin A and the LINC complex act as potential tumor suppressors in Ewing Sarcoma. Cell Death Dis 13:346
198. Bronshtein I et al (2015) Loss of lamin A function increases chromatin dynamics in the nuclear interior. Nat Commun 6:8044
199. Haller F et al (2016) Paediatric and adult soft tissue sarcomas with NTRK1 gene fusions: a subset of spindle cell sarcomas unified by a prominent myopericytic/haemangiopericytic pattern. J Pathol 238:700
200. Tsai JW et al (2022) Adult NTRK-rearranged spindle cell neoplasms of the viscera: with an emphasis on rare locations and heterologous elements. Mod Pathol 35:911

201. Foster CR et al (2011) The role of lamin A in cytoskeleton organization in colorectal cancer cells: A proteomic investigation. Nucleus 2:434
202. Dong CH et al (2021) LMNB2 promotes the progression of colorectal cancer by silencing p21 expression. Cell Death Dis 12:331
203. Ranade D, Koul S, Thompson J, Prasad KB, Sengupta K (2017) Chromosomal aneuploidies induced upon Lamin B2 depletion are mislocalized in the interphase nucleus. Chromosoma 126:223–244
204. Yu ZY et al (2020) Lamin b1 deficiency promotes malignancy and predicts poor prognosis in gastric cancer. Neoplasma 67:1303
205. Ooi WF et al (2020) Integrated paired-end enhancer profiling and whole-genome sequencing reveals recurrent CCNE1 and IGF2 enhancer hijacking in primary gastric adenocarcinoma. Gut 69:1039
206. Pei S et al (2022) Aberrant nuclear lamina contributes to the malignancy of human gliomas. J Genet Genomics 49:132
207. Gatti G et al (2021) Role of lamin A/C as candidate biomarker of aggressiveness and tumorigenicity in glioblastoma multiforme. Biomedicine 9
208. Nardella M et al (2015) Down-regulation of the Lamin A/C in neuroblastoma triggers the expansion of tumor initiating cells. Oncotarget 6:32821
209. Rauschert I et al (2017) Promoter hypermethylation as a mechanism for Lamin A/C silencing in a subset of neuroblastoma cells. PLoS One 12:e0175953
210. Schermelleh L (2003) Dynamic organization of chromosomes in the mammalian cell nucleus. Dissertation, LMU München: Faculty of Biology. LMU München
211. Sorenson L, Fu Y, Hood T, Warren S, McEachron TA (2019) Targeted transcriptional profiling of the tumor microenvironment reveals lymphocyte exclusion and vascular dysfunction in metastatic osteosarcoma. Onco Targets Ther 8
212. Hatch EM, Fischer AH, Deerinck TJ, Hetzer MW (2013) Catastrophic nuclear envelope collapse in cancer cell micronuclei. Cell 154:47
213. de Leeuw R, Gruenbaum Y, Medalia O (2018) Nuclear lamins: thin filaments with major functions. Trends Cell Biol 28

# Chapter 6
# Changes in Chromatin Organisation and Mechanotransduction Mediated by Mutant Lamins in Laminopathies

**Subhradip Nath and Kaushik Sengupta**

**Abstract** The filamentous nuclear lamina network is predominantly made up of lamin proteins, which play pivotal roles in maintaining nuclear architecture, chromatin structure, mechanotransduction and various other nuclear processes. Among the lamin proteins, A type is the principal mechanical component of the nucleus and controls gene expression via direct interactions with chromatin and chromatin modulator proteins. Mutations in the *LMNA* gene cause a plethora of diseases termed laminopathies, including various cardiac and muscular ailments, by disrupting nuclear integrity and mechanotransduction signalling. These mutations also alter chromatin organisation and epigenetic landscape, tantamount to compromised cellular homeostasis. In this chapter, we elaborate on lamin A's molecular structure, assembly dynamics and its role in nuclear mechanotransduction and chromatin maintenance. Additionally, we highlighted the pathological consequences of lamin A dysfunction and discussed emerging approaches aimed at rationalising the cellular- and tissue-specific effects during laminopathies.

**Keywords** Nuclear lamina · Lamin A · Mechanotransduction · Chromatin organisation · Cytoskeleton · Laminopathies

S. Nath
Biophysics & Structural Genomics Division, Saha Institute of Nuclear Physics, Kolkata, West Bengal, India

K. Sengupta (✉)
Homi Bhabha National Institute, Mumbai, Maharashtra, India
e-mail: kaushik.sengupta@saha.ac.in

© The Author(s), under exclusive license to Springer Nature Switzerland AG 2025

K. Sengupta, A. Muchir (eds.), *Lamins in Genome Organization and Diseases*, Subcellular Biochemistry 115, https://doi.org/10.1007/978-3-032-00537-3_6

## Introduction

The nucleus is the major organelle in all metazoan cells containing the chromatin and is responsible for controlling a multitude of cellular processes. The nucleus is a double-membrane-bound organelle having continuity with endoplasmic reticulum and studded with nuclear pores that facilitates the exchange of molecules between the nucleoplasm and cytoplasm. Underlying the inner nuclear membrane (INM) of the nuclear envelope is a protein meshwork termed as lamina which varies between 14 nm and 18 nm in thickness [1–3]. The nuclear lamina provides structural integrity and rigidity to nucleus and also acts as a tether for the heterochromatin via interaction with the lamina-associated domains or LADs.

In retrospect, fibrous protein mesh underlying the inner nuclear membrane was isolated as an ionic detergent insoluble, electron-dense entity as visualised in electron micrographs of vertebrate cells as early as the 1960s [4]. The lamina is constituted primarily of the lamin protein family. The lamins were first identified biochemically as prominent 60–80 kDa proteins serving as an integral component of nucleus providing structural support to the lamina [5] and playing pivotal roles in maintaining nuclear homeostasis [6–8]. Later, characterised as type V intermediate filaments, lamins stand out as the sole nuclear member within the intermediate protein family and are exclusively found in all metazoan cells [7] except plants.

Based on their sequence homologies, expression patterns and biochemical and dynamic properties, lamins can be classified into A and B types, each having distinct characteristics [9], and form segregated meshwork in lamina. In mammalian cells, the *LMNA* gene gives rise to four lamin proteins through alternative splicing: the predominant isoforms lamin A and lamin C, along with minor isoforms LAΔ10 and the germ-cell-specific isoform LC2 [9]. On the other hand, B-type lamins include lamins B1 and B2 which are encoded by LMNB1 and LMNB2 genes, respectively [10–13]. B-type lamins are constitutively expressed from the embryonic stage, whereas the expression of A-type lamins is developmentally regulated and primarily expressed in differentiated tissues [14].

Ecdysozoans, including *Caenorhabditis elegans* and *Drosophila melanogaster*, possess one or more lamin subtypes. Ce-lamin and Dm0 were the first lamin protein types reported in *C. elegans* and *Drosophila*, respectively [15, 16]. Ce-lamin differs from human lamin A in its longer head domain, lacking two heptad repeats in coil 2A and having a relatively short tail domain, yet sharing features with B-type lamins such as farnesylation at the CAAX box [17]. In contrast, Dm0, which is longer than the mammalian lamin B1 and more similar in length to lamin B2, is expressed throughout development [18]. *Drosophila* possesses another LamC which is often regarded as a lamin A variant being regulated developmentally [19]. While LamC shares traits with mammalian lamin C, being classified as an A-type lamin, the classification of *Drosophila*'s lamin Dmo as a B type is challenged by its protein structure. Neither lamin Dmo nor Ce-lamin genes, the first invertebrate lamin genes analysed, are representative of their taxonomic groups. Studies on the early development of *Xenopus* revealed that their oocytes primarily contain one B-type lamin,

lamin B3 (XLB3), along with minor amounts of XLB2 and XLB1 [20, 21]. In mice, type B lamins are expressed throughout early development, while lamins A/C are initially detected in the trophoblast on embryonic day 9, in the embryoblast on day 10 and in myoblasts and other mesenchymal tissues by day 11 [22, 23]. Later findings indicate that lamins B1 and B2, but not lamins A/C, are present in undifferentiated human and mouse embryonic stem cells [14]. Additionally, throughout embryogenesis, a marked increase in lamin A and a corresponding increase in lamina to nucleoplasm localisation ratio is observed with each differentiation step. This increase in lamin A expression not only correlates with zona hardening but also serves to reinforce the nuclear lamina, crucial for withstanding nuclear strain during development and regulating differentiation processes, thereby influencing the primary lineage of embryonic cells [24]. These observations provide a closer correlation between expression of lamin A/C, cellular differentiation and mechanotransduction. The absence of lamins A/C in undifferentiated human ES cells has been proposed as a contributing factor to the high deformability of the nuclei in these cells [25]. Variations in the expression of A-type and B-type lamins have been observed in adult human tissues and during the process of neurogenesis in the adult rat brain.

Since the initial discovery of lamin A/C mutation and its association with dilated cardiomyopathy in the late 1990s [26], a significant number of mutations in lamin A/C have been identified. To date, nearly 500 such mutations have been documented (http://www.umd.be/LMNA/), which leads to various conditions such as muscular dystrophies, lipodystrophies, dilated cardiomyopathy and premature aging syndrome affecting multiple organs [27–29]. Similarly, research has also demonstrated that lamin B1 duplication leads only to autosomal dominant leukodystrophy at that time [11]. Interestingly, recent work by Cristofoli et al. has linked missense mutations in lamin B1 with severe autosomal dominant microcephaly for the first time [30]. These discoveries have spurred significant research efforts focusing primarily on understanding lamin A/C mutations and their repercussions on cells and tissues.

Based on the identification of potential signalling pathways implicated in these mutants, researchers have provided insights that have given rise to two distinct explanatory frameworks: the structural hypothesis and gene regulation hypotheses. The structural hypothesis posits that mutations in lamin A disrupt higher-order assembly, consequently causing aberrant network formation, thereby affecting the nuclear shape and rigidity. This disruption consequently disturbs the transmission of external forces to the nucleus. On the other hand, the gene regulation hypothesis suggests that mutations in lamin A lead to aberrant transcription, splicing and epigenetic modification, thereby adversely regulating the gene expression pattern [31].

## Lamins: Structure and Assembly

As a member of the intermediate protein family, the nuclear lamins also exhibit the typical tripartite structural characteristic of intermediate filaments, comprising a highly α-helical central rod domain flanked by a short globular amino-terminal 'head' domain and a longer carboxyterminal 'tail' domain [8, 32, 33]. For lamin A, the central α-helical or rod domain comprises of four α-helical segments termed 1A (33–70 a.a.), 1B (80–218 a.a.), 2A and 2B (243–383 a.a.), where 1A and 1B are connected by linker L1 and 1B and 2A is connected by linker L12 giving rise to a 351-a.a.-long helical region, 48 amino acids longer than cytoplasmic intermediate filaments [34, 35]. The existence of linker L2, which connects the domains 2A and 2B, is proven, and in many cases it is referred as a stutter region in rod 2 domain, nonetheless providing a flexible link between the two subdomains. The extra region in rod 1 domain which is typically not observed in any other intermediate filaments is due to nuclear lamin A showing a special feature, i.e. its helix 1B contained 42 amino acids, or 'six heptads', more than cytoplasmic IF proteins [32], which partake in lamin homo-polymerisation. The tail domain contains a highly conserved structural motif resembling a type S immunoglobulin fold (Ig fold, 428–549 amino acids) [36, 37] along with a nuclear localisation signal (417 amino acids) necessary for transport into the nucleus, located between the carboxyterminal end of the central rod domain and the Ig fold.

Lamin A is expressed as a premature protein pre-lamin A (664 amino acids) with a CAAX box which is farnesylated by farnesyl-transferase, recruiting lamin A to the nuclear periphery with the help of hydrophobic interaction with the nuclear envelope [38]. Subsequently, AAX motif is cleaved by an endopeptidase RCE1 (Ras-conversating enzyme 1) and/or ZMPSTE24 (zinc metalloprotease related to Ste24p) to yield the mature lamin A having 18 fewer amino acid residues [39–41], whereas B-type lamins retain the farnesyl moiety in their mature form [42], thus enabling it to bind directly to nuclear envelope. The importance of CAAX box cleavage in lamin A is demonstrated by a splice site mutation that results in deletion of the ZMPSTE24 cleavage site, resulting in the accumulation of a permanently isoprenylated lamin A, leading to progeria [43]. Lamin C, on the other hand, lacks the CAAX box and is never isoprenylated, as it possesses a distinct six-amino-acid sequence at its tail end. Irrespective of that, lamin C can be still be found at the nuclear periphery, showing that isoprenylation might not be essential for lamin's localisation, which may be mediated by other proteins [44, 45].

Incorporation of lamin A into the nuclear lamina creates a large portion of lamin proteins in an insoluble pool at the nuclear rim, justifying the fact that A-type lamins are predominantly localised to inner nuclear envelope in fully differentiated cells [2]. It is known that lamin proteins exist in the 54-nm-long coiled-coil dimeric form in solution before forming the filament [46]. The assembly of lamins into functional filaments follows a hierarchical and coordinated sequence of molecular interactions, involving both longitudinal and lateral associations of dimers and higher-order polymeric forms. Lamin A assembly starts with the formation of coiled-coil dimers through the parallel alignment of rod domains, which polymerise longitudinally in a head-to-tail manner ($A_{CN}$ interactions) via electrostatic interactions, regulated by phosphorylation of lamin A head and tail domain (Ser22 and Ser392 phosphorylation by CDC2 and other kinases). Following longitudinal assembly, lateral A11 interactions stabilise the lamin tetramers (72 nm long, with 19-nm-long overhangs), while A22 interactions form four-helix bundles, termed as protofilaments, introducing curvature and flexibility to the lamin filaments through its 2B stutter region (L2). Protofilaments assemble from staggered tetramers and laterally associate into thin, ~3.5-nm-wide filaments, which assembles into the nuclear lamina [2, 46, 47]. Even though the polymerisation of lamin is majorly coordinated by the rod domain, both head and tail domains have been implicated in lamin A assembly [48]. In vitro analysis previously demonstrated that lamin A assembly starts by hydrophobic interactions driven by dimer formation within the helical rod regions, followed by higher-order assembly through parallel head-to-tail interactions and lateral compaction, ultimately forming an ~10 nm filaments [49, 50]. This understanding was later refined by cryo-electron microscopy studies, which revealed that lamins within the nuclear lamina meshwork assemble into tetramers forming filaments approximately 3.5 nm thick [2]. Elasticity and curvature, driven by A22 and $A_{CN}$ interactions, bestow lamin filaments with mechanical resilience, enabling the nuclear lamina to absorb and distribute mechanical stress while safeguarding chromatin. This assembly mechanism not only explains the lamina's structural adaptability but also offers broader insights into the mechanics of intermediate filament proteins and their role in mechanotransduction.

## Lamin A and Mechanotransduction

Mechanotransduction is a highly developing area in cell biology which focuses on the mechanism of sensing the external mechanical stimuli and their conversion to the physicochemical changes inside the cell [51] by complex combination of biochemical conduits. Although mechanotransduction loosely refers to cellular processes involving conversion of mechanical stimuli into biochemical signals, it is often applied more broadly to describe the overall cellular changes like changes in gene expression and structural reorganisation in response to its mechanical environment. The term 'mechanosensing' is largely used to describe the initiation of mechanotransduction events while denoting the downstream signalling and changes in the gene expression as 'mechanotransduction signalling' [52]. This is important for many cellular functions by allowing the cell to maintain its relative state of homeostasis with respect to the properties of the outside environment. Being the principal mechanosensing component of the nucleus, lamin A plays one of the crucial roles in overall mechanotransduction-mediated signalling.

As mentioned before, the mechanical properties of lamin filaments are primarily driven by its structural flexibility provided by its electrostatic interactions. The crystal structures of the lamin A fragments revealed differential bending angles at the linker regions, which partake in stabilising the coiled-coil interactions between adjacent amino acids. The bending at the L1 region supports the left-handed coiled-coil interaction in rod 1A domain, irrespective of the phase shift between coils 1A and 1B, while the L12 region might destabilise coiled-coil interactions within parallel dimers [53] giving rise to its structural flexibility. Structure relaxation simulations confirmed bending angles between 46° and 64° at the linker regions contribute towards the formation of a snake-like zigzag filament model, which adds flexibility to lamin A filaments towards stretching and bending. This is possible due to the compression and sliding within the lamin A coiled-coil rod domain, which enables the filaments to shorten and elongate with applied force. This process is facilitated by electrostatic interactions and flexible linker regions like L1, L2 and L12 [54], allowing small-scale deformations without significant unfolding, enhancing filament elasticity. The unstructured head and tail domains at the N- and C-termini also interact electrostatically with the rod domain, stabilising the tetrameric structure and supporting head-to-tail assembly. Cross-linking data further support a tandem stagger model, where linker regions fold over the coiled-coil termini, enabling rod shortening by up to 10 nm, which enhances the flexibility and adaptability of lamin filaments under mechanical stress. 1B and 2B rod domains of lamin A are essential for higher-order assembly and network formation but also have a significant strain resistance property. Despite their similar structural composition, these two motifs exhibit key differences in their mechanical properties, including modes of unfolding and unfolding forces. Despite the longer length of 1B domain, the inter-helical interactions were comparably stronger for the 2B domain [55], translating to ~200% strain resistance for 1B and 100% strain resistance for 2B domain. Also, through single molecule force spectroscopy studies, it has been found that 1B unzips at slightly lower force (~70 pN) compared to the lamin 2B dimer (~80 pN), a measurement that is backed up by molecular dynamic simulation. Therefore, the elasticity of lamin A protein molecule stems most prominently from the 1B domain which also plays a critical role in the polymerisation of the molecule. However, both of the domains were showing unfolding through several grey pathways, suggesting that they are flexible in nature. This importance of the 1B domain is supported by the observation that 42% of the total rod domain mutations are reported to be in 1B compared to 24% in 2B [55]. Similar to the rod domain, the Ig fold domain of lamin A also demonstrates unique mechanical properties, notably in its unfolding behaviour under force. Compared to the 1B and 2B domains, lamin Ig folds unfold at lower forces, indicating their involvement in providing structural support and flexibility to lamin proteins. The Ig domain of lamin A unfolds during fluid sheer stress in isolated nucleus exposing a previously buried cysteine residue, which was not observed in intact cells [56]. Later, it was found that lamin Ig folds yielded to lower force (~56 pN) compared to the signature elastomeric I27 domain, indicating that these lower unfolding forces are also necessary in detecting minute forces received by the nucleus. Although the tail domain does not have robust mechanical properties, this is compensated by having multiple motifs for its interaction with other participants of mechanotransduction pathway.

During the mechanotransduction process, lamins interact with various components such as SUN1, emerin, actins and many others to conduct the mechanical signal transduction in an effective manner. Most of the mechanical signals like fluid sheer stress, compression and stretching, extracellular matrix (ECM) stiffness and topographical changes are primarily sensed by the integrins, present in the focal adhesion (FA) complexes. Integrins act to anchor the cells, to the ECM, consisting of various matrix proteins such as fibronectin, collagen, laminin, etc., whose composition and concentration is sensed by these integrins and sent as a signal through the receptor tyrosine kinases (RTKs) [57, 58]. One of the fundamental RTKs in the FA complex is focal adhesion kinase or FAK, whose activation by autophosphorylation marks the beginning of intracellular mechanotransduction, by activating downstream signalling cascades. Similarly, paxillin, another member of FA complex, is also phosphorylated by FAK in response to integrin-mediated mechanical signalling [59]. Phospho-paxillin in turn moonlights as a transcription factor and translocates to the nucleus to convey various mechanically induced signalling cascades [60]. Along with such diffusional signalling cascades, integrins primarily transfer the mechanical signals through the cytoskeletal filaments like actins via cooperation between talin and vinculin. Talins are one of the first proteins to get recruited in the juvenile FA complexes. With the help of FAK and paxillin, talins open up in a force-dependent manner to expose vinculin binding sites. This indicates that the accumulation of vinculin in the focal adhesion complex directly reflects the force, vis-a-vis mechanical stimulus encountered by that particular FA. Vinculins in turn have strong actin binding sites [61], more prominent than that of talin, which binds to actin stress fibres, composed predominantly of beta-actin polymers, decorated with cell-line-specific myosins and tropomyosins. The actin cytoskeleton, which generates as well as transmits mechanical forces, appears as a central player in cellular mechanosensing. Through an array of studies on individual actin filaments, it has been proven that mechanical stimuli apprehended by the actin filaments alter their association with the regulatory proteins, affecting their assembly and dynamics [62]. Changes in filament dynamics could be provoked by mechanical stress applied to the filament and therefore provide the structural basis for our understanding of actin mechanotransduction, with the filament itself acting as a mechanosensor [63]. Dynamics and turnover of actin stress filaments are one of the most crucial signals in the entire mechanotransduction pathway. Not only do the actins transfer the mechanical signals from the FA complex to the nucleus, but the polymerisation of actins is also tightly regulated in a stress-dependent manner, starting by FA complex, which acts as a mechanical cue for many other signalling pathways. FAK and paxillin again are major players in actin polymerisation [64, 65] by interacting with various prominent actin regulatory pathway members such as ERK1/2, rhoA, ROCK and many others [66]. Actin stress fibres not only affect the mechanical signal transfer in a direct way, but its turnover, i.e. F/G actin ratio, also bypasses the HIPPO signalling pathway and regulates nuclear sequestration of YAP/TAZ, a key mechanoeffector protein [67]. Taken together, actin remodelling is one of the key parameters in the mechanotransduction pathway which links various pathways such as rhoA/ROCK, MAPK and HIPPO into a common cause.

On the other side of these stress fibres, actins bind to the giant spectrin repeat containing a group of proteins called as nesprins. There are several types of nesprins that are expressed in a tissue-specific manner but more or less have a common functional parameter, binding to terminus of cytoskeletal filaments and tethering them to nucleus. Nesprins are part of the linker of nucleoskeleton and cytoskeleton or LINC complex which, as the name suggests, links the cytoskeleton like microfilaments, microtubules and intermediate filaments to the nucleoskeleton. The nucleoskeleton is primarily composed of lamin intermediate filaments and nuclear actins, thus creating a suspension bridge of mechanotransduction pathway. Nesprins are embedded in outer nuclear membrane from where it interacts with SUN (Sad1p, UNC-84) protein of the LINC complex through its KASH (Klarsicht, ANC-1, Syne homology) domain in the lumen of nuclear envelope [68]. SUN1, another transmembrane protein embedded in the inner nuclear membrane, associates with lamins and helps regulating the mechanotransduction signals [58, 69]. Along with lamins, SUN proteins also help in the organisation of chromosome territories by binding directly to chromatin, chromatin regulators and the nuclear pore complex subunits. It has been observed that disruption of LINC complex proteins by knocking out either or both member of LINC complex reduces nuclear stiffness and disturbs the transmembrane actin-associated nuclear or TAN lines, which are key components in maintaining nuclear homeostasis as well as mechanotransduction [70]. Along the same line, disrupting LINC complex almost completely abolishes nuclear deformation when cells are subjected to high mechanical stimulus, without compromising mechanoresponsive gene activation [71], whereas it has been proven that an intact LINC complex is indeed necessary for the detection and translation of low magnitude mechanical signals [72], stating its crucial role in detecting sensitive mechanical cues from ECM for fine-tuning mechanical signals. Thus, following the LINC complex, mechanical signals finally reach the lamins, which, as discussed earlier, helps in modulating the mechanical signals and translates it to epigenetic signals and differential gene expression.

Besides direct effect in mechanotransduction, it has been observed that lamins also play several key roles indirectly in the transfer of mechanical signals. Studies by several groups have shown that lamin A deficiency and mutations lead to increased activation of ERK1/2 through MEK kinase activity, which is one of the key mechanically activated kinases of MAPK signalling pathway [73–77]. Upon the stimulus-induced ERK activation, two opposing outcomes in cellular force generation can occur: first is the generation of contractile forces through actomyosin contraction, and the other is the generation of extending forces through actin polymerisation [78]. ERK1/2 join forces with rhoA, FAK and formins in a coordinated manner which critically analyses the mechanical signals and regulates actin polymerisation through multi-level regulatory pathways [79]. ERK1/2 activation plays a significant role in the generation of various cellular projections, including lamellipodia, filopodia and invadopodia, thereby promoting cell migration and invasion. These pseudopodia form through nucleation or branching of actins in the cortical region through the activation of Arp2/3 complex [80], the major actin nucleator of the cell. Along with promoting actin polymerisation, ERK1/2 has been recently identified to also interact with one of the major actin depolymerising proteins, cofilin-1 [75], which gets phosphorylated at T25 position by ERK1/2 to promote actin depolymerisation. Phospho-cofilin-1 binds to myocardin-related transcription factor A (MRTF-A) in the cytoplasm, thus preventing the stimulation of serum response factor (SRF) in the nucleus leading to decreased α-tubulin acetylation by reducing the expression of α-tubulin acetyltransferase leading to mislocalisation of Connexin 43 (Cx43), an important component for GAP junctions [81]. Given that SRF's transcriptional activity is dependent on actin cytoskeleton dynamics, which in turn is controlled directly by lamin A through ERK1/2, it is becoming increasingly evident that the three cytoskeletal systems correlate with each other in vital cellular processes and that their purpose and dynamics are often closely interwoven.

## Defects in Mechanotransduction in Lamin A Mutations

First identified in the late 1990s, laminopathies refer to a set of rare diseases arising primarily from mutations in the *LMNA* gene. These disorders affect specific tissues, including striated muscles, peripheral nerves, adipose tissue, etc., causing a plethora of diseases such as cardiomyopathy (DCM), Emery-Dreifuss muscular dystrophy (AD-EDMD), Hutchinson-Gilford progeria syndrome (HGPS), atypical Werner's syndrome (WS), restrictive dermopathy (RD) and mandibuloacral dysplasia (MAD). The primary characteristic of laminopathic cells is altered nuclear homeostasis through abnormal nuclear architecture leading from altered lamin A mechanics. Studies have shown that mutations in lamin A lead to alterations in its secondary and tertiary structure leading to improper polymerising kinetics [82], resulting in changes such as increased lamina mesh size, lamin filament bundling, viscoelastic alterations and aggregation within the nucleus [83, 84]. Certain mutations, particularly those associated with DCM, cause enlarged heart chambers and spongy heart tissues, elevating cardiac tissue stiffness and heightened risk of myocardial infarction. These mutations implicated in the pathogenesis of DCM possess distinct secondary and tertiary conformations compared to the wild type, which was reflected in gross distribution and localisation defects in the protein [82]. One such mutation, E161K, showed drastic changes in its secondary and tertiary structure, whereas K97E exhibited a much higher self-association constant, reflected in larger aggregates in cells ectopically expressing the mutant ex vivo [55]. The mutants K97E, L183P, K260N and Y267C show at least tenfold high self-associating affinities and less stable interaction with lamin B1 compared to wild-type lamin A [84]. The mutations causing DCM lead to a fragile and poorly elastic nuclear envelope, which is unable to bear strain deformation, explaining why such lamin A mutations lead to abnormally shaped fragile nuclei in the first place. In case of AD-EDMD causing mutation R453W, the Ig fold of lamin A shows a much lower unfolding force compared to the wild type and is found to be intrinsically misfolded, thereby rendering the lamin A network weaker and less resilient to force [85]. The same mutation shows a much lower diffusion rate compared to the wild type in mouse myoblasts, which might cause the destabilisation of nuclear lamina through improper dynamic stability of the lamina [34]. These observations suggest that in each type of laminopathic point mutation, based on the type and position of mutation, different biophysical aspects of lamin A protein are altered, which directly results in the change in its viscoelastic properties, interaction and overall dynamics, required for proper functioning of the protein.

Along with structural and mechanical changes in lamin A, it has been seen that in H222P, a widely accepted model mutation for lamin A-mediated DCM, ERK1/2 kinase gets hyperactivated. ERK1/2 hyperactivation leads to the overactivation of cofilin-1, an actin destabiliser protein, which in turn alters the actin stress fibre kinetics in cells by depolymerising the actin filaments [75, 81]. Phosphorylation of FHOD1 and FHOD3 by hyperactivated ERK1/2 in H222P mutation also inhibits their actin-bundling activity, resulting in abnormal nuclear positioning in differentiated cardiac muscles [73]. It has been observed that due to this misregulation of cytoskeletal filament dynamics, the stiffness of the cells also increased along with reduced cellular motility [74]. However, inhibition of MEK kinase, an upstream modulator of ERK1/2 activation, partially prevents DCM phenotypes [86, 87]. In mouse models of H222P lamin A mutants, it has been observed that inhibitors of ERK/JNK pathway blocked increased expression of RNAs encoding natriuretic peptide precursors and proteins involved in sarcomere architecture that occurred in placebo-treated mice and prevented left ventricular end-systolic dilatation, increased ejection fraction and decreased myocardial fibrosis [88].

Due to the elevated levels of ERK1/2 during laminopathies, impaired muscle cell differentiation is also observed in muscle progenitor cells having lamin A mutations [89]. AD-EDMD-causing mutant R453W shows elevated ERK1/2 activation, resulting in incomplete exit from the cell cycle and differentiation into myocytes with a lower level of multi-nucleation. This abnormality of differentiation might be correlated to defective mechanotransduction through YAP/TAZ. It has been observed that cells having lamin A mutation show a heightened mechanosensitivity indicated through elevated nuclear localisation of YAP [90, 91]. Notably, during lamin A H222P mutation, the nuclear-to-cytoplasmic (N/C) ratio of YAP shifted towards higher nuclear YAP for the cell, which counters the anti-proliferative signal needed to initiate myoblast differentiation, despite hippo activation [92, 93]. This observation of nuclear YAP localisation is quite contradictory to what has been observed through the destabilisation of actins in a non-lamin A-dependent manner, i.e. using actin destabilisation drugs such as cytochalasin and latrunculin, where a disturbance in actin polymerisation process in turn reduces the N/C ratio of YAP/TAZ. Although YAP activation and nuclear localisation are primarily regulated by actins and the hippo pathway, it is the nuclear pore complex which is the 'gateway' to the nucleoplasm, and studies showed that NPC remodelling can also regulate the YAP N/C ratio in a force-dependent manner [94, 95], and as lamin A constitutes the matrix for NPC distribution and has proven interaction with the same, it is highly probable that the final YAP nuclear import during laminopathies is mediated by NPC in a lamin A-dependent manner and not solely based on actin remodelling.

Overall, we can conclude that a single point mutation in lamin A protein can mislead the mechanotransduction processes inside cell, as well as disturb the cellular proliferation and differentiation processes. If we look closer into the differentiation of cells, we can appreciate the fact that this involves chromatin reorganisation and change in chromosome territories [25], which sheds light on the role of lamin in safeguarding chromatin architecture and positioning.

## Lamin and Chromatin Interaction

At one point or another, each of the cytoskeletal filament groups interacts and modulates chromatin reorganisation, whether it be microtubule-mediated spindle fibre assembly or nuclear actin-mediated chromatin territory movements. Similarly, it has been well established that lamin A plays a crucial role in chromatin remodelling through mechanotransduction via regulation of various transcription factors and alteration of epigenetic landscape along with maintaining nuclear architecture, preventing chromatin damage. Lamins interact with various histone subunits and regulate the intricate chromatin to form regions known as lamina-associated domains or LADs [96–98]. LADs are a large fraction of the mammalian genome which ranges from 10 megabases down to 100 kilobases, which comprise of ~40% of the partially repressed human genome, by their association to the nuclear lamina. LADs contain characteristic heterochromatic markers, such as H3K9me2/me3 and H3K27me3, that exhibit condensed chromatin structures. LADs are categorised into two major groups: constitutive (cLADs) and facultative LADs (fLADs). While the organisation of cLADs is highly static, fLADs are comparatively more dynamic, often showing attachment to and detachment from the nuclear lamina in response to environmental cues [99]. Nuclear lamins also provide anchoring sites, including site in its tail region, for binding to the LEM (Lap2α, Emerin, Man1; a BAF-binding motif) domain-containing proteins which have overlapping function essential for chromosome segregation and cell division [100], potentially enhancing the role of the lamin A-BAF-LEM complex in organising chromatin architecture [101]. Notably, BAF, LAP2α and various histone subunits which interact with lamin A help in the modulation of epigenetic changes through mechanotransduction pathway. BAF, LAP2α and other proteins responsible for chromatin organisation interact with lamins at the inner nuclear membrane, influencing chromatin organisation based on nuclear envelope dynamics [102]. It facilitates the tethering of chromatin to the nuclear periphery through its interaction with lamins, together with regulating gene expression patterns and nuclear architecture. They are also essential for organismal and cell-type-specific gene regulation [103] and are part of a larger organisation called chromatin territories. Chromatin territories are compartments created by the individual chromatins with their self- and cross-chromatin associations and are necessary to maintain proper gene expression, genome stability and broadly the implementation of a hierarchy of subnuclear functions [104].

Chromatin territories exhibit partial characteristics of liquid-liquid phase separation (LLPS) inside nucleus [105] and maintain the state through the heterochromatin protein 1 or HP1. HP1s are a group of proteins that binds to both hetero- and euchromatin and shows a strong correlation with lamin A. Mammalian HP1 exists in three major isoforms, α, β and γ. HP1α and HP1β are enriched in heterochromatic foci, whereas HP1γ has been detected in both euchromatin and heterochromatin [106, 107]. The interaction between HP1γ and lamin A is well established and can be validated as both the proteins are chromatin associated and regulate the assembly of chromatin territories. Both HP1-gamma and lamin are implicated in DNA repair processes, suggesting their coordinated roles in maintaining genome stability. Disruption of either HP1-gamma or lamin function can lead to defects in DNA repair mechanisms, predisposing cells to genomic instability and disease.

Lamin A also participates in DNA repair, replication, transcription and many other mechanisms associated with the chromatin maintenance. This is evident by the elevated amount of lamin-chromatin interaction in the nuclear interior during DNA replication, as the interaction is proven to be necessary for the maintenance of the double amount of chromatin after the replicative phase. The Ig fold motif located in the lamin A C-terminus binds directly to proliferating cell nuclear antigen or PCNA, a key processivity factor which is required for DNA chain elongation and enables DNA polymerases to synthesise long segments of DNA without detaching from the template [31]. Also, lamin A interacts with RNA polymerase II, which acts to promote the re-association of hundreds of proteins with newly replicated chromatin via pathways that are not observed in steady-state chromatin including ATP-dependent remodellers, transcription factors and histone methyl-transferases [108]. Lamin A also participates in the post-transcriptional modification of mRNAs through the interaction with the spliceosomes, as it is also observed that lamin A has a direct interaction with matrin-3 through its tail domain [109] indicating its possible role in RNA processing and related activities.

Beyond this, the involvement of lamin A extends further into safeguarding genomic architecture, particularly in maintaining chromatin organisation and stability. Along with DNA damage repair, lamin A is one of the major players responsible for modulating telomere stability and also contributes to the spatial organisation of telomeres within the nucleus [110, 111], through its interactions with telomere-repeat binding factor or TRF proteins [112, 113]. TRFs are the key group of proteins involved in telomere maintenance and genome stability and bind specifically to repeated telomeric DNA sequences, forming a complex known as shelterin complex, which protects telomeres and regulates their stability. TRF1 and TRF2 along with proteins like RAP1, TIN2, TPP1 and POT1 consist the shelterin complex proteins required for maintaining telomere integrity [112, 114, 115], among which TRF1, TRF2 and POT1 are the ones that directly bind to DNA. Telomere protection involves the insertion of the 30 overhangs facilitated by TRF2 into telomeric DNA, forming t-loops [114], which in turn is phosphorylated by ERK1/2, a protein known for its correlation with lamin A, in both normal and cancer cells [116]. Similarly, TRF1 is also phosphorylated by ERK1/2, in 17 different residues, and it is seen that inhibiting ERK1/2 causes telomere uncapping in patient-derived glioblastoma models [117]. Thus, we can observe that there is no part of chromatin maintenance which is untouched by lamin A, whether it be chromatin organisation, maintenance or protection.

## Alteration of Chromatin Organisation and Epigenetic Changes in Laminopathies

Our cells depend heavily on the mechanotransduction pathways to translate mechanical signals from the extracellular matrix into biochemical responses to naturally adapt according to the received cues. When these pathway are disrupted, the cell's ability to adapt to mechanical stress and recover from damage diminishes significantly [118]. It has been reported that reduction or complete absence of lamin A disrupts chromatin architecture and its tethering to the nuclear lamina, leading to an extensive chromatin remodelling. Aberrant gene expression patterns observed in laminopathies highlight the roles of both chromatin and epigenetic remodelling factors to this differential gene regulation.

Across laminopathies, whether involving muscular dystrophy or cardiomyopathy, one common feature is the profound alteration of nuclear architecture and chromatin dynamics. Mutations in lamin A disrupt its interactions with chromatin and chromatin-associated proteins, leading to cascading effects on nuclear processes such as transcription, chromatin remodelling and DNA repair. Some specific tail domain lamin A mutations, such as R527P and L530P, which are associated with Emery-Dreifuss muscular dystrophy (EDMD), severely impair the structural dynamics of the Ig fold, the key DNA-binding motif of lamin A. This disruption reduces lamin A's ability to interact with the chromatin, which might lead to alteration of chromatin territories. Along the same line, these mutations also abrogate the interaction between lamin A and nuclear actins, an essential filament for organising chromatin territories [119]. Nuclear actin, a cofactor for transcription, chromatin remodelling and nuclear assembly, is also affected in terms of concentration and function, as the lack of proper interaction with lamin A within nucleus [120]. Similarly, dilated cardiomyopathy (DCM)-associated lamin A mutations also aggravate chromatin disorganisation inside nucleus. It has been observed that these DCM-causing mutations alter heterochromatin markers, such as H3K9me3, which are highly associated with LADs, thereby influencing epigenetic regulation and chromatin compartmentalisation (24153156). Additionally, mutations in the lamin A rod domain deplete heterochromatin proteins HP1$\alpha$ and HP1$\beta$ through their proteolytic degradation, while HP1$\gamma$ are reported to remain unaffected in this case [121]. However, other studies have shown that HP1$\gamma$ can sequester within lamin A mutant aggregates in the nucleus, leading to its eventual degradation alongside other HP1 isoforms [121]. This widespread disruption of heterochromatin proteins contributes to the misregulation of chromatin territory architecture and gene expression [122], particularly in muscle and cardiac tissues affected by EDMD and DCM. For instance, the DCM-causing E161K mutation, which sequesters RNA polymerase II, has been observed to alter the location of chromosome 13, a normally heterochromatic region, towards the nuclear interior as revealed by 3D-FISH studies [123]. These mislocalisations point to the fact that lamin A dysfunction alters chromatin territories and epigenetic regulation.

In progeroid syndromes such as Hutchinson-Gilford progeria syndrome (HGPS), an aberrant form of lamin A, called progerin, is produced due to improper processing of prelamin A. Progeria-associated lamin A mutations also lead to physical distortions in the nucleus, such as the formation of nuclear blebs and micronuclei. These structural anomalies interfere with chromatin segregation and cell cycle progression, ultimately causing widespread nuclear dysfunction [124]. It has been noticed that these telomeres are widely associated in either nuclear interior or it is abundant in the nuclear blebs, characteristic of progeroid mutation. Further, progerin lacks the ability to bind TRF2, a critical shelterin complex protein, leading to telomere instability and subsequent genomic damage [113]. This instability is further worsened by the aberrant localisation of LADs, which causes substantial changes in the spatial organisation of chromatin and the regulation of gene expression [125, 126]. Additionally, chromatin reprogramming in progeroid conditions results in the loss of histone modifications that are essential for the proper maintenance of heterochromatin and euchromatin balance, further contributing to genomic instability. Mouse embryonic fibroblasts expressing progerin or lacking A-type lamins show marked defects in telomere homeostasis and DNA repair, driven by the degradation of 53BP1 through cathepsin-L and reduced expression of BRCA1 and RAD51 [127]. These defects highlight the critical role of lamin A in maintaining genome stability.

In summary, the absence of properly functioning lamin A leads to a cascade of nuclear defects, including telomere shortening, loss of chromatin integrity and defective nuclear assembly. These changes manifest in tissue-specific phenotype, with skeletal muscle, cardiac tissue and rapidly dividing cells such as fibroblasts being most severely affected. In conclusion, laminopathies are caused by mutations in the lamin A/C gene, which disrupts the balance and structure of the nucleus causing a variety of tissue-specific diseases including accelerated aging. Excessive activation of pathways such as ERK1/2, as well as perturbation in mechanosensing, further adversely affects cellular functions like mechanotransduction, chromatin architecture and muscle growth. Furthermore, lamin A mutations procure a significant impact on genome reprogramming as they affect telomere stability, chromatin architecture and gene expression. Understanding these intricate and interlinked pathways is critical for understanding the aberrant physiological processes of laminopathies and discovering prospective therapeutic targets to treat these debilitating diseases.

## References


1. Stiekema M, van Zandvoort MAMJ, Ramaekers FCS, Broers JLV (2020) Structural and mechanical aberrations of the nuclear lamina in disease. Cells 9
2. Turgay Y, Eibauer M, Goldman AE et al (2017) The molecular architecture of lamins in somatic cells. Nature 543:261–264
3. Turgay Y, Medalia O (2017) The structure of Lamin filaments in somatic cells as revealed by cryo-electron tomography. Nucleus 8:475–481
4. Fawcett DW (1966) On the occurrence of a fibrous lamina on the inner aspect of the nuclear envelope in certain cells of vertebrates. Am J Anat 119:129–145
5. Houben F, Ramaekers FCS, Snoeckx LHEH, Broers JLV (2007) Role of nuclear lamina-cytoskeleton interactions in the maintenance of cellular strength. Biochim Biophys Acta, Mol Cell Res 1773:675–686
6. Aaronson RP, Blobel G (1975) Isolation of nuclear pore complexes in association with a lamina. Proc Natl Acad Sci 72:1007–1011
7. Aebi U, Cohn J, Buble L, Gerace L (1986) The nuclear lamina is a meshwork of intermediate-type filaments
8. McKeon FD, Kirschner MW, Caput D (1986) Homologies in both primary and secondary structure between nuclear envelope and intermediate filament proteins. Nature 319:463–468
9. Lin F, Worman HJ (1993) Structural organization of the human gene encoding nuclear Lamin A and nuclear Lamin C. J Biol Chem 268:16321–16326
10. Adam SA (2017) The nucleoskeleton. Cold Spring Harb Perspect Biol 9
11. Bhattacharjee P, Dasgupta D, Sengupta K (2015) Molecular events in Lamin B1 Homopolymerization: a biophysical characterization. J Phys Chem B 119:14014–14021
12. Butin-Israeli V, Adam SA, Goldman AE, Goldman RD (2012) Nuclear Lamin functions and disease. Trends Genet 28:464–471
13. Sengupta D, Sengupta K (2022) Lamin A and telomere maintenance in aging: two to tango. Mutat Res Mol Mech Mutagen 825:111788
14. Constantinescu D, Gray HL, Sammak PJ et al (2006) Lamin A/C expression is a marker of mouse and human embryonic stem cell differentiation. Stem Cells 24:177–185
15. Osman M, Paz M, Landesman Y et al (1990) Molecular analysis of the drosophila nuclear Lamin gene. Genomics 8:217–224
16. Peter A, Stick R (2012) Evolution of the Lamin protein family: what introns can tell. Nucleus 3:44–59
17. Rzepecki R, Gruenbaum Y (2018) Invertebrate models of Lamin diseases. Nucleus 9:227–234
18. Bossie CA, Sanders MM (1993) A cDNA from Drosophila melanogaster encodes a Lamin C-like intermediate filament protein. J Cell Sci 104:1263–1272
19. Riemer D, Stuurman N, Berrios M et al (1995) Expression of drosophila Lamin C is developmentally regulated: analogies with vertebrate A-type lamins. J Cell Sci 108:3189–3198
20. Benavente R, Krohne G, Franke WW (1985) Cell type-specific expression of nuclear lamina proteins during development of Xenopus laevis. Cell 41:177–190
21. Stick R, Hausen P (1985) Changes in the nuclear lamina composition during early development of Xenopus laevis. Cell 41:191–200
22. Rober RA, Weber K, Osborn M (1989) Differential timing of nuclear Lamin A/C expression in the various organs of the mouse embryo and the young animal: a developmental study. Development 105:365–378
23. Stewart C, Burke B (1987) Teratocarcinoma stem cells and early mouse embryos contain only a single major Lamin polypeptide closely resembling Lamin B. Cell 51:383–392
24. Skory RM, Moverley AA, Ardestani G et al (2023) The nuclear lamina couples mechanical forces to cell fate in the preimplantation embryo via actin organization. Nat Commun 14:3101
25. Pajerowski JD, Dahl KN, Zhong FL et al (2007) Physical plasticity of the nucleus in stem cell differentiation. Proc Natl Acad Sci USA 104:15619–15624

26. Fatkin D, MacRae C, Sasaki T et al (1999) Missense mutations in the rod domain of the Lamin A/C gene as causes of dilated cardiomyopathy and conduction-system disease. N Engl J Med 341:1715–1724
27. Dutta S, Das JK, Maganti L et al (2018) Skeletal muscle dystrophy mutant of Lamin A alters the structure and dynamics of the Ig fold domain. Sci Rep 8:1–17
28. Gundersen GG, Worman HJ (2013) Nuclear positioning. Cell 152:1376–1389
29. Worman HJ, Ostlund C, Wang Y (2010) Diseases of the nuclear envelope. Cold Spring Harb Perspect Biol 2
30. Cristofoli F, Moss T, Moore HW et al (2020) De novo variants in LMNB1 cause pronounced syndromic microcephaly and disruption of nuclear envelope integrity. Am J Hum Genet 107:753–762
31. Shumaker DK, Solimando L, Sengupta K et al (2008) The highly conserved nuclear Lamin Ig-fold binds to PCNA: its role in DNA replication. J Cell Biol 181:269–280
32. Herrmann H, Aebi U (2016) Intermediate filaments: structure and assembly. Cold Spring Harb Perspect Biol 8
33. Parry DAD, Conway JF, Steinert PM (1986) Structural studies on Lamin. Similarities and differences between Lamin and intermediate-filament proteins. Biochem J 238:305–308
34. Mukherjee C, Sengupta D, Maganti L et al (2022) Slower diffusion and anomalous association of R453W Lamin A protein alter nuclear architecture in AD-EDMD. RSC Adv 12:32129–32141
35. Strelkov SV, Schumacher J, Burkhard P et al (2004) Crystal structure of the human Lamin a coil 2B dimer: implications for the head-to-tail association of nuclear lamins. J Mol Biol 343:1067–1080
36. Conway JF, Parry DAD (1989) Structure and spatial organisation of intermediate filament and nuclear Lamin. Molecules 140–149
37. Dhe-Paganon S, Werner ED, Chi YI, Shoelson SE (2002) Structure of the globular tail of nuclear Lamin. J Biol Chem 277:17381–17384
38. Hennekes H, Nigg EA (1994) The role of isoprenylation in membrane attachment of nuclear lamins. A single point mutation prevents proteolytic cleavage of the Lamin A precursor and confers membrane binding properties. J Cell Sci 107:1019–1029
39. Corrigan DP, Kuszczak D, Rusinol AE et al (2005) Prelamin a endoproteolytic processing in vitro by recombinant Zmpste24. Biochem J 387:129–138
40. de Leeuw R, Gruenbaum Y, Medalia O (2018) Nuclear Lamins: thin filaments with major functions. Trends Cell Biol 28:34–45
41. Holtz D, Tanaka RA, Hartwig J, McKeon F (1989) The CaaX motif of Lamin A functions in conjunction with the nuclear localization signal to target assembly to the nuclear envelope. Cell 59:969–977
42. Dittmer T, Misteli T (2011) The Lamin protein family. Genome Biol 12:222
43. Broers JLV, Ramaekers FCS, Bonne G et al (2006) Nuclear lamins: Laminopathies and their role in premature ageing. Physiol Rev 86:967–1008
44. Davies BSJ, Coffinier C, Yang SH et al (2011) Investigating the purpose of prelamin a processing. Nucleus 2:4–9
45. Goldberg MW, Fiserova J, Huttenlauch I, Stick R (2008) A new model for nuclear lamina organization. Biochem Soc Trans 36:1339–1343
46. Ahn J, Jo I, Kang S et al (2019) Structural basis for Lamin assembly at the molecular level. Nat Commun 10:1–12
47. Lilina AV, Chernyatina AA, Guzenko D, Strelkov SV (2020) Lateral A11 type tetramerization in lamins. J Struct Biol 209:107404
48. Sasse B, Aebi U, Stuurman N (1998) A tailless drosophila Lamin Dm0 fragment reveals lateral associations of dimers. J Struct Biol 123:56–66
49. Herrmann H, Foisner R (2003) Intermediate filaments: novel assembly models and exciting new functions for nuclear lamins. Cell Mol Life Sci 60:1607–1612

50. Stuurman N, Heins S, Aebi U (1998) Nuclear lamins: their structure, assembly, and interactions. J Struct Biol 122:42–66
51. Ingber DE (2003) Mechanobiology and diseases of mechanotransduction. Ann Med 35:564–577
52. Isermann P, Lammerding J (2013) Nuclear mechanics and mechanotransduction in health and disease. Curr Biol 23:R1113–R1121
53. Ahn J, Jo I, Jeong S et al (2023) Lamin filament assembly derived from the atomic structure of the antiparallel four-helix bundle. Mol Cells 46:309–318
54. Makarov AA, Zou J, Houston DR et al (2019) Lamin A molecular compression and sliding as mechanisms behind nucleoskeleton elasticity. Nat Commun 10:1–17
55. Bera M, Ainavarapu SRK, Sengupta K (2016) Significance of 1B and 2B domains in modulating elastic properties of Lamin A. Sci Rep 6:27879
56. Swift J, Ivanovska IL, Buxboim A et al (2013) Nuclear Lamin-a scales with tissue stiffness and enhances matrix-directed differentiation. Science 341:1358–1375
57. Boudreau NJ, Jones PL (1999) Extracellular matrix and integrin signalling: the shape of things to come. Biochem J 339:481–488
58. Martino F, Perestrelo AR, Vinarský V et al (2018) Cellular mechanotransduction: from tension to function. Front Physiol 9:1–21
59. Burridge K, Fath K, Kelly T et al (1988) Focal adhesions: transmembrane junctions between the extracellular matrix and the cytoskeleton. Annu Rev Cell Biol 4:487–525
60. Zhang Y, Kishi H, Morita T, Kobayashi S (2021) Paxillin controls actin stress fiber formation and migration of vascular smooth muscle cells by directly binding to the active Fyn. FASEB J 35:1–20
61. Humphries JD, Wang P, Streuli C et al (2007) Vinculin controls focal adhesion formation by direct interactions with Talin and actin. J Cell Biol 179:1043–1057
62. Michelot A, Drubin DG (2011) Building distinct actin filament networks in a common cytoplasm. Curr Biol 21:560–569
63. Galkin VE, Orlova A, Egelman EH (2012) Actin filaments as tension sensors. Curr Biol 22:R96–R101
64. Case LB, Waterman CM (2015) Integration of actin dynamics and cell adhesion by a three-dimensional, mechanosensitive molecular clutch. Nat Cell Biol 17:955–963
65. Wilson OJ, Bradley H, Shaw CS, Wagenmakers AJM (2014) Paxillin and focal adhesion kinase colocalise in human skeletal muscle and its associated microvasculature. Histochem Cell Biol 142:245–256
66. Xu B, Ju Y, Song G (2014) Role of p38, ERK1/2, focal adhesion kinase, RhoA/ROCK and cytoskeleton in the adipogenesis of human mesenchymal stem cells. J Biosci Bioeng 117:624–631
67. Seo J, Kim J (2018) Regulation of hippo signaling by actin remodeling. BMB Rep 51:151–156
68. Haque F, Lloyd DJ, Smallwood DT et al (2006) SUN1 interacts with nuclear Lamin A and cytoplasmic Nesprins to provide a physical connection between the nuclear lamina and the cytoskeleton. Mol Cell Biol 26:3738–3751
69. Wolfenson H, Yang B, Sheetz MP (2019) Steps in Mechanotransduction pathways that control cell morphology. Annu Rev Physiol 81:585–605
70. Gant Luxton GW, Gomes ER, Folker ES et al (2011) TAN lines: a novel nuclear envelope structure involved in nuclear positioning. Nucleus 2:173–181
71. Lombardi ML, Jaalouk DE, Shanahan CM et al (2011) The interaction between nesprins and sun proteins at the nuclear envelope is critical for force transmission between the nucleus and cytoskeleton. J Biol Chem 286:26743–26753
72. Uzer G, Thompson WR, Sen B et al (2015) Cell Mechanosensitivity to extremely low-magnitude signals is enabled by a LINCed nucleus. Stem Cells 33:2063–2076
73. Antoku S, Wu W, Joseph LC et al (2019) ERK1/2 phosphorylation of FHOD connects signaling and nuclear positioning alternations in cardiac Laminopathy. Dev Cell 51:602–616.e12

74. Chatzifrangkeskou M, Kah D, Lange JR et al (2020) Mutated Lamin A modulates stiffness in muscle cells. Biochem Biophys Res Commun 529:861–867
75. Chatzifrangkeskou M, Yadin D, Marais T et al (2018) Cofilin-1 phosphorylation catalyzed by ERK1/2 alters cardiac actin dynamics in dilated cardiomyopathy caused by Lamin A/C gene mutation. Hum Mol Genet 27:3060–3078
76. Emerson LJ, Holt MR, Wheeler MA et al (2009) Defects in cell spreading and ERK1/2 activation in fibroblasts with Lamin A/C mutations. Biochim Biophys Acta Mol basis Dis 1792:810–821
77. Muchir A, Wu W, Worman HJ (2009) Reduced expression of A-type lamins and emerin activates extracellular signal-regulated kinase in cultured cells. Biochim Biophys Acta Mol basis Dis 1792:75–81
78. Nayak AN, Hirashima T (2023) Tug-of-war via ERK signaling pathway for tissue organization – ERK activation to force generation. Curr Opin Cell Biol 85:102249
79. Selig M, Lauer JC, Hart ML, Rolauffs B (2020) Mechanotransduction and stiffness-sensing: mechanisms and opportunities to control multiple molecular aspects of cell phenotype as a design cornerstone of cell-instructive biomaterials for articular cartilage repair. Int J Mol Sci 21:5399
80. Tanimura S, Takeda K (2017) ERK signalling as a regulator of cell motility. J Biochem 162:145–154
81. Le Dour C, Chatzifrangkeskou M, Macquart C et al (2022) Actin-microtubule cytoskeletal interplay mediated by MRTF-A/SRF signaling promotes dilated cardiomyopathy caused by LMNA mutations. Nat Commun 13:7886
82. Bhattacharjee P, Banerjee A, Banerjee A et al (2013) Structural alterations of Lamin A protein in dilated cardiomyopathy. Biochemistry 52:4229–4241
83. Banerjee A, Rathee V, Krishnaswamy R et al (2013) Viscoelastic behavior of human Lamin A proteins in the context of dilated cardiomyopathy. PLoS One 8:e83410
84. Bhattacharjee P, Dasgupta D, Sengupta K (2017) DCM associated LMNA mutations cause distortions in lamina structure and assembly. Biochim Biophys Acta - Gen Subj 1861:2598–2608
85. Bera M, Kotamarthi HC, Dutta S et al (2014) Characterization of unfolding mechanism of human Lamin A Ig fold by single-molecule force spectroscopy-implications in EDMD. Biochemistry 53:7247–7258
86. Favreau C, Delbarre E, Courvalin JC, Buendia B (2008) Differentiation of C2C12 myoblasts expressing Lamin A mutated at a site responsible for Emery-Dreifuss muscular dystrophy is improved by inhibition of the MEK-ERK pathway and stimulation of the PI3-kinase pathway. Exp Cell Res 314:1392–1405
87. Wu W, Iwata S, Homma S et al (2014) Depletion of extracellular signal-regulated kinase 1 in mice with cardiomyopathy caused by Lamin A/C gene mutation partially prevents pathology before isoenzyme activation. Hum Mol Genet 23:1–11
88. Wu W, Muchir A, Shan J et al (2011) Mitogen-activated protein kinase inhibitors improve heart function and prevent fibrosis in cardiomyopathy caused by mutation in Lamin A/C gene. Circulation 123:53–61
89. Hah J, Kim D-H (2019) Deciphering nuclear Mechanobiology in Laminopathy. Cells 8:231
90. Donnaloja F, Carnevali F, Jacchetti E, Raimondi MT (2020) Lamin A/C Mechanotransduction in Laminopathies. Cells 9
91. Owens DJ, Fischer M, Jabre S et al (2020) Lamin mutations cause increased YAP nuclear entry in muscle stem cells. Cells 9:1–14
92. Bertrand AT, Ziaei S, Ehret C et al (2014) Cellular microenvironments reveal defective mechanosensing responses and elevated YAP signaling in LMNA-mutated muscle precursors. J Cell Sci 127:2873–2884
93. Owens DJ, Messéant J, Moog S et al (2021) Lamin-related congenital muscular dystrophy alters mechanical signaling and skeletal muscle growth. Int J Mol Sci 22:1–22

94. Elosegui-Artola A, Andreu I, Beedle AEM et al (2017) Force triggers YAP nuclear entry by regulating transport across nuclear pores. Cell 171:1397–1410
95. Kofler M, Kapus A (2023) Nuclear import and export of YAP and TAZ. Cancers (Basel) 15:7–11
96. Carollo PS, Barra V (2023) Chromatin epigenetics and nuclear lamina keep the nucleus in shape: examples from natural and accelerated aging. Biol Cell 115:1–12
97. Smith ER, Leal J, Amaya C et al (2021) Nuclear Lamin A/C expression is a key determinant of paclitaxel sensitivity. Mol Cell Biol 41:1–15
98. Vahabikashi A, Adam SA, Medalia O, Goldman RD (2022) Nuclear lamins: structure and function in mechanobiology. APL Bioeng 6:011503
99. Wang B, Luo Q, Medalia O (2024) Lamins and chromatin join forces. Adv Biol Regul 101059
100. Samson C, Petitalot A, Celli F et al (2018) Structural analysis of the ternary complex between Lamin A/C, BAF and emerin identifies an interface disrupted in autosomal recessive progeroid diseases. Nucleic Acids Res 46:10460–10473
101. Shevelyov YY, Ulianov SV (2019) The nuclear lamina as an organizer of chromosome architecture. Cells 8:1–13
102. Foisner R, Gerace L (1993) Integral membrane proteins of the nuclear envelope interact with lamins and chromosomes, and binding is modulated by mitotic phosphorylation. Cell 73:1267–1279
103. Cheedipudi SM, Matkovich SJ, Coarfa C, Hu X, Robertson MJ, Sweet M et al (2019) Genomic reorganization of Lamin-associated domains in cardiac myocytes is associated with differential gene expression and DNA methylation in human dilated cardiomyopathy. Circ Res 124(8):1198–1213
104. Cremer T, Cremer M, Hübner B et al (2020) The Interchromatin compartment participates in the structural and functional Organization of the Cell Nucleus. BioEssays 42:1–18
105. Wang B, Zhang L, Dai T et al (2021) Liquid–liquid phase separation in human health and diseases. Signal Transduct Target Ther 6:290
106. Hediger F, Gasser SM (2006) Heterochromatin protein 1: Don't judge the book by its cover! Curr Opin Genet Dev 16:143–150
107. Hiragami K, Festenstein R (2005) Heterochromatin protein 1: a pervasive controlling influence. Cell Mol Life Sci 62:2711–2726
108. Bandau S, Alvarez V, Jiang H et al (2024) RNA polymerase II promotes the organization of chromatin following DNA replication. EMBO Rep 25(3):1387
109. Depreux FF, Puckelwartz MJ, Augustynowicz A et al (2015) Disruption of the Lamin A and matrin-3 interaction by myopathic LMNA mutations. Hum Mol Genet 24:4284–4295
110. Gonzalez-Suarez I, Gonzalo S (2010) Nurturing the genome: A-type lamins preserve genomic stability. Nucleus 1:129–135
111. Makhija E, Jokhun DS, Shivashankar GV (2016) Nuclear deformability and telomere dynamics are regulated by cell geometric constraints. Proc Natl Acad Sci USA 113:E32–E40
112. Giraud-Panis MJ, Pisano S, Benarroch-Popivker D et al (2013) One identity or more for telomeres? Front Oncol 3:1–16
113. Wood AM, Danielsen JMR, Lucas CA et al (2014) TRF2 and Lamin A/C interact to facilitate the functional organization of chromosome ends. Nat Commun 5:1–9
114. De Lange T (2005) Shelterin: the protein complex that shapes and safeguards human telomeres. Genes Dev 19:2100–2110
115. Lu W, Zhang Y, Liu D et al (2011) Telomere- strucutre, function, and regulation. Exp Cell Res 4:133–141
116. Picco V, Coste I, Giraud-Panis MJ et al (2016) ERK1/2/MAPK pathway-dependent regulation of the telomeric factor TRF2. Oncotarget 7:46615–46627
117. Cherfils-Vicini J, Gilson E (2019) Inhibiting TRF 1 upstream signaling pathways to target telomeres in cancer cells. EMBO Mol Med 11:10–11
118. Laghmach R, Di Pierro M, Potoyan DA (2021) The interplay of chromatin phase separation and lamina interactions in nuclear organization. Biophys J 120:5005–5017

119. Olave IA, Reck-Peterson SL, Crabtree GR (2002) Nuclear actin and actin-related proteins in chromatin remodeling. Annu Rev Biochem 71:755–781
120. Simon DN, Zastrow MS, Wilson KL (2010) Direct actin binding to A- and B-type Lamin tails and actin filament bundling by the Lamin A tail. Nucleus 1:264–272
121. Chaturvedi P, Parnaik VK (2010) Lamin a rod domain mutants target heterochromatin protein 1α and β for proteasomal degradation by activation of F-box protein, FBXW10. PLoS One 5:e10620
122. Sengupta D, Sengupta K (2024) Lamin A K97E leads to NF-κB-mediated dysfunction of inflammatory responses in dilated cardiomyopathy. Biol Cell 1–21:e2300094
123. Mewborn SK, Puckelwartz MJ, Abuisneineh F et al (2010) Altered chromosomal positioning, compaction, and gene expression with a Lamin A/C gene mutation. PLoS One 5:e14342
124. Taimen P, Pfleghaar K, Shimi T et al (2009) A progeria mutation reveals functions for Lamin A in nuclear assembly, architecture, and chromosome organization. Proc Natl Acad Sci USA 106:20788–20793
125. Kubben N, Adriaens M, Meuleman W et al (2012) Mapping of Lamin A- and progerin-interacting genome regions. Chromosoma 121:447–464
126. Puckelwartz MJ, Depreux FFS, McNally EM (2011) Gene expression, chromosome position and Lamin A/C mutations. Nucleus 2:162–167
127. Das A, Grotsky DA, Neumann MA et al (2013) Lamin A Δexon9 mutation leads to telomere and chromatin defects but not genomic instability. Nucl (United States) 4:410–419

# Chapter 7
# ECM-Lamin Crosstalk in the Regulation of Genomic Stability

**Bipasa Mandal, Indrakshi Banerjee, Md Wasim Akram Ddoza Hazari, Atanu Mondal, and Chandrima Das**

**Abstract** The extracellular matrix (ECM) and nuclear lamina are fundamental components of cellular architecture, playing pivotal roles in mechanotransduction and gene regulation. The dynamic interplay between the ECM and lamin proteins profoundly influences chromatin organization, epigenetic modifications, and genomic stability. Variations in ECM stiffness and composition can induce significant alterations in nuclear architecture, including changes in nuclear morphology and lamin A/C expression levels. These structural changes, in turn, modulate histone modifications, DNA methylation patterns, and chromatin compaction, all of which are critical for regulating gene expression. Disruption of this intricate ECM-lamin interaction can lead to aberrant gene expression, increased genomic instability, and the progression of various diseases, particularly cancer. A comprehensive understanding of ECM-lamin dynamics offers valuable insights into the mechanisms underlying epigenetic regulation and genome maintenance, with potential implications in developing novel therapeutic strategies against diseases associated with dysregulation of mechanotransduction.

**Keywords** Extracellular matrix · Lamin · Nuclear architecture · Gene expression · Mechanotransduction · Histone methylation · Genomic instability · Exosomes · Cell-free DNA

B. Mandal · I. Banerjee · M. W. A. D. Hazari · A. Mondal · C. Das (✉)
Biophysics and Structural Genomics Division, Saha Institute of Nuclear Physics, Kolkata, West Bengal, India

Homi Bhabha National Institute, BARC Training School Complex, Mumbai, Maharashtra, India
e-mail: chandrima.das@saha.ac.in

© The Author(s), under exclusive license to Springer Nature Switzerland AG 2025 
K. Sengupta, A. Muchir (eds.), *Lamins in Genome Organization and Diseases*, Subcellular Biochemistry 115, https://doi.org/10.1007/978-3-032-00537-3_7

## Introduction

The nucleus is a highly organized structure whose architectural integrity is essential for maintaining genomic stability and regulating gene expression. This organization is governed by distinct chromosomal territories (CTs), where gene-rich regions are typically localized toward the core of the nucleus, while sequences having poor gene abundance are positioned at the periphery [1, 2]. The spatial arrangement of chromatin within these territories is further regulated by a complex network of proteins and RNA, collectively referred to as the nuclear matrix or scaffold [3]. This nuclear substructure not only organizes CTs but also facilitates DNA expression by housing chromatin-modifying enzymes and metabolic activities, many of which are compartmentalized into specific nuclear domains [3].

A critical aspect of nuclear organization is its dynamic interplay with the ECM and nuclear lamins, which are fundamental components of cellular architecture. The ECM and lamins play pivotal roles in mechanotransduction and gene regulation, with their interaction profoundly influencing chromatin organization, epigenetic modifications, and genomic stability [4–6]. Variations in ECM stiffness and composition can induce significant changes in nuclear architecture, including alterations in nuclear morphology and expression level of lamin A/C genes [7]. These structural modifications, in turn, modulate histone modifications, DNA methylation patterns, and chromatin compaction, all of which are critical for proper gene expression [7]. Disruption of this intricate ECM-lamin crosstalk can lead to aberrant gene expression, increased genomic instability, and the progression of diseases, particularly cancer.

One of the key consequences of disrupted ECM-nuclear communication is the release of unstable genomic regions into the extracellular space as cell-free DNA (cfDNA). Initially identified by Mandel et al. in 1948 [8], cfDNA refers to extracellular DNA molecules found in body fluids such as serum and plasma [9]. These molecules include host DNA from the nuclear and mitochondrial genomes [10, 11], as well as foreign DNA from pathogens such as bacteria or viruses [12]. CfDNA typically exists as double-stranded DNA (dsDNA) fragments ranging from 100 to 200 base pairs (bp) [13], although larger fragments spanning several kilobase pairs (Kbp) have also been reported [14, 15].

Exosomes, a subset of extracellular vesicles (EVs) ranging from 30 nm to 200 nm in size, play a critical role in the transport of nucleic acids, including cfDNA, cfRNA, and miRNAs. These vesicles are enriched with genetic material, which can be either embedded on their surface or encapsulated within their lumen. Notably, exosomes are implicated in tumor progression and metastasis [16], with cancer patients exhibiting elevated levels of these vesicles, often referred to as oncosomes. The cfDNA within exosomes often reflects the mutational landscape of the parental tumor cells [17], underscoring their potential as noninvasive biomarkers for early cancer detection.

The release of exosomes is regulated by various mechanisms, including the stiffness of the extracellular matrix. Emerging evidence suggests a strong correlation between ECM stiffness and cancer progression [18–20]. In this chapter, our cynosure is on the critical role of ECM-lamin crosstalk in regulating genomic stability through epigenetic mechanisms and the release of cfDNA. By exploring the dynamic interplay between the ECM and nuclear lamins, we aim to provide insights into the mechanisms underlying epigenetic regulation and genome maintenance. This understanding has significant connotations for the development of novel therapeutic strategies against diseases associated with mechanotransduction dysregulation, particularly cancer.

## ECM and Nuclear Crosstalk: Mechanotransduction and Cellular Signaling

The ECM is far more than a static structural scaffold; it is a dynamic and highly interactive signaling platform that orchestrates cellular behavior through intricate molecular mechanisms. The ECM influences nuclear function, gene expression, epigenetic regulation, and genomic stability playing a pivotal role in both physiological homeostasis and pathological conditions such as cancer.

### *Mechanotransduction and Nuclear Crosstalk: Bridging the ECM and the Nucleus*

The ECM and the nucleus engage in a bidirectional communication network that is essential for maintaining cellular homeostasis and enabling adaptive responses to environmental changes. This crosstalk is mediated through mechanotransduction, a process where mechanical stimuli from the ECM get translated into biochemical signals capable of influencing nuclear function and gene expression. This process can affect various cellular activities, including differentiation, proliferation, ECM deposition, and angiogenesis [21, 22]. Mechanosensory proteins, particularly integrins, serve as primary mediators in this complex signaling cascade [23–25]. Integrins are among the most extensively researched mechanosensors [26]. They consist of alpha and beta subunits, which include extracellular, transmembrane, and intracellular domains. Integrins detect extracellular stimuli and trigger various signaling cascades [27]. Additionally, they play a crucial role in forming focal adhesions—protein complexes located at the interface between the ECM and the cytoskeleton. Focal adhesions are made up of a diverse array of proteins, comprising of talin, focal adhesion kinase (FAK), and Src, among others [28].

G proteins, extracellular signal-regulated kinases (ERKs), stretch-activated ion channels, and receptor tyrosine kinases (RTKs) are key intracellular elements involved in mechanotransduction. G proteins are located at focal adhesion sites and induce a conformational change in response to ECM stiffness, which ultimately influences cell growth [29]. RTKs are activated downstream of G proteins and undergo dimerization when stimulated. They can be activated by external signals, such as growth factors, triggering a downstream signaling cascade [30]. This activation of RTKs can lead to the stimulation of ERKs, which play a crucial role in gene expression and protein synthesis [31]. Additionally, there are stretch-activated ion channels that regulate the entry and exit of ions, and it has been observed that intracellular calcium ion concentrations increase during cell stretching [32].

These molecular transducers play a significant role in detecting subtle changes in mechanical properties of the microenvironment, initiating conformational modifications that propagate signals directly to the nuclear interior through focal adhesion complexes [21, 33]. The cytoskeleton plays a pivotal role in this transmission, dynamically reorganizing its structural components—actins, microtubules, and microfilaments—to counteract and respond to external mechanical forces. It has been reported that stress can induce the expression of α-smooth muscle actin (α-SMA), which is crucial for maintaining cytoskeletal contractility [34]. The cytoskeleton carries signals from the ECM to the nucleus by interacting with focal adhesion complexes (FAC), which are formed in response to both intracellular and extracellular signals [35]. These FACs engage with Klarsicht, ANC-1, and Syne homology (KASH) proteins situated on the outer membrane of the nuclear envelope. KASH then interacts with SUN-domain proteins, such as Sad1p and UNC-84, located on the inner nuclear membrane to form the linker of nucleoskeleton and cytoskeleton (LINC) complex. This further connects to lamin, chromosomes, and other nuclear components [36, 37]. The SUN-domain-containing proteins interact with nesprins through their carboxy-terminal regions, presumably within the perinuclear space [38]. Additionally, nesprins engage with actin filaments via their amino terminus [39]. This integrated mechanism demonstrates how the ECM communicates with the nucleus through the cellular cytoskeleton to influence nuclear shape, chromatin organization, and transcriptional activity.

The diversity of mechanical force origins is remarkable, ranging from ECM rigidity and fluidic dynamics to tissue-specific mechanical characteristics [40]. Cellular responses to these stimuli involve sophisticated secretory mechanisms, including the release of critical signaling molecules such as transforming growth factor-β (TGF-β) and fibroblast growth factor (FGF) [41]. These biochemical mediators facilitate ECM remodeling and contribute to cellular adaptation and phenotypic modulation. Molecular signaling components orchestrate the intricate translation of mechanical signals. These molecular machines undergo complex conformational changes, triggering downstream signaling cascades that ultimately influence gene expression, protein synthesis, and cellular behavior.

## *Regulation of Nuclear Lamina by ECM*

The nuclear lamina, a meshwork of lamin proteins beneath the inner nuclear membrane, plays a critical role in mechanotransduction. These lamin proteins act as a mechanosensor that bridges external mechanical cues and nuclear structural organization. Its plasticity enables it to undergo conformational changes in response to mechanical forces, influencing nuclear stiffness, chromatin organization, and transcriptional activity. Tissue-specific variations in lamin A/C expression further highlight its role in cellular adaptation, with high levels in bone tissue contributing to nuclear stiffness and lower levels in adipose tissue allowing for greater nuclear flexibility [7, 42]. In response to mechanical stress, the electrostatic interactions between the positively charged Ig-tail domain and negatively charged regions of the nearby lamin filaments are disrupted, leading to lamina reorganization [43]. Lamin A/C is key in determining tissue-specific cell differentiation. Mechanical stimuli originating from the ECM can direct lamin to stabilize the genome properly in response to mechanical stress and facilitate gene expression during cell differentiation in tissue-specific manner. These processes are essential for maintaining nuclear shape and preventing DNA damage [44].

In pathological contexts such as cancer metastasis, mechanotransduction assumes profound significance. The altered tumor microenvironment, characterized by increased ECM stiffness and enhanced protein crosslinking [45, 46], creates a unique mechanical landscape that facilitates cellular migration and invasion. Matrix metalloproteinases (MMPs) play a crucial role in this process, reorganizing ECM architecture to support cancer cell dissemination [47]. Compression has been shown to enhance cell adherence to the substratum and promote the release of MMPs, which in turn remodel the ECM, further enabling invasion [48, 49]. The ECM in solid tumors is stiffer and activates numerous mechano-signaling pathways. Lysyl oxidase, a crosslinker of ECM proteins, contributes to ECM stiffness, induces fibrosis, and activates integrin signaling to promote tumorigenesis [19]. The MMPs organize the ECM in a manner that facilitates the spread of cancer cells. At the tumor-stroma interface, collagen is rearranged to enable more effective invasion and subsequent migration of these cells [50]. Research indicates that fluid flow increases the number and velocity of migratory cells in a three-dimensional (3D) culture model with tumor cells suspended in the extracellular matrix. Additionally, asymmetrical matrix adhesion stresses from fluid flow induce cancer cells to migrate upstream, creating a gradient of integrin activation [51].

The extraordinary complexity of ECM-nuclear crosstalk underscores the dynamic nature of cellular communication. By translating mechanical information into biochemical and genetic responses, cells maintain a sophisticated mechanism of environmental sensing and adaptive behavior, with implications spanning physiological homeostasis and pathological transformation in cancer.

## ECM and Epigenetic Alterations Leading to Changes in the Gene Expression: Molecular Mechanisms of Cellular Transformation

The ECM is a dynamic signaling hub that profoundly influences gene expression and cellular behavior. Through a combination of mechanical and biochemical signaling pathways, the ECM communicates with the nucleus to regulate transcriptional activity, chromatin organization, and epigenetic modifications. Epigenetic regulation represents a sophisticated molecular mechanism through which ECM interactions dynamically modulate cellular gene expression and phenotypic characteristics. The cytoskeleton serves as a critical conduit for transmitting ECM-derived signals to the nucleus, enabling cells to integrate environmental cues with genetic responses. Focal adhesion complexes play a central role in relaying ECM signals, recruiting and activating signaling molecules that propagate mechanical and biochemical cues to the nuclear interior [35].

### *Mechanical Interactions Between ECM Impacting Nuclear Architecture and Gene Expression*

Integrin-mediated mechanical signals can physically distort nuclear shape, redistribute nucleoli [52], and trigger cascading epigenetic modifications. Notably, laminin, a major component of ECM, emerges as a crucial signaling molecule, particularly in mammary epithelial cell differentiation, demonstrating the ability to modulate gene expression through sophisticated mechanisms including histone acetylation, chromatin remodeling, and recruitment of transcriptional complexes [53].

The nuclear lamina serves as a critical interface between ECM signals and gene expression, with lamin proteins interacting with heterochromatin regions, transcription factors, and specific genomic domains to facilitate gene silencing and chromosomal organization. High-resolution mapping has identified approximately 1300 large DNA domains that bind to the nuclear lamina, known as lamina-associated domains (LADs), which are predominantly located in gene-poor regions and characterized by distinct epigenetic markers such as H3K27me3 and insulator proteins [54]. Notably, a majority of these domains were found in gene-poor regions; specifically, 75% of large gene deserts interacted with lamin B1. The levels of active transcription markers were remarkably low in the promoters of genes located within the domains associated with lamin B. These domains were typically well defined by the presence of H3K27me3, the insulator protein CCCTC-binding factor (CTCF), suggesting these genomic regions are isolated from others so that heterochromatin formation and gene silencing is not continued [55]. LAD bodies are enriched with constitutive heterochromatin marks H3K9Me2/3 and DNA methylation marks [56].

Lamins A/C and B influence the chromosomal organization, mostly associated with gene silencing, and have been reported to interact with heterochromatin regions [57]. Research has shown that specific proteins, such as germ cell-less 1 (GCL1), which is essential for chromatin condensation [58], and barrier-to-autointegration factor (BAF), which inhibits gene expression by obstructing transcription activators [59], interact with the lamin-binding, LAP2-Emerin-MAN1 (LEM) domain proteins which are also present in the nuclear lamina. Lamin A/C and lamin B have been found to bind and colocalize with transcription factors MOK2 and Oct-1, both of which are known to suppress certain genes [60, 61]. It is suggested that these transcription factors are associated with genes at the lamina to maintain sustained silencing, such as Oct-1 repressing the collagenase gene which localizes in the nuclear periphery along with lamin B [61]. As Oct-1 dissociates from the nuclear periphery, the expression of the collagenase gene increases, further confirming the colocalization, whereas the effect of localization of MOK2 on the nuclear periphery in gene silencing is yet to be ventured [61, 62].

Epigenetic modifications are emerging as a key mechanism through which gene expression is regulated by ECM. Integrin signaling has been shown to epigenetically activate specific genes by promoting histone acetylation, for instance, α6-integrin signaling promotes histone acetylation and facilitates transcription factor binding, demonstrating the nuanced posttranslational regulatory processes [63]. The dynamic nature of these interactions is particularly evident in developmental contexts, where gene cluster interactions with the nuclear lamina can correlate with precise upregulation or downregulation of specific genetic elements [64].

## *Epigenetic Regulation by ECM: Molecular Dynamics of Chromatin Modification*

The mechanical properties of the ECM, such as stiffness and elasticity, profoundly influence histone methylation states through mechanotransduction pathways. Increased ECM stiffness in the tumor microenvironment, for example, has been shown to promote H3K27me3-mediated gene silencing, which in other reports has been shown to contribute to cancer progression [65, 66]. Mechanical forces from the ECM are transmitted to the nucleus via the cytoskeleton and nuclear lamina, activating signaling pathways that modulate histone methylation patterns[124]. The biochemical composition of the ECM, including its protein and growth factor content, also plays a critical role in regulating histone methylation. ECM components such as fibronectin, collagen, and laminin induce intracellular signaling cascades that influence the activity of histone-modifying enzymes. Among the growth factors (such as TGF-β and FGF) that are sequestered in the ECM, TGF-β can modulate DNA methylation by activating downstream signaling pathways [125, 68].

The dynamic interplay between the ECM and histone methylation enables cells to exhibit remarkable plasticity and maintain epigenetic memory. Cells can rapidly alter their histone methylation patterns in response to changes in ECM composition or mechanics, facilitating adaptive gene expression programs. Histone methylation marks can be stably inherited through cell divisions, contributing to the maintenance of cellular identity and memory of past environmental exposures [69]. Dysregulation of ECM-driven epigenetic mechanisms is implicated in various diseases, including cancer and fibrosis, highlighting the potential of targeting epigenetic pathways for therapeutic applications.

TGF-β1 is recognized for its ability to induce the expression of ECM-related genes, including connective tissue growth factor (CTGF) [70, 71], collagen-α1, and plasminogen activator inhibitor-1 (PAI-1) [72]. Research has shown that TGF-β1 stimulation enhances activating histone marks, such as H3K4me1, H3K4me2, and H3K4me3, at the promoters of these genes while concurrently diminishing repressive marks like H3K9me2 and H3K9me3 [73]. These modifications foster a chromatin environment that is conducive to transcriptional activation. This process is further supported by the recruitment of histone methyltransferases, including SET domain-containing protein 7/9 (SET7/9), to these promoters. Importantly, inhibiting SET7/9 significantly impairs TGF-β1-induced ECM gene expression, underscoring its critical role in this regulatory mechanism (G. [73]).

In corneal fibroblasts, the expression of TGF-β1 protein and ECM genes in response to TGF-β1 is linked to specific histone methylation patterns [126,75]. TGF-β1 stimulation leads to an increase in H3K4 methylation levels as mentioned above, but interestingly a decrease in H3K27me3 at the promoters of these genes was observed [127]. This is accompanied by enhanced recruitment of H3K4 methyltransferases mixed lineage leukemia protein 1 (MLL1) and SET7/9, alongside increased binding of Smad3 to these promoters [77–80]. Collectively, these observations illustrate that histone methylation, particularly the addition or removal of methyl groups on specific lysine residues of histone H3, plays a vital role in modulating ECM gene expression. The dynamic interplay between activating and repressive histone marks ultimately shapes the transcriptional outcomes of ECM-related genes, thereby influencing various physiological and pathological conditions [81].

## The Extracellular Matrix: A Critical Regulator of Genomic Stability

Recent investigations reveal a sophisticated mechanism by which ECM interactions directly influence nuclear architecture, chromatin organization, and DNA damage repair processes. The intricate relationship between the ECM and genomic stability, with a particular emphasis on the role of lamins in mediating this crosstalk, is emerging as a key area of research in the context of pathological states like cancer (Fig. 7.1).

## *ECM Stiffness, DNA Repair Mechanisms, and Lamin Dysfunction*

Lamins, especially lamin A/C, are essential for the homeostatic maintenance of nuclear stiffness and shape, which are critical for proper chromatin organization and gene expression. Mutations in lamin genes, such as LMNA, disrupt nuclear envelope integrity, leading to increased chromosomal aberrations, telomere dysfunction, and accelerated cellular senescence [82]. These alterations can precipitate complex genomic rearrangements, chromosomal missegregation, and potentially malignant transformations [83]. The interplay between ECM stiffness and lamin-mediated nuclear mechanics underscores the importance of the ECM in regulating genomic stability through its influence on nuclear architecture.

The mechanical environment critically influences DNA damage repair efficiency, with lamins playing a key role in this process. Low ECM stiffness compromises double-strand break (DSB) repair by activating specific signaling pathways that inhibit critical repair protein recruitment. The RAP2-MAP4K4/6/7 signaling pathway, for instance, modifies ubiquitin phosphorylation and impedes the formation of essential repair protein complexes like BRCA1 and 53BP1 [84]. Lamins are intimately involved in this process, as they regulate the spatial organization of chromatin and the accessibility of DNA repair machinery to damaged sites. Low stiffness not only hinders DSB repair but also fosters apoptosis through the activation of the Hippo-LATS pathway [84]. This phenomenon serves as an evolutionary strategy that upholds genomic stability by curtailing the proliferation of cells experiencing stress due to DNA damage. In the context of cancer, it underscores the significance of mechanosensing and highlights how tumor stiffness can serve as a potential biomarker. Therefore, therapeutic approaches may focus on modulating ECM stiffness or targeting associated signaling pathways to enhance the efficacy of cancer treatments. Depending on the ECM stiffness, the nuclear lamina exhibits a variety of morphological traits. Nuclei flatten and display a taut lamina on rigid substrates, whereas nuclei on soft substrates are often rounder with wrinkled lamina [85, 86]. Thus, alterations in ECM stiffness translate into changes in nuclear morphology.

Lamin dysfunction can exacerbate genomic instability by impairing DNA damage response and repair processes. For example, mutations in LMNA disrupt the nuclear lamina's ability to maintain chromatin organization, leading to the mislocalization of repair proteins and inefficient DNA repair [87]. This compromised repair capacity increases the risk of chromosomal aberrations and genomic instability, which are hallmarks of cancer progression. The ECM's mechanical properties, therefore, play a critical role in modulating lamin function and, consequently, genomic stability.

## *Histone Modifications, DNA Damage Response, and Lamins*

Histone ubiquitination emerges as a sophisticated molecular mechanism for DNA damage signaling, with lamins playing a regulatory role in this process. Specific ubiquitination marks on histones, particularly H2A at residues K13, K15, and K119, serve as critical signals for DNA damage response [88]. These modifications orchestrate the recruitment of repair proteins and influence the selection between homologous recombination and nonhomologous end-joining repair pathways [89]. Proteins like SMARCAD1 work with ubiquitinated histones to promote chromatin remodeling required for end resection during HR [90]. By initiating the activation either of the DNA repair pathways, HR (BRCA1, RAD18, RNF169) and NHEJ (53BP1), H2AK13/15ub plays a crucial role in the DDR pathway [90]. Lamins, through their interactions with chromatin, help maintain the spatial organization of histone modifications, ensuring efficient DNA damage signaling and repair.

Lamin dysfunction can disrupt these interactions, leading to aberrant histone modifications and impaired DNA damage response. For instance, mutations in LMNA can alter the distribution of histone marks, such as H3K27me3, which are critical for gene silencing and chromatin compaction [91]. The disruption of lamin can lead to an aggravated response to DNA damage, increasing the risk of genomic instability and cancer progression. The ECM, through its influence on lamin function, therefore plays a critical role in regulating histone modifications and DNA damage response.

## *ECM, Tumor Microenvironment, and Lamin-Mediated Genomic Instability*

The tumor microenvironment represents a complex ecosystem where ECM composition, mechanical properties, and lamin-mediated nuclear modifications converge to influence genomic instability. These modifications compromise DNA repair mechanisms, promote chromosomal aberrations, and facilitate cellular transformation. For example, increased ECM stiffness in the tumor microenvironment can lead to nuclear envelope rupture [92] and chromatin reorganization [93], which might be responsible for the release of cell-free DNA (cfDNA), a biomarker of genomic instability.

Inflammatory cytokines and mechanical stress within the tumor microenvironment further exacerbate lamin dysfunction, leading to increased genomic instability [94, 95]. Hypoxia, another common feature of the tumor microenvironment, can induce lamin modifications that disrupt nuclear mechanics and chromatin organization [96]. These changes create a permissive environment for cancer progression, highlighting the critical role of the ECM and lamin in maintaining genomic stability.

## *Genomic Instability and Cell-Free DNA: The Role of Lamins*

Genetic instability is a prevalent characteristic in many cancers, though the extent to which it escalates remains unclear. The mechanisms underlying these instabilities, whether they occur randomly or are driven by specific processes, are still not fully understood. Numerous clinical observations have identified genetic instability as a potential biological marker. One of the common forms is chromosomal instability (CIN) which refers to the high rates of changes in chromosomal numbers and structure observed in cancer cells compared to normal cells. While some mutations are present in certain cells within the tumor, others are only found in a limited number of cells [97]. This suggests that cancer cells are the progenitors of single mutations but, over time, continue to acquire additional mutations and instances of chromosomal instabilities [98]. In addition to CIN, which is the most prevalent type of genetic instability, another significant form is microsatellite instability (MIN) [97] which is the lengthening or shortening of oligonucleotide sequence repeats within microsatellite sequences [99]. An increased number of base pair mutations in these sequences characterizes MIN [100]. The genetic instability has been linked to DNA damage repair genes. Germline mutations in several genes associated with breast cancer susceptibility, such as BRCA1, BRCA2, PALB2, BRIP1, RAD50, NBN (Nijmegen breakage syndrome protein 1), WRN (Werner syndrome helicase), BLM (Bloom syndrome helicase), RECQL4 (RecQ protein-like 4), and the Fanconi anemia genes, have all been connected to the mechanisms involved in repairing DNA double-strand breaks (DSBs) and crosslinks between corresponding DNA strands. These mutations increase the risk of developing various cancers, including breast cancers, ovarian cancers, and leukemia and lymphomas [101–103].

Cell-free DNA (cfDNA) has emerged as a crucial biomarker reflecting genomic instability, with lamin playing a significant role in its generation and characterization. The release of cfDNA is intimately connected to nuclear structural changes, DNA damage, and cellular stress responses. Lamins, through their role in maintaining nuclear integrity, influence the release of cfDNA by regulating nuclear envelope stability and chromatin organization.

Lamin dysfunction can lead to nuclear envelope rupture and DNA fragmentation, probably contributing to the release of cfDNA [104]. Additionally, altered nuclear mechanical properties resulting from lamin mutations can facilitate cfDNA release through mechanisms involving chromatin reorganization and compromised DNA repair processes. These alterations provide valuable insights into cancer progression and have potential role in therapeutic monitoring. The analysis of cfDNA in liquid biopsies, for example, offers a noninvasive method for detecting tumor-specific mutations and monitoring treatment responses [105, 106].

## *ECM and cfDNA in Cancer: Molecular Dynamics and Signaling*

The identification of cell-free nucleic acids (cfNAs) in the bloodstream began in 1948 when Mandel and Metais first discovered them [8]. However, it wasn't until 1993 that interest in this area surged within the scientific community, following the detection of the K-Ras gene sequence in circulating tumor DNA [17]. Notably, mutations in the K-Ras gene were found in pancreatic carcinoma patients [107]. Over the past two decades, extensive research has been conducted on circulating cell-free nucleic acids (cfNAs), which include circulating cfDNAs, circulating cell-free RNAs (cfRNAs), and microRNAs. These nucleic acids are found in high concentrations in body fluids, particularly in blood, and their levels can increase under disease conditions [108]. As a result, they have emerged as important biomarkers in liquid biopsies, offering a noninvasive method for disease detection and monitoring [105]. The quantification of circulating free DNA (cfDNA) can be accomplished through various therapeutic interventions that do not necessitate tissue biopsies [109]. Elevated levels of cfDNA have been observed in a range of pathological conditions, including but not limited to noncancerous diseases. The levels are elevated in benign lesions, tissue injuries, infections, and inflammatory diseases as well [110].

Cell-free DNA (cfDNA) has emerged as a critical biomarker in cancer biology, offering insights into tumor dynamics, progression, and therapeutic responses. The release of cfDNA into the extracellular space is a complex process influenced by both cellular mechanisms and the ECM. Passive mechanisms of cfDNA release, such as apoptosis and necrosis, generate distinct DNA fragmentation patterns. Apoptotic cell death produces cfDNA fragments of approximately 167 base pairs, associated with nucleosomal histones, while necrotic cell death results in larger, irregularly sized DNA fragments [111]. This process is marked by membrane disruption and organelle dysfunction, which expose the cellular components to degradative enzymes. Furthermore, macrophage-mediated clearance of necrotic cells significantly contributes to cfDNA release, with the presence of long cfDNA fragments potentially indicating an overload in the phagocytic system. Circulating tumor cells (CTCs) contribute minimally to cfDNA release, but chromosomal instability in cancer cells can lead to the formation of micronuclei and extrachromosomal DNA fragments [17, 112]. Circulating tumor cells (CTCs) also play a minor role in cfDNA release. Upon entering circulation, CTCs encounter various biophysical forces, including the swift flow of blood, collisions with diverse blood elements, and hemodynamic forces, which may result in their rupture and the subsequent release of cfDNA. However, the contribution of CTCs to cfDNA is minimal, as evidence of cfDNA presence has been detected even in instances where CTCs are absent, suggesting that the correlation between these two factors is not significant [112]. Chromosomal instability in cancer can give rise to the formation of micronuclei and double minutes (DMs). Micronuclei, which contain aberrantly segregated chromosomal DNA, may release cfDNA into the extracellular space [113, 114]. Likewise, DMs, which are extrachromosomal DNA fragments, can also contribute to cfDNA [115]. These fragments may exit the nucleus and be released in the form of micronuclei or may be degraded via autophagy, leading to further release of cfDNA into the extracellular environment [110, 116].

Active cfDNA release mechanisms involve extracellular vesicles (EVs) and exosomes, which play a pivotal role in intercellular communication and cancer progression. Cancer-associated exosomes, or oncosomes, reflect the genomic and mutational profile of parental tumor cells and correlate with tumor aggressiveness. ECM stiffness significantly influences exosome release [18–20], with increased rigidity activating signaling pathways such as Akt, PI3K, and Rab8, enhancing exosome secretion. It also corroborates with the fact that the expression of phosphatase and TENsin homolog deleted on chromosome 10 (PTEN) which is upstream of PI3K and Akt and counteracts their function decreases [19, 117–119]. Increased stiffness of ECM produces more exosomes, and inhibiting focal adhesion kinase (FAK) has been shown to decrease Akt activation and reduce exosome release from cells on a stiff ECM. Conversely, increased ECM stiffness activates Rab8, and its inhibition leads to a decrease in exosome secretion. Higher levels of Rab8 on softer ECMs, however, enhance exosome release, demonstrating that Rab8 is essential for this process. Akt phosphorylates Rabin8 at Serine 149, alleviating its self-inhibition, which enables Rab8 to bind GTP, thereby promoting exosome secretion and establishing a connection between ECM stiffness and exosome release [120]. Furthermore, the influence of exosomes derived from stiff ECMs on tumor growth has been noted [121]. Research indicates that exosomes from rigid ECMs contribute to greater tumor cell growth compared to those from softer ECMs, suggesting that ECM stiffness plays a significant role in influencing both tumor cells and their exosomes [122]. Additionally, exosomes from stiff ECMs likely affect other cell types within the tumor microenvironment, including immune cells and fibroblasts [120].

The nuclear lamina also plays a critical role in cfDNA release, with lamin proteins maintaining nuclear integrity and organizing chromatin. Deficiency of lamin may contribute to the generation and release of cfDNA as it has a role in disruption of DDR, genomic integrity maintenance, and nuclear rupture [87, 123]. The interplay between ECM mechanics, nuclear integrity, and cfDNA release might have significant implications for cancer diagnostics and treatment strategies. The analysis of cfDNA from liquid biopsies devises a noninvasive method for detecting tumor-specific mutations, monitoring treatment responses, and predicting disease progression [105]. Understanding the molecular mechanisms of cfDNA release opens new avenues for therapeutic intervention, with potential targets including ECM stiffness, exosome secretion, and lamin-associated pathways (Fig. 7.1).

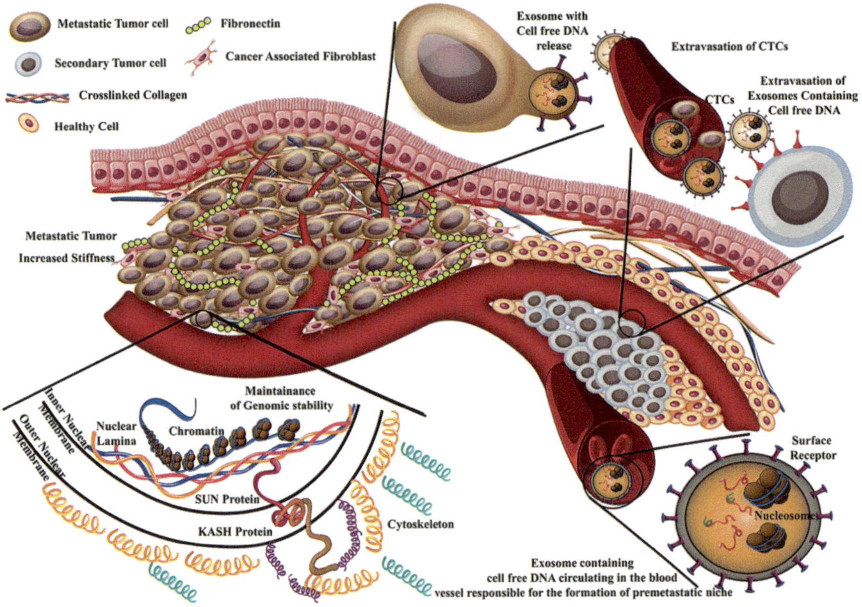


**Fig. 7.1** ECM stiffness-lamin crosstalk maintains genomic stability and the release of cfDNA-containing exosomes, establishing a pre-metastatic niche at the secondary site. The ECM stiffness is sensed by integrins, which serve as pivotal mechanotransduction mediators. Through cytoskeletal proteins, integrins activate KASH and SUN proteins (known as the LINC complex) located in the outer and inner nuclear membrane, respectively. KASH proteins extend into the nuclear interior and anchor to lamins in the nuclear lamina. This structural connection facilitates chromatin organization, thereby influencing gene expression and ensuring genomic stability. Mutations in the lamin genes can lead to the mislocalization of DDR proteins, and this causes genomic instability. In cancer, heightened genomic instability leads to the accumulation and release of extranuclear DNA, known as cell-free DNA (cfDNA). Among its subpopulations, cfDNA actively secreted via exosomes plays a crucial role, and its release is closely linked to ECM stiffness. A stiffer ECM enhances exosome secretion, thereby increasing the release of this active pool of cfDNA. These exosomes, along with circulating tumor cells, are transported through angiogenic blood vessels. They are hypothesized to reach distant sites, where they interact with immune cells, priming the microenvironment for the establishment of a pre-metastatic niche. Thus, metastatic tumors not only disseminate cells that seed secondary sites but also release molecular signals that facilitate invasion, immune modulation, and tumor proliferation, orchestrating a favorable environment for metastatic progression

## Conclusion and Future Perspectives

The ECM is not merely a passive architectural component; rather, it functions as an active and dynamic regulator of cellular genetic stability. Its complex interactions with lamins and the nuclear architecture emphasize the intricate relationships between the mechanical environment and genomic integrity. Lamins serve as critical regulators of nuclear mechanics and chromatin organization, playing a vital role in preserving genomic stability and inhibiting cancer progression. Dysregulation of lamin function, often triggered by alterations in ECM stiffness and composition, can result in heightened genomic instability, thereby underscoring the significance of the ECM-lamin axis in cellular adaptation and pathological transformation.

Gaining insight into the molecular mechanisms that govern ECM-mediated genomic regulation presents promising opportunities for cancer research. The mechanical properties of the ECM may serve as potential biomarkers, while therapeutic strategies could focus on ECM stiffness or related signaling pathways to enhance the efficacy of cancer treatments. By elucidating the role of lamins in this context, researchers can uncover new perspectives on cellular adaptation and disease mechanisms, paving the way for innovative therapeutic interventions. The dynamic interplay between the ECM and lamins highlights the importance of the microenvironment in shaping cellular behavior and underscores the potential of targeting ECM-related pathways for both diagnostic and therapeutic applications.

Several unanswered questions remain in this fascinating area of lamin/ECM crosstalk in the maintenance of nuclear architecture and genomic stability:

- The conversion of biophysical stimuli into biochemical signals is a fundamental process in cellular mechanotransduction, enabling cells to respond to their mechanical environment. How do cells detect biophysical stimuli and convert them into biochemical signals to initiate these downstream cellular processes?
- Cell surface proteins, such as integrins and mechanosensors, play a critical role in sensing mechanical forces and ECM stiffness, which influence cellular behavior and signaling pathways. What molecular mechanisms allow surface proteins to detect and respond to mechanical forces and variations in extracellular matrix stiffness?
- Epigenetic regulators are traditionally known to function within the nucleus. Do they impact the cell-free DNA release mechanisms in some direct/indirect manner? Can the cell-free DNA pools impact the intercellular communication mechanism in an active/passive manner?
- Circulating tumor cells (CTCs) are key players in cancer metastasis, but their survival dynamics and functional roles in the circulatory system are not fully understood. What are the functional roles of circulating tumor cells (CTCs), and what factors determine their survival and persistence in the bloodstream?
- The release of cell-free DNA (cfDNA) is often associated with cellular stress or death, but whether this process is entirely random or regulated remains unclear. Is the release of cell-free DNA (cfDNA) from cells a regulated process, or does it occur purely through random cellular events?

In summary, advancing research into the dynamic interplay between lamin/ECM crosstalk and epigenetic mechanisms holds immense promise for uncovering novel therapeutic targets, biomarkers, and innovative treatment strategies. A deeper understanding of these interactions in the regulation of gene expression and genomic stability could revolutionize cancer management, paving the way for more effective therapies and significantly improving patient outcomes.

**Acknowledgments** CD acknowledges funding from Core Research Grant (CRG/2022/005052) by the Department of Science and Technology, the Department of Biotechnology (BT/PR52628/MED/30/2553/2024), and Basic and Applied Research in Biophysics and Material Science (RSI 4002) by the Department of Atomic Energy (DAE), Govt. of India.

## References

1. Cremer T, Cremer M, Dietzel S, Müller S, Solovei I, Fakan S (2006) Chromosome territories – a functional nuclear landscape. Curr Opin Cell Biol 18(3):307–316
2. Mahy NL, Perry PE, Bickmore WA (2002) Gene density and transcription influence the localization of chromatin outside of chromosome territories detectable by FISH. J Cell Biol 159(5):753
3. Nickerson JA (2001) Experimental observations of a nuclear matrix. J Cell Sci 114(Pt 3):463–474
4. Bissell MJ, Hall HG, Parry G (1982) How does the extracellular matrix direct gene expression? J Theor Biol 99(1):31–68
5. Plachot C, Lelièvre SA (2004) DNA methylation control of tissue polarity and cellular differentiation in the mammary epithelium. Exp Cell Res 298(1):122–132
6. Zoubiane GS, Valentijn A, Lowe ET, Akhtar N, Bagley S, Gilmore AP, Streuli CH (2004) A role for the cytoskeleton in prolactin-dependent mammary epithelial cell differentiation. J Cell Sci 117(Pt 2):271–280
7. Swift J, Ivanovska IL, Buxboim A, Harada T, Dingal PCDP, Pinter J, Pajerowski JD, Spinler KR, Shin JW, Tewari M, Rehfeldt F, Speicher DW, Discher DE (2013) Nuclear Lamin-A scales with tissue stiffness and enhances matrix-directed differentiation. Science 341(6149)
8. Mandel P, Metais P (1948) Nuclear acids in human blood plasma. C R Seances Soc Biol Fil 142(3–4):241–243
9. Hui L, Maron JL, Gahan PB (2015) Other body fluids as non-invasive sources of cell-free DNA/RNA. In: Advances in predictive, preventive and personalised medicine, vol 5, pp 295–323
10. Kohler C, Radpour R, Barekati Z, Asadollahi R, Bitzer J, Wight E, Bürki N, Diesch C, Holzgreve W, Zhong XY (2009) Levels of plasma circulating cell free nuclear and mitochondrial DNA as potential biomarkers for breast tumors. Mol Cancer 8:105
11. Zachariah RR, Schmid S, Buerki N, Radpour R, Holzgreve W, Zhong X (2008) Levels of circulating cell-free nuclear and mitochondrial DNA in benign and malignant ovarian tumors. Obstet Gynecol 112(4):843–850
12. Pietrzak B, Kawacka I, Olejnik-Schmidt A, Schmidt M (2023) Circulating microbial cell-free DNA in health and disease. Int J Mol Sci 24(3):3051
13. Stroun M, Anker P, Lyautey J, Lederrey C, Maurice PA (1987) Isolation and characterization of DNA from the plasma of cancer patients. Eur J Cancer Clin Oncol 23(6):707–712
14. Giacona MB, Ruben GC, Iczkowski KA, Roos TB, Porter DM, Sorenson GD (1998) Cell-free DNA in human blood plasma: length measurements in patients with pancreatic cancer and healthy controls. Pancreas 17(1):89–97
15. Thierry AR, El Messaoudi S, Gahan PB, Anker P, Stroun M (2016) Origins, structures, and functions of circulating DNA in oncology. Cancer Metastasis Rev 35(3):347–376
16. Patel NJ, Ashraf A, Chung EJ (2023) Extracellular vesicles as regulators of the extracellular matrix. Bioengineering (Basel, Switzerland) 10(2)
17. Grabuschnig S, Bronkhorst AJ, Holdenrieder S, Rodriguez IR, Schliep KP, Schwendenwein D, Ungerer V, Sensen CW (2020) Putative origins of cell-free DNA in humans: a review of active and passive nucleic acid release mechanisms. Int J Mol Sci 21(21):1–24

18. Laklai H, Miroshnikova YA, Pickup MW, Collisson EA, Kim GE, Barrett AS, Hill RC, Lakins JN, Schlaepfer DD, Mouw JK, LeBleu VS, Roy N, Novitskiy SV, Johansen JS, Poli V, Kalluri R, Iacobuzio-Donahue CA, Wood LD, Hebrok M et al (2016) Genotype tunes pancreatic ductal adenocarcinoma tissue tension to induce matricellular-fibrosis and tumor progression. Nat Med 22(5):497
19. Levental KR, Yu H, Kass L, Lakins JN, Egeblad M, Erler JT, Fong SFT, Csiszar K, Giaccia A, Weninger W, Yamauchi M, Gasser DL, Weaver VM (2009) Matrix crosslinking forces tumor progression by enhancing integrin signaling. Cell 139(5):891–906
20. Thomas D, Radhakrishnan P (2019) Tumor-stromal crosstalk in pancreatic cancer and tissue fibrosis. Mol Cancer 18(1):14
21. Orr AW, Helmke BP, Blackman BR, Schwartz MA (2006) Mechanisms of mechanotransduction. Dev Cell 10(1):11–20
22. Vogel V, Sheetz M (2006) Local force and geometry sensing regulate cell functions. Nat Rev Mol Cell Biol 7(4):265–275
23. Friedland JC, Lee MH, Boettiger D (2009) Mechanically activated integrin switch controls alpha5beta1 function. *Science* 323(5914):642–644
24. Ma Z, Finkel TH (2010) T cell receptor triggering by force. Trends Immunol 31(1):1–6
25. Paszek MJ, Boettiger D, Weaver VM, Hammer DA (2009) Integrin clustering is driven by mechanical resistance from the Glycocalyx and the substrate. PLoS Comput Biol 5(12):e1000604
26. Guo W, Giancotti FG (2004) Integrin signalling during tumour progression. Nat Rev Mol Cell Biol 5(10):816–826
27. Urbich C, Dernbach E, Reissner A, Vasa M, Zeiher AM, Dimmeler S (2002) Shear stress-induced endothelial cell migration involves integrin signaling via the fibronectin receptor subunits alpha(5) and beta(1). Arterioscler Thromb Vasc Biol 22(1):69–75
28. Seong J, Wang N, Wang Y (2013) Mechanotransduction at focal adhesions: from physiology to cancer development. J Cell Mol Med 17(5):597–604
29. Govey PM, Jacobs JM, Tilton SC, Loiselle AE, Zhang Y, Freeman WM, Waters KM, Karin NJ, Donahue HJ (2014) Integrative transcriptomic and proteomic analysis of osteocytic cells exposed to fluid flow reveals novel mechano-sensitive signaling pathways. J Biomech 47(8):1838–1845
30. Ullrich A, Schlessinger J (1990) Signal transduction by receptors with tyrosine kinase activity. Cell 61(2):203–212
31. Jo H, Sipos K, Go YM, Law R, Rong J, McDonald JM (1997) Differential effect of shear stress on extracellular signal-regulated kinase and N-terminal Jun kinase in endothelial cells. Gi2- and Gbeta/gamma-dependent signaling pathways. J Biol Chem 272(2):1395–1401
32. Pommerenke H, Schmidt C, Dürr F, Nebe B, Lüthen F, Müller P, Rychly J (2002) The mode of mechanical integrin stressing controls intracellular signaling in osteoblasts. J Bone Miner Res 17(4):603–611
33. Ingber DE (2006) Cellular mechanotransduction: putting all the pieces together again. FASEB J 20(7):811–827
34. Wang J, Su M, Fan J, Seth A, McCulloch CA (2002) Transcriptional regulation of a contractile gene by mechanical forces applied through integrins in osteoblasts. J Biol Chem 277(25):22889–22895
35. Sheetz MP (2001) Cell control by membrane-cytoskeleton adhesion. Nat Rev Mol Cell Biol 2(5):392–396
36. Fridkin A, Penkner A, Jantsch V, Gruenbaum Y (2009) SUN-domain and KASH-domain proteins during development, meiosis and disease. Cell Mol Life Sci 66(9):1518–1533
37. Starr DA, Fischer JA (2005) KASH 'n Karry: the KASH domain family of cargo-specific cytoskeletal adaptor proteins. BioEssays 27:1136–1146
38. Crisp M, Liu Q, Roux K, Rattner JB, Shanahan C, Burke B, Stahl PD, Hodzic D (2006) Coupling of the nucleus and cytoplasm: role of the LINC complex. J Cell Biol 172(1):41–53

39. Zhang Q, Skepper JN, Yang F, Davies JD, Hegyi L, Roberts RG, Weissberg PL, Ellis JA, Shanahan CM (2001) Nesprins: a novel family of spectrin-repeat-containing proteins that localize to the nuclear membrane in multiple tissues. J Cell Sci 114(Pt 24):4485–4498
40. Wang JHC, Thampatty BP (2006) An introductory review of cell mechanobiology. Biomech Model Mechanobiol 5(1):1–16
41. Wells RG, Discher DE (2008) Matrix elasticity, cytoskeletal tension, and TGF-beta: the insoluble and soluble meet. Sci Signal 1(10)
42. Alcorta-Sevillano N, Macías I, Rodríguez CI, Infante A (2020) Crucial role of Lamin A/C in the migration and differentiation of MSCs in bone. Cells 9(6)
43. Makarov AA, Zou J, Houston DR, Spanos C, Solovyova AS, Cardenal-Peralta C, Rappsilber J, Schirmer EC (2019) Lamin A molecular compression and sliding as mechanisms behind nucleoskeleton elasticity. Nat Commun 10(1):1–17
44. Malashicheva A, Perepelina K (2021) Diversity of nuclear Lamin A/C action as a key to tissue-specific regulation of cellular identity in health and disease. Front Cell Dev Biol 9
45. Provenzano PP, Inman DR, Eliceiri KW, Knittel JG, Yan L, Rueden CT, White JG, Keely PJ (2008) Collagen density promotes mammary tumor initiation and progression. BMC Med 6(1):1–15
46. Winkler J, Abisoye-Ogunniyan A, Metcalf KJ, Werb Z (2020) Concepts of extracellular matrix remodelling in tumour progression and metastasis. Nat Commun 11(1):1–19
47. Blavier L, Lazaryev A, Shi XH, Dorey FJ, Shackleford GM, DeClerck YA (2010) Stromelysin-1 (MMP-3) is a target and a regulator of Wnt1-induced epithelial-mesenchymal transition (EMT). Cancer Biol Ther 10(2):198–208
48. Renò F, Grazianetti P, Stella M, Magliacani G, Pezzuto C, Cannas M (2002) Release and activation of matrix metalloproteinase-9 during in vitro mechanical compression in hypertrophic scars. Arch Dermatol 138(4):475–478
49. Tse JM, Cheng G, Tyrrell JA, Wilcox-Adelman SA, Boucher Y, Jain RK, Munn LL (2012) Mechanical compression drives cancer cells toward invasive phenotype. Proc Natl Acad Sci USA 109(3):911–916
50. Provenzano PP, Eliceiri KW, Campbell JM, Inman DR, White JG, Keely PJ (2006) Collagen reorganization at the tumor-stromal interface facilitates local invasion. BMC Med 4(1)
51. Polacheck WJ, German AE, Mammoto A, Ingber DE, Kamm RD (2014) Mechanotransduction of fluid stresses governs 3D cell migration. Proc Natl Acad Sci USA 111(7):2447–2452
52. Maniotis AJ, Chen CS, Ingber DE (1997) Demonstration of mechanical connections between integrins, cytoskeletal filaments, and nucleoplasm that stabilize nuclear structure. Proc Natl Acad Sci USA 94(3):849–854
53. Streuli CH, Schmidhauser C, Bailey N, Yurchenco P, Skubitz APN, Roskelley C, Bissell MJ (1995) Laminin mediates tissue-specific gene expression in mammary epithelia. J Cell Biol 129(3):591
54. Guelen L, Pagie L, Brasset E, Meuleman W, Faza MB, Talhout W, Eussen BH, De Klein A, Wessels L, De Laat W, Van Steensel B (2008) Domain organization of human chromosomes revealed by mapping of nuclear lamina interactions. Nature 453(7197):948–951
55. Chang L, Li M, Shao S, Li C, Ai S, Xue B, Hou Y, Zhang Y, Li R, Fan X, He A, Li C, Sun Y (2020) Nuclear peripheral chromatin-Lamin B1 interaction is required for global integrity of chromatin architecture and dynamics in human cells. Protein Cell 13(4):258
56. Sun C, Zhao Y, Guo L, Qiu J, Peng Q (2025) The interplay between histone modifications and nuclear lamina in genome regulation. J Genet Genomics 52(1):24–38
57. Akhtar A, Gasser SM (2007) The nuclear envelope and transcriptional control. Nat Rev Genet 8(7):507–517
58. Kimura T, Ito C, Watanabe S, Takahashi T, Ikawa M, Yomogida K, Fujita Y, Ikeuchi M, Asada N, Matsumiya K, Okuyama A, Okabe M, Toshimori K, Nakano T (2003) Mouse germ cell-less as an essential component for nuclear integrity. Mol Cell Biol 23(4):1304–1315

59. Segura-Totten M, Kowalski AK, Craigie R, Wilson KL (2002) Barrier-to-autointegration factor: major roles in chromatin decondensation and nuclear assembly. J Cell Biol 158(3):475–485
60. Dreuillet C, Tillit J, Kress M, Ernoult-Lange M (2002) In vivo and in vitro interaction between human transcription factor MOK2 and nuclear Lamin A/C. Nucleic Acids Res 30(21):4634
61. Imai SI, Nishibayashi S, Takao K, Tomifuji M, Fujino T, Hasegawa M, Takano T (1997) Dissociation of Oct-1 from the nuclear peripheral structure induces the cellular aging-associated collagenase gene expression. Mol Biol Cell 8(12):2407–2419
62. Mattout-Drubezki A, Gruenbaum Y (2003) Dynamic interactions of nuclear lamina proteins with chromatin and transcriptional machinery. Cell Mol Life Sci 60(10):2053–2063
63. Jolivet G, Pantano T, Houdebine LM (2005) Regulation by the extracellular matrix (ECM) of prolactin-induced alpha s1-casein gene expression in rabbit primary mammary cells: role of STAT5, C/EBP, and chromatin structure. J Cell Biochem 95(2):313–327
64. Rozario T, DeSimone DW (2009) The extracellular matrix in development and morphogenesis: a dynamic view. Dev Biol 341(1):126
65. Adhikari S, Singh V, Nandi S, Ghosal M, Raj NS, Khanna J, Bhattacharya A, Kabiraj A, Mondal A, Vasudevan M, Senapati D, Roy H, Sengupta K, Notani D, Das C (2024) UBR7 in concert with EZH2 inhibits the TGF-β signaling leading to extracellular matrix remodeling. Cell Rep 43(7):114394
66. Cozzolino AM, Noce V, Battistelli C, Marchetti A, Grassi G, Cicchini C, Tripodi M, Amicone L (2016) Modulating the substrate stiffness to manipulate differentiation of resident liver stem cells and to improve the differentiation state of hepatocytes. Stem Cells Int 2016
67. Liu R, Jin Y, Tang WH, Qin L, Zhang X, Tellides G, Hwa J, Yu J, Martin KA (2013) Ten-eleven translocation-2 (TET2) is a master regulator of smooth muscle cell plasticity. Circulation 128(18):2047–2057
68. Gloe T, Sohn HY, Meininger GA, Pohl U (2002) Shear stress-induced release of basic fibroblast growth factor from endothelial cells is mediated by matrix interaction via integrin alpha(v)beta3. J Biol Chem 277(26):23453–23458
69. Almouzni G, Cedar H (2016) Maintenance of epigenetic information. Cold Spring Harb Perspect Biol 8(5):a019372
70. Denton CP, Abraham DJ (2001) Transforming growth factor-beta and connective tissue growth factor: key cytokines in scleroderma pathogenesis. Curr Opin Rheumatol 13(6):505–511
71. Guha M, Xu Z, Tung D, Lanting L, Natarajan R (2007) Specific down-regulation of connective tissue growth factor attenuates progression of nephropathy in mouse models of type 1 and type 2 diabetes. FASEB J 21(12):3355–3368
72. Baricos WH, Shah SV (1984) Increased cathepsin D-like activity in cortex, tubules, and glomeruli isolated from rats with experimental nephrotic syndrome. Biochem J 223(2):393–399
73. Sun G, Reddy MA, Yuan H, Lanting L, Kato M, Natarajan R (2010) Epigenetic histone methylation modulates fibrotic gene expression. J Am Soc Nephrol 21(12):2069–2080
74. Jeon ES, Kim JH, Ryu H, Kim EK (2012) Lysophosphatidic acid activates TGFBIp expression in human corneal fibroblasts through a TGF-β1-dependent pathway. Cell Signal 24(6):1241–1250
75. Yam GHF, Wang K, Jhanji V, Choy KW, Baum L, Pang CP (2012) In vitro amyloid aggregate forming ability of TGFBI mutants that cause corneal dystrophies. Invest Ophthalmol Vis Sci 53(9):5890–5898
76. Yellore VS, Rayner SA, Aldave AJ (2011) TGFB1-induced extracellular expression of TGFBIp and inhibition of TGFBIp expression by RNA interference in a human corneal epithelial cell line. Invest Ophthalmol Vis Sci 52(2):757
77. Martin C, Zhang Y (2005) The diverse functions of histone lysine methylation. Nat Rev Mol Cell Biol 6(11):838–849
78. Ruthenburg AJ, Allis CD, Wysocka J (2007) Methylation of lysine 4 on histone H3: intricacy of writing and reading a single epigenetic mark. Mol Cell 25(1):15–30

79. Schnaper HW, Hayashida T, Hubchak SC, Poncelet AC (2003) TGF-beta signal transduction and mesangial cell fibrogenesis. Am J Physiol Renal Physiol 284(2):F243
80. Shilatifard A (2006) Chromatin modifications by methylation and ubiquitination: implications in the regulation of gene expression. Annu Rev Biochem 75:243–269
81. Maeng YS, Lee GH, Choi SI, Kim KS, Kim EK (2015) Histone methylation levels correlate with TGFBIp and extracellular matrix gene expression in normal and granular corneal dystrophy type 2 corneal fibroblasts. BMC Med Genet 8(1):74
82. Martins F, Sousa J, Pereira CD et al (2020) Nuclear envelope dysfunction and its contribution to the aging process. Aging Cell 19(5):e13143
83. Zuela N, Bar DZ, Gruenbaum Y (2012) Lamins in development, tissue maintenance and stress. EMBO Rep 13(12):1070–1078
84. Deng M, Lin J, Nowsheen S, Liu T, Zhao Y, Villalta PW, Sicard D, Tschumperlin DJ, Lee SB, Kim JJ, Lou Z (2020) Extracellular matrix stiffness determines DNA repair efficiency and cellular sensitivity to genotoxic agents. Sci Adv 6(37)
85. Wang TC, Abolghasemzade S, McKee BP, Singh I, Pendyala K, Mohajeri M, Patel H, Shaji A, Kersey AL, Harsh K, Kaur S, Dollahon CR, Chukkapalli S, Lele PP, Conway DE, Gaharwar AK, Dickinson RB, Lele TP (2024) Matrix stiffness drives drop like nuclear deformation and Lamin A/C tension-dependent YAP nuclear localization. Nat Commun 15(1):1–17
86. Xia Y, Cho S, Vashisth M, Ivanovska IL, Dingal PCDP, Discher DE (2019) Manipulating the mechanics of extracellular matrix to study effects on the nucleus and its structure. Methods (San Diego, Calif) 157:3–14
87. Gonzalo S (2014) DNA damage and Lamins. Adv Exp Med Biol 773:377
88. Mattiroli F, Penengo L (2021) Histone ubiquitination: an integrative signaling platform in genome stability. Trends Genet 37(6):566–581
89. Horn V, Uckelmann M, Zhang H, Eerland J, Aarsman I, le Paige UB, Davidovich C, Sixma TK, van Ingen H (2019) Structural basis of specific H2A K13/K15 ubiquitination by RNF168. Nat Commun 10(1):1–12
90. Chakraborty S, Pandita RK, Hambarde S, Mattoo AR, Charaka V, Ahmed KM, Iyer SP, Hunt CR, Pandita TK (2018) SMARCAD1 phosphorylation and ubiquitination are required for resection during DNA double-strand break repair. IScience 2:123
91. Salvarani N, Crasto S, Miragoli M, Bertero A, Paulis M, Kunderfranco P, Serio S, Forni A, Lucarelli C, Dal Ferro M, Larcher V, Sinagra G, Vezzoni P, Murry CE, Faggian G, Condorelli G, Di Pasquale E (2019) The K219T-Lamin mutation induces conduction defects through epigenetic inhibition of SCN5A in human cardiac laminopathy. Nat Commun 10(1):1–16
92. Denais CM, Gilbert RM, Isermann P, McGregor AL, Te Lindert M, Weigelin B, Davidson PM, Friedl P, Wolf K, Lammerding J (2016) Nuclear envelope rupture and repair during cancer cell migration. *Science* 352(6283):353
93. Stowers RS, Shcherbina A, Israeli J, Gruber JJ, Chang J, Nam S, Rabiee A, Teruel MN, Snyder MP, Kundaje A, Chaudhuri O (2019) Matrix stiffness induces a tumorigenic phenotype in mammary epithelium through changes in chromatin accessibility. Nat Biomed Eng 3(12):1009
94. Aivaliotis IL, Pateras IS, Papaioannou M, Glytsou C, Kontzoglou K, Johnson EO, Zoumpourlis V (2012) How do cytokines trigger genomic instability? J Biomed Biotechnol 2012:536761
95. Willaume S, Rass E, Paulafontanilla-Ramirez P, Moussa A, Wanschoor P, Bertrand P (2021) A link between replicative stress, Lamin proteins, and inflammation. Genes 12(4):552
96. Li Y, Zhao L, Li XF (2021) Hypoxia and the tumor microenvironment. Technol Cancer Res Treat 20:15330338211036304
97. Negrini S, Gorgoulis VG, Halazonetis TD (2010) Genomic instability – an evolving hallmark of cancer. Nat Rev Mol Cell Biol 11(3):220–228
98. Sansregret L, Vanhaesebroeck B, Swanton C (2018) Determinants and clinical implications of chromosomal instability in cancer. Nat Rev Clin Oncol 15(3):139–150
99. Leach FS, Nicolaides NC, Papadopoulos N, Liu B, Jen J, Parsons R, Peltomäki P, Sistonen P, Aaltonen LA, Nyström-Lahti M, Guan XY, Zhang J, Meltzer PS, Yu JW, Ten Kao F, Chen

DJ, Cerosaletti KM, Fournier REK, Todd S et al (1993) Mutations of a mutS homolog in hereditary nonpolyposis colorectal cancer. Cell 75(6):1215–1225

100. Al-Tassan N, Chmiel NH, Maynard J, Fleming N, Livingston AL, Williams GT, Hodges AK, Davies DR, David SS, Sampson JR, Cheadle JP (2002) Inherited variants of MYH associated with somatic G:C—>T:a mutations in colorectal tumors. Nat Genet 30(2):227–232
101. Bachrati CZ, Hickson ID (2003) RecQ helicases: suppressors of tumorigenesis and premature aging. Biochem J 374(Pt 3):577–606
102. Kennedy RD, D'Andrea AD (2006) DNA repair pathways in clinical practice: lessons from pediatric cancer susceptibility syndromes. J Clin Oncol 24(23):3799–3808
103. Ripperger T, Gadzicki D, Meindl A, Schlegelberger B (2008) Breast cancer susceptibility: current knowledge and implications for genetic counselling. Eur J Hum Genet 17(6):722
104. Earle AJ, Kirby TJ, Fedorchak GR, Isermann P, Patel J, Iruvanti S, Moore SA, Bonne G, Wallrath LL, Lammerding J (2020) Mutant lamins cause nuclear envelope rupture and DNA damage in skeletal muscle cells. Nat Mater 19(4):464–473
105. Adhit KK, Wanjari A, Menon S, K, S. (2023) Liquid biopsy: an evolving paradigm for non-invasive disease diagnosis and monitoring in medicine. Cureus 15(12):e50176
106. Jung K, Fleischhacker M, Rabien A (2010) Cell-free DNA in the blood as a solid tumor biomarker – a critical appraisal of the literature. Clin Chim Acta 411(21–22):1611–1624
107. Zvereva M, Roberti G, Durand G, Voegele C, Nguyen MD, Delhomme TM, Chopard P, Fabianova E, Adamcakova Z, Holcatova I, Foretova L, Janout V, Brennan P, Foll M, Byrnes GB, McKay JD, Scelo G, Le Calvez-Kelm F (2020) Circulating tumour-derived KRAS mutations in pancreatic cancer cases are predominantly carried by very short fragments of cell-free DNA. EBioMedicine 55:102462
108. Bruno DCF, Donatti A, Martin M, Almeida VS, Geraldis JC, Oliveira FS, Dogini DB, Lopes-Cendes I (2020) Circulating nucleic acids in the plasma and serum as potential biomarkers in neurological disorders. Braz J Med Biol Res 53(10):e9881
109. García-Pardo M, Makarem M, Li JJN, Kelly D, Leighl NB (2022) Integrating circulating-free DNA (cfDNA) analysis into clinical practice: opportunities and challenges. Br J Cancer 127(4):592
110. Stejskal P, Goodarzi H, Srovnal J, Hajdúch M et al (2023) Circulating tumor nucleic acids: biology, release mechanisms, and clinical relevance. Mol Cancer 22(1):1–21
111. Kustanovich A, Schwartz R, Peretz T, Grinshpun A (2019) Life and death of circulating cell-free DNA. Cancer Biol Ther 20(8):1057–1067
112. Papadopoulos N (2020) Pathophysiology of ctDNA release into the circulation and its characteristics: what is important for clinical applications. Tumor Liquid Biopsies 215:163–180
113. Bach DH, Zhang W, Sood AK (2019) Chromosomal instability in tumor initiation and development. Cancer Res 79(16):3995–4002
114. Schütz E, Akbari MR, Beck J, Urnovitz H, Zhang WW, Bornemann-Kolatzki K, Mitchell WM, Nam RK, Narod SA (2015) Chromosomal instability in cell-free DNA is a serum biomarker for prostate cancer. Clin Chem 61(1):239–248
115. Xu K, Ding L, Chang TC, Shao Y, Chiang J, Mulder H, Wang S, Shaw TI, Wen J, Hover L, McLeod C, Wang YD, Easton J, Rusch M, Dalton J, Downing JR, Ellison DW, Zhang J, Baker SJ, Wu G (2019) Structure and evolution of double minutes in diagnosis and relapse brain tumors. Acta Neuropathol 137(1):123–137
116. Gekara NO (2017) DNA damage-induced immune response: micronuclei provide key platform. J Cell Biol 216(10):2999–3001
117. Mouw JK, Yui Y, Damiano L, Bainer RO, Lakins JN, Acerbi I, Ou G, Wijekoon AC, Levental KR, Gilbert PM, Hwang ES, Chen YY, Weaver VM (2014) Tissue mechanics modulate microRNA-dependent PTEN expression to regulate malignant progression. Nat Med 20(4):360–367
118. Rubashkin MG, Cassereau L, Bainer R, DuFort CC, Yui Y, Ou G, Paszek MJ, Davidson MW, Chen YY, Weaver VM (2014) Force engages vinculin and promotes tumor progres-

sion by enhancing PI3K activation of phosphatidylinositol (3,4,5)-triphosphate. Cancer Res 74(17):4597–4611
119. Schrader J, Gordon-Walker TT, Aucott RL, van Deemter M, Quaas A, Walsh S, Benten D, Forbes SJ, Wells RG, Iredale JP (2011) Matrix stiffness modulates proliferation, chemotherapeutic response, and dormancy in hepatocellular carcinoma cells. Hepatology (Baltimore, Md) 53(4):1192–1205
120. Wu B, Liu DA, Guan L, Myint PK, Chin LK, Dang H, Xu Y, Ren J, Li T, Yu Z, Jabban S, Mills GB, Nukpezah J, Chen YH, Furth EE, Gimotty PA, Wells RG, Weaver VM, Radhakrishnan R et al (2023) Stiff matrix induces exosome secretion to promote tumour growth. Nat Cell Biol 25(3):415–424
121. Yang E, Wang X, Gong Z, Yu M, Wu H, Zhang D (2020) Exosome-mediated metabolic reprogramming: the emerging role in tumor microenvironment remodeling and its influence on cancer progression. Sig Transduct Target Ther 5(1):1–13
122. Patwardhan S, Mahadik P, Shetty O, Sen S (2021) ECM stiffness-tuned exosomes drive breast cancer motility through thrombospondin-1. Biomaterials 279:121185
123. Cho S, Vashisth M, Abbas A, Majkut S, Vogel K, Xia Y, Ivanovska IL, Irianto J, Tewari M, Zhu K, Tichy ED, Mourkioti F, Tang HY, Greenberg RA, Prosser BL, Discher DE (2019) Mechanosensing by the lamina protects against nuclear rupture, DNA damage, and cell-cycle arrest. Dev Cell 49(6):920–935
124. Laklai H, Miroshnikova YA, Pickup MW, Collisson EA, Kim GE, Barrett AS, Hill RC, Lakins JN, Schlaepfer DD, Mouw JK, LeBleu VS, Roy N, Novitskiy SV, Johansen JS, Poli V, Kalluri R, Iacobuzio-Donahue CA, Wood LD, Hebrok M, Weaver VM (2016). Genotype tunes pancreatic ductal adenocarcinoma tissue tension to induce matricellular-fibrosis and tumor progression. Nature Medicine, 22(5):497
125. Bai J, Xi Q (2018). Crosstalk between TGF-β signaling and epigenome. Acta Biochimica et Biophysica Sinica, 50(1):60–67
126. Maeng YS, Lee GH, Choi SIl, Kim KS, Kim EK (2015). Histone methylation levels correlate with TGFBIp and extracellular matrix gene expression in normal and granular corneal dystrophy type 2 corneal fibroblasts. BMC Medical Genomics, 8(1):74
127. Jia Y, Reddy MA, Das S, Oh HJ, Abdollahi M, Yuan H, Zhang E, Lanting L, Wang M, Natarajan R (2019). Dysregulation of histone H3 lysine 27 trimethylation in transforming growth factor-β1–induced gene expression in mesangial cells and diabetic kidney. The Journal of Biological Chemistry, 294(34):12695

# Chapter 8
# The Strange Case of the Noncanonical Lamina: Deep Divisions in Nuclear Organisation?

**Mark C. Field and Michael P. Rout**

**Abstract** The nuclear envelope is subtended in most eukaryotes by a proteinaceous lamina, a network that has been recognised since the 1950s. Originally considered as a simple structural support, it is now clear that the lamina can be highly dynamic and participates in a multitude of functions, including transcriptional and epigenetic regulation, definition of chromatin domains, genome stability and the positioning of nuclear pore complexes. The major protein components of the lamina in metazoans are ~60 kDa lamins, which assemble to form fibres and a network and are regulated by multiple mechanisms. Despite a broad taxonomic distribution beyond Metazoa, lamins are not universal and, in at least three major lineages, are absent, specifically fungi, plants and kinetoplastid protists. The latter two possess lineage-specific lamin analogues, the CRWN and NUP-1/NUP-2 proteins, respectively. Here we discuss and compare the kinetoplastid and plant lamina, their origins, components and functions and spectacular examples of convergent evolution.

**Keywords** Nucleus · Trypanosomes · Plants · CRWN proteins · Lamina · Evolution · Gene expression · Nuclear pore complex · Eukaryogenesis

M. C. Field (✉)
School of Life Sciences, University of Dundee, Dundee, Scotland, UK

Biology Centre, Institute of Parasitology, Czech Academy of Sciences, České Budějovice, Czech Republic
e-mail: mcfield@dundee.ac.uk

M. P. Rout
Laboratory of Cellular and Structural Biology, The Rockefeller University, New York, NY, USA

© The Author(s), under exclusive license to Springer Nature Switzerland AG 2025 
K. Sengupta, A. Muchir (eds.), *Lamins in Genome Organization and Diseases*, Subcellular Biochemistry 115, https://doi.org/10.1007/978-3-032-00537-3_8

## Introduction

The origin of eukaryotic cells dates to over $10^9$ years ago [1], and it is generally accepted that the last common eukaryotic ancestor (LECA) possessed a complex cellular architecture including the mitochondrion, an endomembrane system, a flagellum and the nucleus [2]. The latter was likely broadly similar in many respects to the nuclei of extant organisms, with DNA packaged into nucleosomes, possessed of a nucleolus, Cajal bodies (and likely additional subnuclear bodies) and nuclear pore complexes transversing the nuclear envelope to mediate nucleocytoplasmic communication, together with a lamina system [3, 4]. This is very much a contemporary view of the nucleus and only revealed following the advent of widescale genome sequencing and subsequent efforts towards phylogenetic reconstructions of the LECA.

The nuclear lamina was described in the 1960s based on electron micrographs as a "complex supporting layer of fine filaments… observed on the inner aspect of the nuclear envelope" (Fawcett 1966), but identification of the lamin proteins, the stoichiometry of isoforms and finally their sequences were not ascertained until the 1970s and 1980s [5, 6]. Due to their position at the periphery of the nucleus, lamins were originally proposed to function chiefly to support the nuclear envelope, but their dynamic nature (especially in metazoan organisms with an open mitosis) and multiple interactions with other intranuclear assemblies suggested a more sophisticated integration with nuclear events [7].

We are now aware that lamins are related to type V intermediate filaments on account of possession of heptad repeats [8, 9] and that they have a plethora of protein-protein interactions and post-translational modifications, including phosphorylation, prenylation and proteolytic processing [10]. Multiple isoforms are recognised in vertebrates, which assemble into coiled-coil fibres, essentially as dimers that are further assembled into head-to-tail tetrameric units [11, 12]. Type A lamins are expressed from the *LMNA* gene, and alternate splicing generates lamins A and C, while type B lamins are expressed from two separate genes, *LMNB1* and *LMNB2* [9]. Significantly, the complete tertiary structure of a lamin protein remains to be reported.

While originally considered metazoan specific, in part due to the absence of lamin genes from fungal model systems, this is not the case, and lamin genes are present across a broad range of eukaryotes, including social amoeba (antisocial ones having no truck with them), the SAR supergroup and Metazoa [4, 13]. The distribution of lamins is sufficiently deep and broad across the eukaryotes to robustly support the conclusion that a primordial lamin B was present in the LECA [4].

However, it has emerged that there are more complex levels of gene retention and loss. Specifically, there are numerous examples of closely related taxa where one retains a lamin gene and the other does not [4]. This retention behaviour also seems to scale, so that we have the fungal phylum lacking lamins while their sister lineage, the Metazoa, retains them, and within the Amoebozoa only the slime-mould amoebae have lamins while remaining lineages do not (Figs. 8.1 and 8.4). In the case of

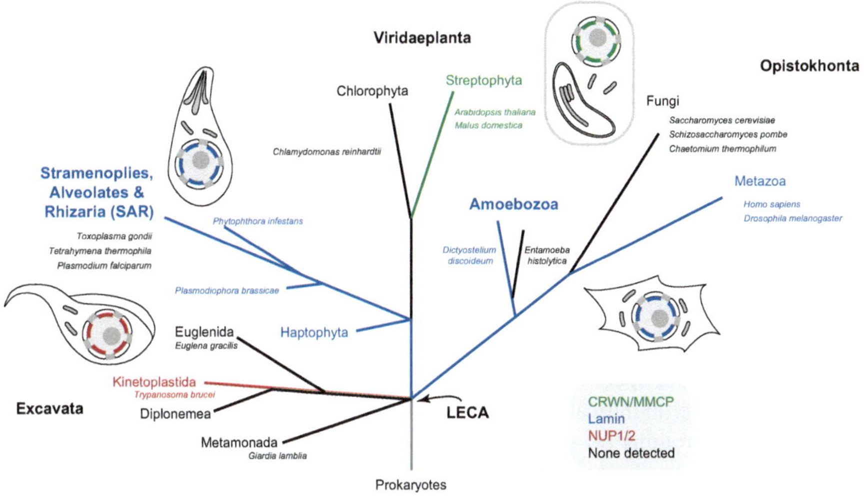


**Fig. 8.1** Distribution of lamina systems across eukaryotes. A simplified eukaryote phylogeny is shown, with lineages and branches colourised according to possession of one of three presently known lamina systems. Apparent loss of the lamin system post the LECA is clear for the fungi, most Excavata and branches of the Viridiplantae and Amoebozoa. Note that for the SAR supergroup lamins are widespread, but major disease or model organisms are examples of taxa lacking the lamin system

plants, none of the lineages possess lamins, although the CRWN (crowded nuclei) family of coiled-coil proteins provide a lamin analogue in the Streptophyta. Finally, the diverse Excavata supergroup also lack lamin genes, and though most lineages have no obvious analogue, the exception is the kinetoplastida, members of which possess the NUP-1/2 system.

The pattern of lamin distribution across the eukarya seems to suggest an almost tenuous grip on retention, with quite frequent secondary loss. The result is that a great many lineages lack identifiable lamina components, including important model organisms such as *Saccharomyces cerevisiae* and *Chlamydomonas reinhardtii*, which begs the question of what, if anything, replaces the lamins. Importantly, we know of two analogue systems, CRWN and NUP-1/2, but there are a great many lineages that lack an obvious lamin candidate. Given the obvious central roles of any of these lamina systems, and the role of lamin defects in a multitude of diseases [14–16], it is difficult to envisage that organisms where none of these systems have been identified do not also possess an analogue that remains to be identified, *albeit* that functionality may be provided in a cryptic manner. For example, Esc1 in *S. cerevisiae* lacks any obvious architectural similarity to other lamina systems but does participate, in a lamin-like manner, in defining regions of heterochromatin [17]. Here we consider the origins, similarities and differences between the trypanosomatid and plant laminas.

## CRWN (or NMCP) Proteins Constitute a Plant Lamina

Crowded nuclei proteins are essential architectural components of land plant nuclei and were first identified as nuclear matrix constituent proteins (NMCPs, renamed to CRWN [18]). CRWNs play significant roles in maintaining nuclear structure, regulating chromatin organisation and influencing gene expression, functions mirroring the lamins [19]. CRWN proteins are categorised into two main groups based on phylogeny: Group 1 includes CRWN1, CRWN2 and CRWN3, while group 2 contains CRWN4 orthologs. Within each clade, monocot sequences group independently from the dicot proteins as CRWN1- and CRWN4-related paralogues and are distinguished by the absence of a C-terminal domain present in dicot orthologs. Further, some dicot species, such as *A. thaliana*, possess multiple CRWN1 paralogues, which overall indicates ongoing evolution of the CRWN lamina system [18].

All CRWN proteins possess a coiled-coil region at their N-terminal, similar to lamins, which is required to maintain nuclear morphology [20] (Fig. 8.2). CRWN1 and CRWN4 localise predominantly at the nuclear periphery. Significantly, there is redundancy such that, depending on the paralogue, single knockouts can result in a mild nuclear morphological defect and near wild-type growth in *A. thaliana*, which becomes increasingly severe with double and triple CRWN knockouts, *albeit* that CRWN2/3 double knockouts are minimally affected, suggesting redundancy with CRWN1. Knockout of all CRWN genes is lethal [18]. More severe CRWN knockout mutants exhibit small, round nuclei, contrasting with the elongated nuclei typical of wild-type plants as well as ploidy abnormalities, again with severity varying according to the specific depleted CRWN paralogues.

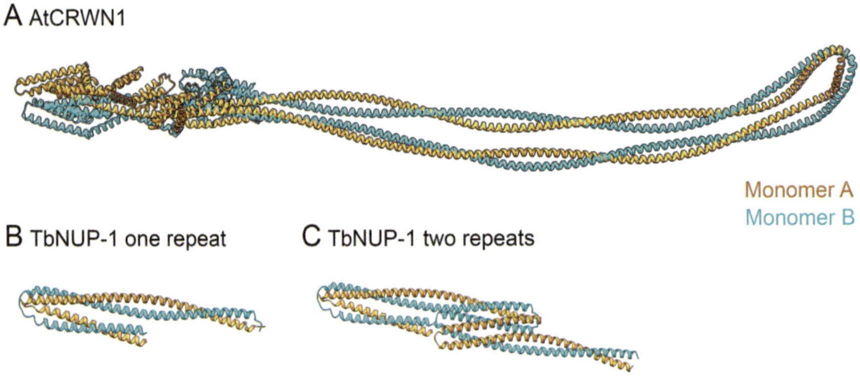


**Fig. 8.2** AlphaFold 3 predictions for lamina protein tertiary structures. Panel A: *A. thaliana* CRWN1, to demonstrate the predominant α-helical topology. Panel B: A single repeat of *T. brucei* NUP-1. Panel C: Two repeats of *T. brucei* NUP-1. Each sequence was modelled as a homodimer (gold and blue for each monomer) using AlphaFold 3 and visualised using PhyMol. Models have been scaled to approximate volumetric parity. The full foldbacks of the helices are likely artifactual but do indicate reliable regions breaking the coiled-coil structure

## CRWN Proteins Constitute a Flexible Network

The apparent genetic redundancy is supported by near identical expression and localisation patterns [21], with CRWN1, CRWN2 and CRWN3 all localising to the nuclear periphery. The cohort of CRWN proteins also appear to form filaments. Consistent with this, CRWN1, CRWN2 and CRWN3 interact with each other, while CRWN1 also interacts with CRWN4. This latter interaction appears rather specific as CRWN4 does not interact with CRWN2 or CRWN3, while homomeric interactions can be detected for all four CRWN proteins ([21], Fig. 8.2). The tighter associations between CRWN1, CRWN2 and CRWN3 may reflect closer evolutionary relationships and, together with homomeric association, suggest a rather more flexible and diverse system than that constructed by lamins. However, in common with both the lamin and NUP-1/2 systems, the configuration of CRWN proteins when assembled into fibres remains to be reported.

CRWN proteins are clearly involved in transcriptional regulation. CRWN1/2 and CRWN1/4 double knockouts trigger ectopic defence responses against pathogens, suggesting activity as negative regulators of immune responses. These include upregulation of salicylic acid biosynthesis genes and lamina disassembly under stress conditions [22, 23]. CRWN1 and CRWN2 play crucial roles in maintaining the repressive histone mark $H_3K_{27}me^3$ at many gene loci [14]. In CRWN1/2 double knockouts, there is reduction in $H_3K_{27}me^3$ levels at the PR1 (pathogenesis-related 1) locus, leading to elevated expression [24]. CRWN proteins also interact with the polycomb repressive complex 2 (PRC2), which is responsible for establishing the $H_3K_{27}me^3$ mark. Transcriptomes are extensively altered in CRWN mutants, with more than 2000 differentially expressed genes in the CRWN1/4 double knockout, including significant alterations in expression of copper-associated genes present as a cluster in chromosome V and critical for copper tolerance [21]. It is also likely that CRWN proteins are part of the mechanism that directs formation of lamina-associated domains (LADs), which are regions characterised by low transcriptional activity and high levels of repressive histone marks such as $H_3K_{27}me^3$ [21, 25].

CRWN1 can bind the plant-specific nucleoporins AtNup82 and AtNup136, which are components of the NPC basket. These nucleoporins possess conserved motifs shared with the plant lamina-associated protein KAKU4 that also facilitates their interaction with CRWN proteins [26]. This interaction suggests that CRWNs are both structural components of the nuclear lamina and connectors of the nucleoskeleton to the NPC. Significantly, KAKU4, AtNup82 and AtNup136 are related at the sequence level, and all three are specific to the Amborellaceae and angiosperms, which is an identical distribution to the expanded CRWN protein cohort [27]. Finally, a phase-separated large complex associated with the plant NPC basket that interacts with PWWP-domain interactor of polycombs 1 (PWO1), a histone reader that recruits PRC2 and interacts with CRWN1, indicates a bridge between the structural functions for CRWN proteins, tethering to the nuclear envelope and the heterochromatin machinery.

Overall, we can reconstruct a hypothetical pathway for the origins of the land plant lamina. The last common land plant ancestor possessed a single CRWN protein and an ancestral version of KAKU4/AtNup82/AtNup136. The ancestral system then diversified into two CRWN clades, with CRWN1–3 as one and CRWN4 as the other, while KAKU4, AtNup82 and AtNup136 also differentiated into three distinct genes. Critically, this allows continued flexibility between CRWN and the NPC together with combinatorial interactions between the CRWN proteins themselves.

## The Trypanosome Lamina

*T. brucei* is a member of the kinetoplastids, a lineage that diverged early from the main eukaryotic lineage post the LECA as members of the Excavata supergroup. The structural and functional divergence of central nuclear activities between trypanosomatids, animals, fungi and plants has been appreciated for decades, including the presence of highly divergent histones, polycistronic transcription coupled to *trans*-splicing, transcription of some protein-coding sequences by RNA polymerase I and the presence of variant surface protein genes important to immune evasion and infectivity at telomere proximal loci [28–30].

NUP-1, the first identified component of the trypanosome nuclear lamina, was described in 2001 [31]. Using a monoclonal antibody (Mab) with nuclear rim staining, but unknown protein antigen [32, 33], the NUP-1 protein was revealed as a band by SDS-PAGE migrating at about 450 kDa and strongly enriched in nuclear fractions [31]. Further, the anti-NUP-1 MAb localises NUP-1 to the inner face of the nuclear envelope under immunoEM, *albeit* being excluded from the region directly beneath the nuclear pore complex. Furthermore, peptide sequencing identified a partial open reading frame containing near-perfect 144 amino acid repeats, with a predicted coiled-coil architecture and short spacers of lower coiled-coil propensity.

## NUP-1 Is Dynamic, with Ordered Assembly

*T. brucei* NUP-1 is a protein of 3647 amino acids, with a central region composed of near prefect repeats. There are orthologs across kinetoplastids, all of which are smaller than *T. brucei* and with varying repeat patterns, but all retain a prediction of essentially a coiled-coil core with periodic breaks, together with distinct α-helical N- and C-terminal domains [34]. If fully extended, this would indicate a protein spanning some ~400 nm; given that the diameter of the trypanosome nucleus is under 2 μm, this suggests that an individual TbNUP-1 protein can access much of the nuclear volume, in stark contrasts to CRWN or lamin proteins [35].

When visualised via epitope tagging of the N- or C-terminus, TbNUP-1 appears as nuclear rim puncta, suggesting a concentration of termini at the nuclear periphery

[34]. This localisation is retained throughout the cell cycle, consistent with closed mitosis in trypanosomes with no evidence for nuclear envelope breakdown. However, persistence of the lamina does not preclude dynamic behaviour, and indeed there is both an apparent less compact spacing during mitosis compared with interphase and a polarised location during anaphase where NUP-1 termini partially accumulate at the distal poles of the separating nuclei [34].

Dual staining using both a C-terminal GFP-fusion protein and an antibody against the repeats indicates considerable spatial differentiation between the termini and the core repeat region, suggesting that at least some TbNUP-1 proteins have an extended configuration [34]. More extensive analysis using cells harbouring differentially tagged TbNUP-1 at both termini and stained using an antibody against the amino acid repeats revealed complex dynamics during the cell cycle ([29], Fig. 8.3). During interphase, all three regions of the protein are predominantly at the nuclear periphery, but during S-phase and through to chromosomal segregation in anaphase, the repeat region migrates deeper within the nucleus, suggesting that TbNUP-1 forms part of the chromosomal segregation machinery. At nuclear division, the protein migrates to the distal poles, with this being more pronounced for the N-terminus. This complexity indicates a central role for TbNUP-1 in nuclear and chromosomal organisation.

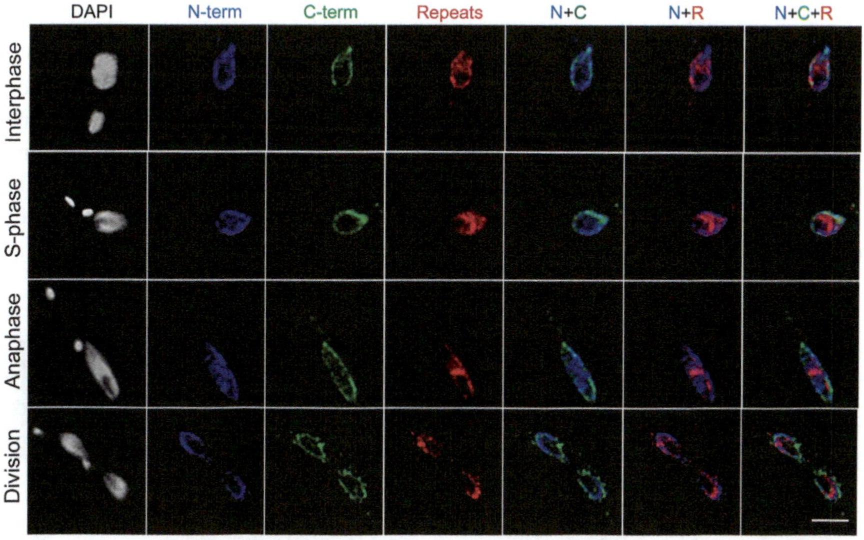


**Fig. 8.3** NUP-1 domains relocate during the cell cycle. Cells expressing a double-tagged version of NUP-1 were imaged by confocal microscopy. DAPI was used to visualise DNA (pseudo-colour white). The N-terminal (N-term) and the C-terminal (C-term) domains were tagged with HA (pseudo-colour blue) and GFP (pseudo-colour green), respectively. To visualise the repeat region of NUP-1 (repeats), affinity-purified rabbit antibodies raised against the repeat were used (pseudo-colour red). The typical distribution of NUP-1 at the nuclear periphery is clear throughout the cell cycle stages as indicated (see text for details). Scale bar: 2 μm. (Figure is reproduced and adapted from Ref. [29]. Copyright retained by the authors as article open access)

An AlphaFold 3 [36] prediction of TbNUP-1 structure offers a possible mechanism for this dynamic behaviour (Fig. 8.2). The prediction highlights the coiled-coil repeats but also the links between these repeats which in the AlphaFold 3 representation are collapsed upon each other, *albeit* the collapse likely an artefact. However, this alternating rod and linker suggests that the repeats can contract and extend, in a manner similar to a Hoberman sphere [37]. This does also suggest the possible formation of TbNUP-1 fibres that are not in register, allowing for more complex branching and network formation than for the lamins.

Assembly of TbNUP-1 has been examined by expression of fragments *in cellulo*, specifically the N- and C-terminal domains alone together with a construct where the repeats have been deleted [29]. All three produce regular spherical assemblies, indicating that both the N- and C-termini are able to oligomerise. These analyses suggest that the N-terminus is more dominant in terms of assembly and anchoring to the nuclear envelope. The domain constructs also recruit endogenous TbNUP-1, indicating that these domains mediate bona fide protein-protein interactions, while migration of the construct polymers to distal positions within dividing nuclei indicates retention of endogenous protein domain targeting. We propose that TbNUP-1 likely assembles via a hub-and-spoke arrangement and with clear homotypic (i.e. N- to N-terminal, C- to C-terminal) and possibly heterotypic (N- to C-terminal) interactions, facilitating great flexibility within the lamina structure, once more an echo of the flexible CRWN system.

TbNUP-1 is a phosphoprotein, with multiple phosphorylation sites within the N- and C-termini and likely a trypanosome-specific CDK substrate (Koreny et al., in preparation). Specifically, mutation of all phosphorylation sites (ser/thr to ala) leads to the appearance of an umbilicus between separating nuclei (Fig. 8.4), which is phenocopied by knockdown of the CDK, CRK3. CRK3 impacts the cell cycle and is highly expressed during cell division [38]. We suggest that phosphorylation (and possibly other post-translational modifications) is responsible for regulating packing and unpacking of TbNUP-1 and specifically during nuclear division and hence an altered conformation and localisation across the cell cycle.

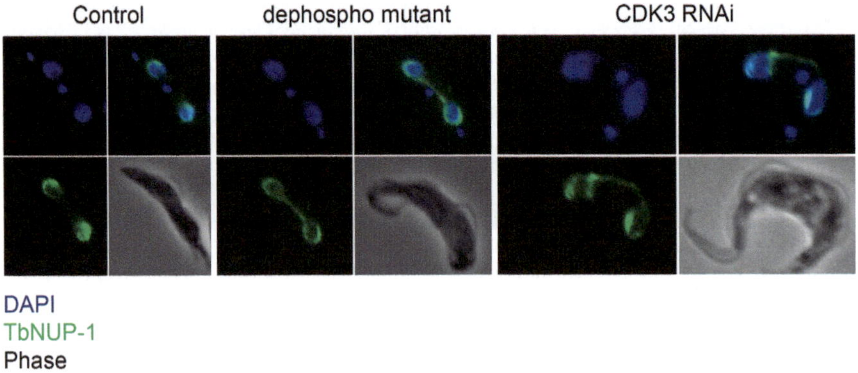


**Fig. 8.4** Phosphorylation mediates TbNUP-1 relocation during nuclear division. Examples of cells possessing wild-type TbNUP-1 (control), a TbNUP-1 mutant with all identified phosphorylation sites mutated to alanine (dephospho-mutant) and cells with CDK3 silenced (CDK3 RNAi). All examples are TbNUP-1::GFP (green) and insect stage parasites, with DAPI stain in blue

## NUP-1 Functions as a Lamin Analogue

Gene silencing of TbNUP-1 reveals functions which align with those reported for lamins in metazoan cells. These impacts include loss of normal nuclear morphology and the emergence of nuclear blebs. The distribution of NPCs is also impacted with the regular spacing of these structures being lost and resulting in pseudocrystalline arrays [34]. NUP-1 also functions in chromatin organisation. The *T. brucei* nuclear genome is organised as 22 megabase chromosomes, several intermediate-sized chromosomes of two to eight hundred kilo- base pairs and ~100 minichromosomes. Originally considered to segregate with distinct timing [33], all classes of chromosome segregate together using the same kinetochore and other machinery [39]. Telomeres during interphase are arranged at the nuclear periphery but become polarised during chromosomal segregation and closely follow TbNUP-1, but this peripheral organisation is, as expected, lost in TbNUP-1 silenced cells [34].

Many megabase chromosome telomeres contain a specialised expression site (ES) for the variant surface glycoprotein, a critical component for antigenic variation [40]. The ES contains a cohort of additional protein-coding genes, collectively referred to as ES-associated genes or ESAGs. All ES genes are, unusually, transcribed by RNA polymerase I as a polycistronic unit, and only a single ES is active at any time; all others are repressed. The single active ES locates to a nuclear interior position, distinct from the inactive ES which are positioned closer to the nuclear periphery [41]. A second protein-coding locus is also transcribed by RNA PolI which encodes the insect stage surface procyclin proteins together with a small number of procyclin-associated genes (PAGs), also transcribed in a polycistronic manner, and repressed in bloodstage parasites [42]. Significantly, both silent VSG ESs and the procyclin locus are derepressed in TbNUP-1 knockdown cells, concordant with the loss of a peripheral location of telomeres [34]. Hence, TbNUP-1 has a very specific impact on maintaining the correct expression profile of a select group of RNA PolI-transcribed developmentally regulated surface protein encoding genes.

## NUP-1 Interacts with the NPC and NUP-2

Protein-protein interactions are central to understanding mammalian lamin biology as they underpin the plethora of processes that lamins mediate: The same is true for TbNUP-1. Direct pullouts of TbNUP-1 identify a large number of nucleoporins, including components of the NPC core scaffold, parts of the transport machinery, TbRan, TbRanBP1 and TbTBC-RootA, a putative RanGAP [43, 44], and components of the nuclear basket [45]. It is likely that many of these represent secondary interactions, for example, the full complement of the trypanosome equivalent of the Nup85 Y-shaped complex is recovered and the entire Ran machinery, the latter a GTPase mediating directional coupling of transport to a GTP/GDP gradient. FG-repeat-containing TbNup98 interacts with multiple regions of the NPC including other FG Nups, the nuclear basket and the scaffold rings and is likely a direct TbNUP-1 interaction partner [29, 43, 44]. TbNUP-1 immunoprecipitations also identify a second large coiled-coil protein, which we designated as TbNUP-2 [43].

An interaction between TbNUP-1 and TbNup98 was also observed by proteomic analysis of the N- and C-terminal domain and N + C-terminal domain constructs [29]. TbNup98 localises as puncta at the NE, consistent with being an NPC component [34, 46], but this is disrupted by overexpression of the N- and C-terminal domain or N + C-terminal domain constructs. Specifically, at short times of overexpression, TbNup98 colocalises with the C-terminal domain, but not N- or N + C domains, while after more protracted induction periods TbNup98 colocalises with all three NUP-1 domain constructs and implies that the TbNUP-1 C-terminus is more dominant in interacting with TbNup98.

Furthermore, silencing TbNup98 disrupts the classical distribution of TbNUP-1 at the nuclear periphery and TbNUP-1 clusters as puncta at the NE. Nuclei in TbNup98-depleted cells are deformed, with evident defects in morphology and aberrant chromatin distribution. DNA content is also disrupted with an increase in >4n cells in the population, with these silenced cells also possessing extranuclear chromatin-containing bodies. Overall, these data suggest a model for TbNUP-1 binding the NPC via TbNup98, and potentially with a preference for the C-terminus, with this interaction being critical to coordinating nuclear division. However, the ability to anchor both termini at the NPC, independent of the central repeats, is fully consistent with the nuclear peripheral localisation of TbNUP-1 N- and C-termini and the more nuclear internal repeat location during chromosomal replication and segregation.

## NUP-2 Is a Second Lamina Component

TbNUP-2 is a 170 kDa protein, predominantly α-helical with several predicted coiled-coil regions, and a phosphoprotein [47]. Orthologs are present across the kinetoplastids and extend to the free-living *Bodo saltans* and maintain a similar architecture in all cases. The absence of a repeat structure differentiates TbNUP-2 from TbNUP-1. The protein localises to the nuclear periphery and, while proximal to TbNUP-1, does not fully colocalise but does concentrate at the distal poles during nuclear division, similar to TbNUP-1. Having both lamina proteins concentrate at the same position in the nucleus where the chromosomes segregate most likely provides additional mechanical support to the spindle.

Similar to TbNUP-1, silencing of TbNUP-2 leads to nuclear blebbing, defective progression through the cell cycle, accumulation of cells with abnormal DNA content and mislocalisation of NPCs. Furthermore, the positioning of telomeres is abnormal, with telomeres frequently accumulating within nuclear blebs. Similar defects are present for the minichromosomes, which overall mirror the impact of TbNUP-1 on chromosomal organisation. Likely in consequence there is a prominent increase in ɣH2A-like marker signal, which indicates double-strand DNA breaks, probably a result of disorganisation of the nucleus. Significantly, silenced cells demonstrate a highly specific altered transcriptome, whereby both the procyclin and silent VSG-ES RNA PolI-transcribed loci are derepressed. The remarkable overall similarity between TbNUP-1 and TbNUP-2 phenotypes is almost certainly due to their codependence, such that silencing of either leads to mislocalisation of the other [43].

Immunoisolation of TbNUP-2 identified robust interaction with TbNup110, a component of the nuclear basket and a suite of seven additional proteins. Subsequent analysis reveals that several are part of an extensive trypanosome NPC nuclear basket [45], and hence while overall structural and functional similarities between TbNUP-1 and TbNUP-2 are significant, it is certainly the case that these proteins are nevertheless nonidentical in their roles.

Of the cohort of proteins identified from TbNUP-1 or TbNUP-2 immunoisolations that are neither NPC or transport factors, only two are conserved across eukaryotes. TbLAP59 is an ortholog of a sphingomyelinase implicated in mediating NPC insertion [45], and TbLAP173 is orthologous to Sac3, the sole component of the TREX-2 complex identified to date in trypanosomes. Of the remainder, TbLAP333 provides a considerable surprise: Lacking any sequence similarity to organisms beyond the kinetoplastida, this protein is predicted by AlphaFold 3 to contain 11 Ig-fold repeats as well as a cluster of transmembrane domains at the C-terminus and hence is structurally related to the NPC membrane proteins yeast Pom152 and human Nup210. The remaining four proteins are predicted to be extensively coiled coil and, based on their interactions, robustly placed as nuclear basket proteins, revealing an amalgam of conserved and lineage-specific components associated with the trypanosome NPC and lamina.

## Evolution of NUP-1 Repeats

*T. brucei* NUP-1 contains multiple coiled-coil segments with linkers separating them. In contrast, lamin rod domains are non-repetitive and consist of unique coiled-coil segments of various lengths separated by short linker regions, which are highly conserved [4] and more similar to the CRWN proteins in this regard [20]. The TbNUP-1 rod domain is actually comprised of two repetitive elements, designated type A and B. Sixteen type A and 15 type B repeats form a coiled-coil rich span with shorter coiled coils separated by multiple non-coiled-coil linker sequences. The repeat core starts and ends with an A-type element ([48], Fig. 8.5).

Analysis of repeat variation and coiled-coil predictions across kinetoplastids identified three groups of sequences; those related to the regular A|B TbNUP-1 architecture, those closely related to *L. major* NUP-1 and a group with more divergent characteristics. *Leishmania* NUP-1 orthologs all possess head and tail domains of very similar lengths and nearly identical sequence. Their rod domains consist of just two types of repeats arranged in the $(A\text{-}B)_n$-A pattern. Unlike the trypanosome

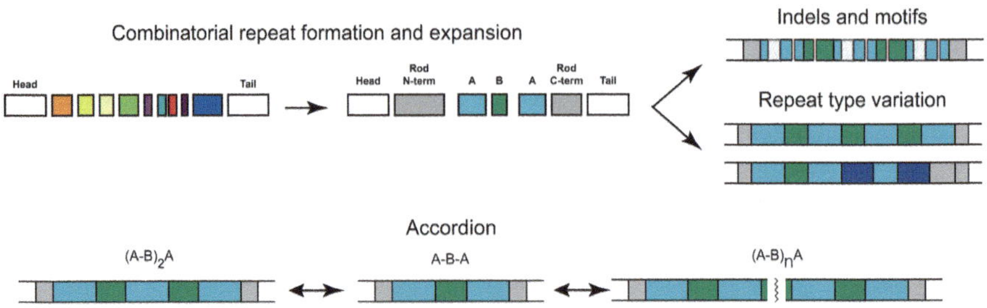


**Fig. 8.5** Modes of expansion and variation of NUP-1 rod domain. We suggest that the highly regular repeats present in most kinetoplastids have evolved from a less regular progenitor, as exemplified by the NUP-1 ortholog in *C. fasciculata* (rainbow repeats top left). This resolved into a regular repeat with A (light blue) and B (green) elements, which could be diversified by indels or other types of sequence variation. Further, we suggest that repeats can evolve through variation in the number of repeats, allowing increases or decreases, similar to an accordion

group, the *Leishmania* A- and B-type repeats are conserved both within and between species, and the major difference between orthologs is the number of repeats comprising the rod. Considering that the number of repeats in *Leishmania* varies in a similar way in trypanosomes but shares relatively little sequence similarity, this suggests an "accordion" effect where, in species with two types of repeats, repeats likely multiplied different numbers of times via domain expansion while maintaining an (A-B)$_n$-A pattern arrangement (Fig. 8.5).

In the monoxenous parasite, *Crithidia fasciculata*, which is considered a basal kinetoplastid form, the regular repeat structure is not present. The syntenic CfNUP-1 gene retains the overall head-rod-tail architecture with the rod consisting of multiple short elements that have a complex repeat organisation. However, despite this, the rod remains predicted as predominantly coiled coil. Hence, we suggest that the highly regular repeats in the trypanosomatids/*Leishmania* are the evolutionary result of a simplification of repeat elements, followed by independent expansion and variation within each lineage [48]. This simplification may allow the interaction between repeats to be flexible and hence allow modification of the network.

## Conclusions

The nucleus originated prior to the LECA, with all reconstructions predicting that the nucleus in the LECA possessed a lamina (most probably related to lamin B), NPCs and other systems that are represented in the vast majority of described extant species. That said, it is also apparent that the period subsequent to the LECA provided ample opportunity for evolution to diversify these systems and that the lamina provides an excellent example. There are at least three systems presently described, lamins, CRWN and NUP-1/2, which have striking similarities despite no evidence for vertical descent and together are a potent example of convergent evolution (Fig. 8.6). We speculate that additional systems may well remain to be uncovered.

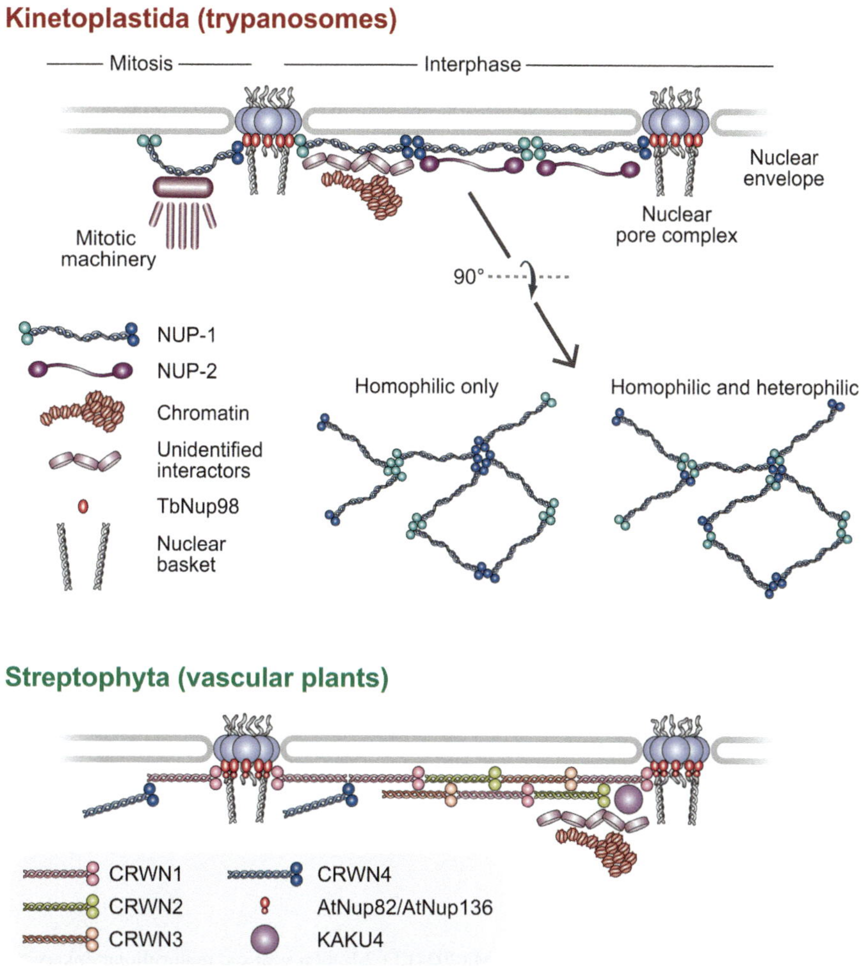


**Fig. 8.6** Organisation of the lamina systems in trypanosomes and plants. Schematics of the lamina systems discussed in the text. Top panel the NUP-1/2 system in kinetoplastida. In this representation, it is assumed that NUP-1 forms a homodimer but that interactions between dinners are primarily mediated by the terminal domains. Lower panel represents the CRWN proteins in vascular plants. Note that the stoichiometry is unknown, but all CRWN proteins can self-interact as well as form promiscuous heterophilic connections. Note that a number of factors that connect the lamina and chromatin remain unknown in both systems, although for plants some components of the polycomb repressive complex 2 (PRC2), responsible for establishing the $H_3K_{27}me^3$ mark, are associated with CRWN isoforms. Top panel is redrawn and expanded from Ref. [29]. Note that protein/complex sizes are not drawn to scale

Each lamina possesses a predominant coiled-coil element, which leads to a common architecture and which is likely modulated by post-translational mechanisms. While NUP-1/2 are only present in organisms with a closed mitosis, lamins are present in organisms with both open and closed mechanisms, while the vast majority of land plants utilise open mitosis. We suggest that the large trypanosome NUP proteins are unsuitable for open mitosis where the lamina has to disassemble, *albeit* that there remains flexibility within the NUP-1/2 network facilitating dynamic reorganisation during the cell cycle. This does lead to diversification in the manner of oligomerisation, with NUP-1 primarily assembling via the non-repetitive N- and C-terminal domains, whereas lamins form fibres that are in register as head-to-tail oligomers. Furthermore, all lamina systems interact with the NPC and mediate spacing of the pores and control gene expression, *albeit* by divergent mechanisms. It remains to be determined how any lamina system is fully built and for the structure of a full monomer to be reported.

The evolutionary distribution of classical lamins is clearly the broadest, and the patchy retention across most eukaryotic lineages contrasts with the CRWN and NUP-1/2 systems, which are clear examples of lineage-specific evolution. The multiple losses of lamins suggest that circumstances can arise comparatively frequently where lamin function is non-essential, and at least in the case of the euglenids, it is clear that lamins were lost prior to their replacement by NUP-1/2.

**Acknowledgements** Many thanks to Chris Caffalette (Rockefeller) for help with the AlphaFold 3 predictions. MCF and MPR would like to thank the many people who have worked on this programme of research for more than two decades, including Kelly duBois, Luke Maishman, Alex Makarov, Ludek Koreny, Norma Padilla and most importantly Sam Obado; our huge gratitude to all.

**Compliance with Ethical Standards** The authors have no conflicts of interest to declare. Work from the authors' laboratories discussed in this article has been funded by the Wellcome Trust, the Medical Research Council (UK) and the National Institutes of Health, which is gratefully acknowledged.

## References

1. Miao L, Yin Z, Knoll AH, Qu Y, Zhu M (2024) 1.63-billion-year-old multicellular eukaryotes from the Chuanlinggou formation in North China. Sci Adv 10:eadk3208
2. Koumandou VL, Wickstead B, Ginger ML, van der Giezen M, Dacks JB, Field MC (2013) Molecular paleontology and complexity in the last eukaryotic common ancestor. Crit Rev Biochem Mol Biol 248:373–396
3. Obado SO, Field MC, Rout MP (2017) Comparative interactomics provides evidence for functional specialization of the nuclear pore complex. Nucleus 8:340–352
4. Koreny L, Field MC (2016) Ancient eukaryotic origin and evolutionary plasticity of nuclear lamina. Genome Biol Evol 8:2663–2671
5. Aaronson RP, Blobel G (1975) Isolation of nuclear pore complexes in association with a lamina. Proc Natl Acad Sci USA 72:1007–1011
6. McKeon FD, Kirschner MW, Caput D (1986) Homologies in both primary and secondary structure between nuclear envelope and intermediate filament proteins. Nature 319(6053):463–468
7. Gerace L, Blobel G (1980) The nuclear envelope lamina is reversibly depolymerized during mitosis. Cell 19:277–287
8. Goldman RD, Gruenbaum Y, Moir RD, Shumaker DK, Spann TP (2002) Nuclear lamins: building blocks of nuclear architecture. Genes Dev 16:533–547

9. Eriksson JE, Dechat T, Grin B, Helfand B, Mendez M, Pallari HM, Goldman RD (2009) Introducing intermediate filaments: from discovery to disease. J Clin Invest 119:1763–1771
10. Adam SA, Butin-Israeli V, Cleland MM, Shimi T, Goldman RD (2013) Disruption of Lamin B1 and Lamin B2 processing and localization by farnesyltransferase inhibitors. Nucleus 4:142–150
11. Makarov AA, Zou J, Houston DR, Spanos C, Solovyova AS, Cardenal-Peralta C, Rappsilber J, Schirmer EC (2019) Lamin a molecular compression and sliding as mechanisms behind nucleoskeleton elasticity. Nat Commun 10:3056
12. Ahn J, Jo I, Jeong S, Lee J, Ha NC (2023) Lamin filament assembly derived from the atomic structure of the antiparallel four-helix bundle. Mol Cells 46:309–318
13. Grafe M, Batsios P, Meyer I, Lisin D, Baumann O, Goldberg MW, Gräf R (2019) Supramolecular structures of the Dictyostelium lamin NE81. Cells 8:162
14. Choi J, Gehring M (2024) CRWN nuclear lamina components maintain the H3K27me3 landscape and promote successful reproduction in Arabidopsis. New Phytol 243:213–228
15. Shin JY, Worman HJ (2022) Molecular pathology of laminopathies. Annu Rev Pathol 17:159–180
16. Wong X, Stewart CL (2020) The laminopathies and the insights they provide into the structural and functional organization of the nucleus. Annu Rev Genomics Hum Genet 21:263–288
17. Faure G, Jézéquel K, Roisné-Hamelin F, Bitard-Feildel T, Lamiable A, Marcand S, Callebaut I (2019) Discovery and evolution of new domains in yeast heterochromatin factor Sir4 and its partner Esc1. Genome Biol Evol 11:572–585
18. Wang H, Dittmer TA, Richards EJ (2013) Arabidopsis CROWDED NUCLEI (CRWN) proteins are required for nuclear size control and heterochromatin organization BMC. Plant Biol 13:200
19. Choi J, Richards EJ (2020) The role of CRWN nuclear proteins in chromatin-based regulation of stress response genes. Plant Signal Behav 15:1694224
20. Yin C, Wang Y, Wang P, Chen G, Sun A, Fang Y (2024) The N-terminal coiled-coil domain of Arabidopsis CROWDED NUCLEI 1 is required for nuclear morphology maintenance. Planta 260:62
21. Sakamoto Y, Sato M, Sato Y, Harada A, Suzuki T, Goto C, Tamura K, Toyooka K, Kimura H, Ohkawa Y, Hara-Nishimura I, Takagi S, Matsunaga S (2020) Subnuclear gene positioning through lamina association affects copper tolerance. Nat Commun 11:5914
22. Wang N, Wang Z, Tzourtzou S, Wang X, Bi X, Leimeister J, Xu L, Sakamoto T, Matsunaga S, Schaller A, Jiang H, Liu C (2023) The plant nuclear lamina disassembles to regulate genome folding in stress conditions. Nat Plants 9:1081–1093
23. Choi J, Strickler SR, Richards EJ (2019) Loss of CRWN nuclear proteins induces cell death and salicylic acid defense signaling. Plant Physiol 179:1315–1329
24. Han Z, Xiong D, Schneiter R, Tian C (2023) The function of plant PR1 and other members of the CAP protein superfamily in plant-pathogen interactions. Mol Plant Pathol 24:651–668
25. Cao Y, Yan H, Sheng M, Liu Y, Yu X, Li Z, Xu W, Su Z (2024) Nuclear lamina component KAKU4 regulates chromatin states and transcriptional regulation in the Arabidopsis genome. BMC Biol 22:80
26. Mermet S, Voisin M, Mordier J, Dubos T, Tutois S, Tuffery P, Baroux C, Tamura K, Probst AV, Vanrobays E, Tatout C (2023) Evolutionarily conserved protein motifs drive interactions between the plant nucleoskeleton and nuclear pores. Plant Cell 35:4284–4303
27. Ciska M, Hikida R, Masuda K, Díaz M, de la Espina S (2019) Evolutionary history and structure of nuclear matrix constituent proteins, the plant analogues of lamins. J Exp Bot 70:2651–2664
28. Faria J, Briggs EM, Black JA, McCulloch R (2022) Emergence and adaptation of the cellular machinery directing antigenic variation in the African trypanosome. Curr Opin Microbiol 70:102209
29. Padilla-Mejia NE, Koreny L, Holden J, Vancová M, Lukeš J, Zoltner M, Field MC (2021) A hub-and-spoke nuclear lamina architecture in trypanosomes. J Cell Sci 134:jcs251264
30. Marcianò G, Ishii M, Nerusheva OO, Akiyoshi B (2021) Kinetoplastid kinetochore proteins KKT2 and KKT3 have unique centromere localization domains. J Cell Biol 220:e202101022

31. Trenaman A, Tinti M, Wall RJ, Horn D (2024) Post-transcriptional reprogramming by thousands of mRNA untranslated regions in trypanosomes. Nat Commun 15:8113
32. Rout MP, Field MC (2001) Isolation and characterization of subnuclear compartments from Trypanosoma brucei. identification of a major repetitive nuclear lamina component. J Biol Chem 276:38261–38271
33. Ogbadoyi E, Ersfeld K, Robinson D, Sherwin T, Gull K (2000) Architecture of the *Trypanosoma brucei* nucleus during interphase and mitosis. Chromosoma 108:501–513
34. Ersfeld K, Gull K (1997) Partitioning of large and minichromosomes in *Trypanosoma brucei*. Science 276:611–614
35. DuBois KN, Alsford S, Holden JM, Buisson J, Swiderski M, Bart JM, Ratushny AV, Wan Y, Bastin P, Barry JD, Navarro M, Horn D, Aitchison JD, Rout MP, Field MC (2012) NUP-1 is a large coiled-coil nucleoskeletal protein in trypanosomes with Lamin-like functions. PLoS Biol 10:e1001287
36. Field MC, Adung'a V, Obado S, Chait BT, Rout MP (2012) Proteomics on the rims: insights into the biology of the nuclear envelope and flagellar pocket of trypanosomes. Parasitology 139:1158–1167
37. Abramson J, Adler J, Dunger J, Evans R, Green T, Pritzel A, Ronneberger O, Willmore L, Ballard AJ, Bambrick J, Bodenstein SW, Evans DA, Hung CC, O'Neill M, Reiman D, Tunyasuvunakool K, Wu Z, Žemgulytė A, Arvaniti E, Beattie C, Bertolli O, Bridgland A, Cherepanov A, Congreve M, Cowen-Rivers AI, Cowie A, Figurnov M, Fuchs FB, Gladman H, Jain R, Khan YA, Low CMR, Perlin K, Potapenko A, Savy P, Singh S, Stecula A, Thillaisundaram A, Tong C, Yakneen S, Zhong ED, Zielinski M, Žídek A, Bapst V, Kohli P, Jaderberg M, Hassabis D, Jumper JM (2024) Accurate structure prediction of biomolecular interactions with AlphaFold 3. Nature 630:493–500
38. Hoberman C (1991) Radial expansion/retraction truss structures. US patent 5,024,031 18 Jun 1991
39. Hammarton TC, Clark J, Douglas F, Boshart M, Mottram JC (2003) Stage-specific differences in cell cycle control in *Trypanosoma brucei* revealed by RNA interference of a mitotic cyclin. J Biol Chem 278:22877–22886
40. Müller LSM, Cosentino RO, Förstner KU, Guizetti J, Wedel C, Kaplan N, Janzen CJ, Arampatzi P, Vogel J, Steinbiss S, Otto TD, Saliba AE, Sebra RP, Siegel TN (2018) Genome organization and DNA accessibility control antigenic variation in trypanosomes. Nature 563:121–125
41. Navarro M, Gull K (2001) A pol I transcriptional body associated with VSG mono-allelic expression in *Trypanosoma brucei*. Nature 414:759–763
42. Haenni S, Studer E, Burkard GS, Roditi I (2009) Bidirectional silencing of RNA polymerase I transcription by a strand switch region in *Trypanosoma brucei*. Nucleic Acids Res 37:5007–5018
43. Maishman L, Obado SO, Alsford S, Bart JM, Chen WM, Ratushny AV, Navarro M, Horn D, Aitchison JD, Chait BT, Rout MP, Field MC (2016) Co-dependence between trypanosome nuclear lamina components in nuclear stability and control of gene expression. Nucleic Acids Res 44:10554–10570
44. Obado SO, Brillantes M, Uryu K, Zhang W, Ketaren NE, Chait BT, Field MC, Rout MP (2016) Interactome mapping reveals the evolutionary history of the nuclear pore complex. PLoS Biol 14:e1002365
45. Butterfield ER, Obado SO, Scutts SR, Zhang W, Chait BT, Rout MP, Field MC (2024) A lineage-specific protein network at the trypanosome nuclear envelope. Nucleus 15:2310452
46. DeGrasse JA, DuBois KN, Devos D, Siegel TN, Sali A, Field MC, Rout MP, Chait BT (2009) Evidence for a shared nuclear pore complex architecture that is conserved from the last common eukaryotic ancestor. Mol Cell Proteomics 8:2119–2130
47. Urbaniak MD, Martin DM, Ferguson MA (2013) Global quantitative SILAC phosphoproteomics reveals differential phosphorylation is widespread between the procyclic and bloodstream form lifecycle stages of *Trypanosoma brucei*. J Proteome Res 12:2233–2244
48. Makarov A, Butenkov A, Lukeš J, Field MC (2025) Origins and evolution of the kinetoplastida nuclear lamina protein NUP-1. Nucleus. (in preparation)

# Chapter 9
# The Role of Lamins in Genome Organisation: A Modelling Perspective

**Shuvadip Dutta and Mithun K. Mitra**

**Abstract** Lamin proteins, predominantly located near the nuclear lamina, are known to play an important role in maintaining large-scale genome organisation. Disruptions in lamin assembly have been implicated in various diseases exhibiting altered chromatin organisation, nuclear stiffness and chromosome dynamics. Theoretical and computational modelling is essential for understanding these diverse observations through the lens of physical forces and interactions. This review focuses on modelling efforts aimed at elucidating the multifaceted role of lamins in genome organisation and how disruptions in these interactions contribute to the pathologies observed in different cell types, disease conditions and lamin mutants. We categorise the major modelling approaches used to study lamin-mediated chromatin organisation, summarise key findings within each category and highlight future directions in developing a comprehensive understanding of the role of lamins in the spatial organisation of the genome.

**Keywords** Lamin · Chromatin · Genome organisation · Polymer models · Coarse-grained models of lamin · Chromosome positioning · Laminopathies

## Introduction

The spatial organisation of interphase chromosomes is a complex, hierarchical phenomenon crucial for nuclear function [1]. A hallmark of chromatin organisation is the hierarchical folding of chromatin into distinct structural domains, from topologically associating domains (TADs) at scales of hundreds of kilobases, to megabase-scale A/B compartments which segregate chromatin into active (A) and

S. Dutta · M. K. Mitra (✉)
Department of Physics, Indian Institute of Technology Bombay, Mumbai, Maharashtra, India
e-mail: mkmitra@iitb.ac.in

© The Author(s), under exclusive license to Springer Nature Switzerland AG 2025

K. Sengupta, A. Muchir (eds.), *Lamins in Genome Organization and Diseases*, Subcellular Biochemistry 115, https://doi.org/10.1007/978-3-032-00537-3_9

inactive (B) compartments and, finally, to chromosome territories at the scales of chromosomes [2] (Fig. 9.1). TADs, which are genomic regions characterised by high self-interaction, are thought to form through cohesin-mediated loop extrusion, bringing distant loci into proximity for regulatory interactions like enhancer–promoter communication [3–5]. In contrast, A and B compartments, segregating active and inactive chromatin [6], are thought to arise from polymer–polymer phase separation of chromatin segments with similar epigenetic marks, which influence both chromatin structure and the transcriptional status of these genomic regions. Chromosome territories, where each chromosome occupies distinct nuclear regions, result from slow interphase chromatin dynamics and architectural constraints, limiting interchromosomal mixing [7].

A key mediator of chromatin organisation is believed to be chromatin interactions with lamin proteins present in the nuclear lamina (NL) [8, 9]. A fibrous meshwork of nuclear lamins (type V intermediate filaments) forms just beneath the inner nuclear membrane (INM) in metazoans and constitutes the main component of the NL [10, 11]. In somatic mammalian cells, A-type lamins, specifically lamin A and lamin C, are encoded by the single *LMNA* gene and undergo alternative splicing. B-type lamins, namely, lamin B1 and lamin B2, are encoded by separate genes, *LMNB1* and *LMNB2*, respectively [12]. While both A- and B-type lamins are expressed in most differentiated cells, only A-type lamins are developmentally regulated [8]. Their expression is minimal in early embryonic stages and pluripotent stem cells, but it increases as cells differentiate. In contrast, B-type lamins are expressed more consistently [13, 14]. In invertebrates like *Drosophila melanogaster*, two lamin genes are known. Lamin C is expressed only late during embryonic development, making it an analogue of vertebrate A-type lamins. Similar to vertebrate lamin C, *Drosophila* lamin C lacks the COOH-terminal isoprenylation motif, while the early-expressed, ubiquitously present Dm0 lamin gene contains the CaaX motif and is considered equivalent to vertebrate B-type lamins [15].

At the NL, the meshwork of lamin filaments is anchored to the INM through interactions with INM proteins and the nuclear pore complexes (NPCs) [16]. Cryo-electron tomography (cryo-ET) in mouse embryonic fibroblasts revealed that lamins assemble into 3.5 nm thick filaments [17], which collectively form a 14–16 nm

**Fig. 9.1** (continued) distinct chromatin interaction preferences based on epigenetic signatures. Additionally, each chromosome occupies a specific spatial region in the nucleus, forming distinct chromosome territories. Chromatin models may either use a bead-spring representation of a polymer or a continuum field theoretic representation. (Bottom left) Illustration of the phase-field model of chromatin organisation within the nucleus. In this model, chromatin is represented as a series of territories consisting of euchromatin (coloured red) and heterochromatin (coloured blue), which may be further subdivided into facultative and constitutive heterochromatin. A schematic representation of the phase-field model-derived nuclear architecture is shown. (Middle right) Representation of the bead-spring polymer model for chromatin. The polymer model is composed of beads that are classified according to experimental data. The coarse graining scale determines the features used to represent beads. (Bottom right) Two primary simulation methods are highlighted: simulation of short genomic segments, which models specific nuclear regions, and the whole-genome simulation, which examines genome-wide structural and functional properties

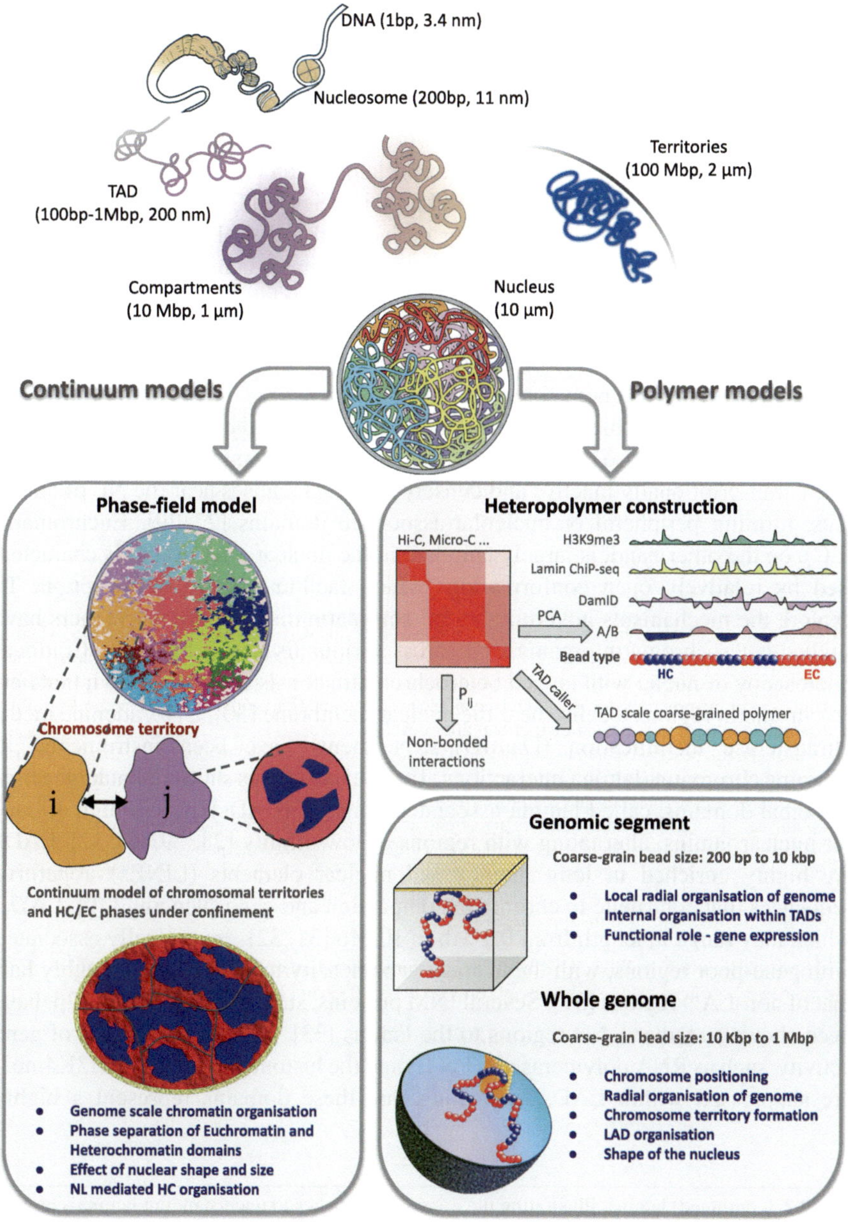


**Fig. 9.1** Approaches to chromatin modelling. (Top) Schematic illustrating the hierarchical organisation of the eukaryotic genome. The chromatin structure is shown at multiple levels of folding within the nucleus, starting from nucleosome (~200 bp, 11 nm) resolution. Depending on gene activity, chromatin density varies and folds into higher-order domains called topologically associating domains (TADs) on the sub-megabase scale. At larger (micron) scales, chromatin is partitioned into active ('A') and repressive ('B') compartments, with each compartment reflecting

thick NL meshwork attached to the inner nuclear membrane [17]. A-type and B-type lamins are believed to form separate but highly interconnected filamentous networks beneath the INM [18] (Fig. 9.2). These filaments are highly variable in length and also exhibit high flexibility [19]. The structural properties of the NL, such as lamin filament length distribution, meshwork density and NL thickness, may vary depending on the cell type [13]. The nuclear lamina can regulate chromatin organisation through interactions with the INM proteins. Several INM proteins are known to interact with lamins, either directly or indirectly [20]. The lamin B receptor (LBR) contains a binding site for HP1 and thus facilitates lamin B interactions with chromatin [21]. Similarly, emerin, which interacts with lamin A, contains a binding site for the chromatin-bridging protein BAF (barrier-to-autointegration factor) [20]. These INM proteins thus connect chromatin to the lamina, pointing to the role of lamins in large-scale genome organisation [22–26].

Beyond the hierarchical organisation of the genome, the nucleus also exhibits radial organisation based on biochemical and functional properties. Heterochromatin (HC), transcriptionally inactive and densely packed, localises near the NL or nucleolus, forming peripheral or nucleolar-associated domains [27–29]. Euchromatin (EC), on the other hand, is largely enriched in the nuclear interior and is characterised by relatively open conformations which facilitate active transcription. To explore the mechanisms governing radial chromatin distribution, researchers have studied heterochromatin organisation across various tissues and species. Confocal microscopy of nuclei with tagged heterochromatin consistently has shown that heterochromatin regions localise near the nuclear membrane [30]. DNA adenine methyltransferase identification (DamID) experiments have been instrumental in mapping chromatin–lamina interactions. These experiments showed that large chromosomal domains, called lamina-associated domains (LADs), have a high affinity for nuclear lamins, alternating with regions of low affinity [28, 29, 31, 32]. LADs are highly enriched in long interspersed nuclear elements (LINEs), repetitive sequences that contribute to chromatin compaction and gene silencing [29]. LADs, which may range in length from 0.1 Mb to 10 Mb [31, 32], are typically associated with gene-poor regions, with the average gene density in LADs being roughly half that of non-LAD regions [31]. Several INM proteins, such as LBR and LAP2β, have been shown to tether LAD regions to the lamins [33]. Moreover, markers of gene activity, such as RNA polymerase II (Pol II) and the histone modification H3K4me2, are suppressed within LADs, indicating that these domains represent a highly

**Fig. 9.2** (continued) lamina, illustrating the diverse approaches to understanding lamina–chromatin interactions. (**a**) The static surface model depicts the lamina as a rigid, static surface that interacts with chromatin. This model assumes no flexibility in the lamina structure and focuses on its role in chromatin anchoring at the nuclear periphery. (**b**) In order to account for nuclear shape fluctuations, fluctuating surface models were developed and can result in different chromatin architectures. (**c**) In contrast to these continuum surface models, coarse-grained particle models of chromatin allow for precise one-on-one interactions between lamin and chromatin domains. (**d**) These particle models can also incorporate dynamic lamin, which can result in lamin aggregates in the nuclear interior. Each class of model offers unique insights and serves to answer distinct questions related to the mechanical properties, dynamics and biological significance of the nuclear lamina

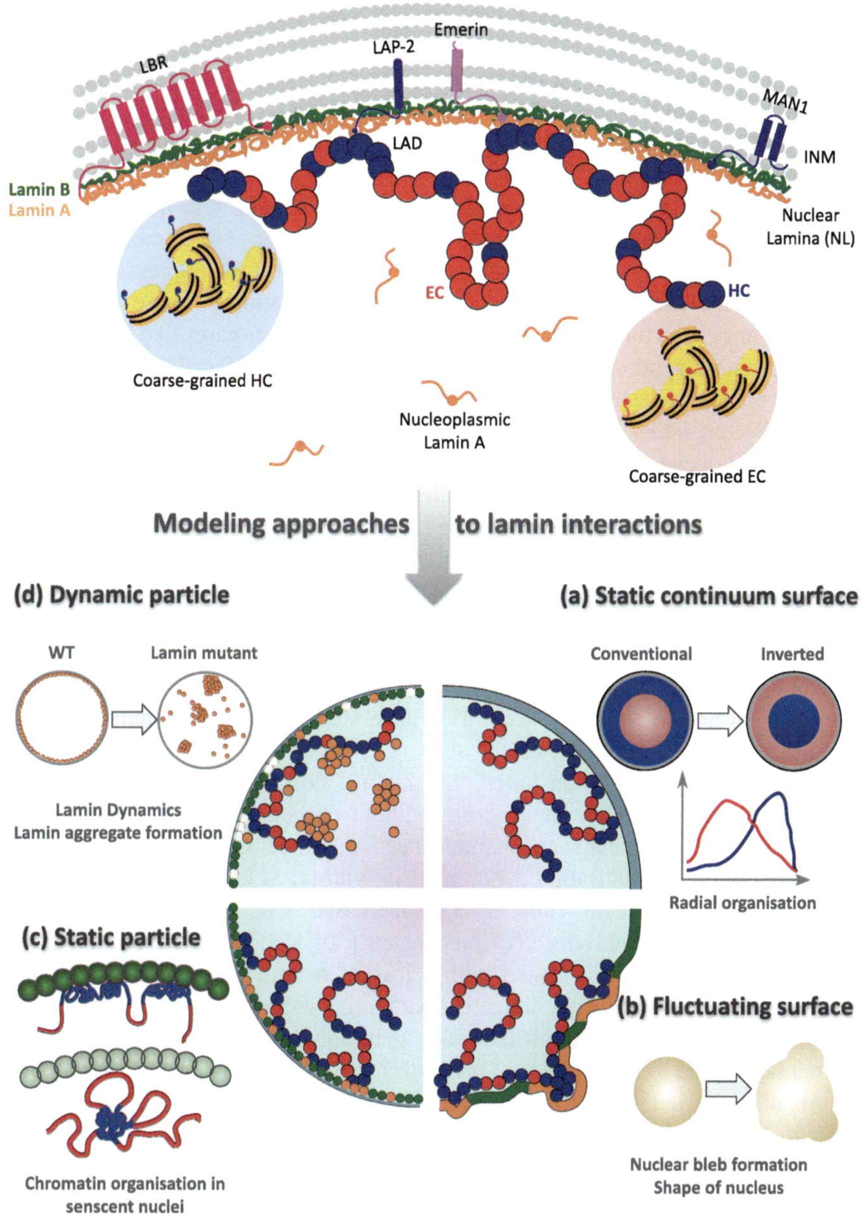


**Fig. 9.2** Modelling approaches to incorporate lamin–chromatin interactions. The top panel shows a schematic illustration of the nuclear periphery, highlighting the organisation of lamin filaments, lamin-binding proteins and chromatin. Lamin filaments form a meshwork at the inner nuclear membrane, interacting with various chromatin regions through protein complexes. These interactions are crucial for the positioning of chromatin at the nuclear periphery and influence nuclear structure and function. The chromatin anchored to the inner membrane creates lamina-associated domain (LAD). The bottom panels show a categorisation of existing models for coarse-grained

repressive chromatin environment [28, 29, 31]. Nearly 40% of the human genome is composed of LADs [31].

The role of the nuclear lamina in this process has been specifically examined through targeted mutations in *LBR* and *LMNA*, helping to distinguish the distinct roles of LBR- and lamin A/C-mediated tethering. Notably, the simultaneous absence of both LBR and lamin A/C ($LBR^{-/-}$, $LMNA^{-/-}$) led to the loss of peripheral heterochromatin and a striking inversion of nuclear architecture, with heterochromatin relocating from the nuclear envelope to the nuclear interior [34]. In most cell types, deletion of lamin A/C alone was compensated by ectopic LBR expression, thereby maintaining the conventional radial chromatin organisation [34]. Several *LMNA* mutations have been linked to a variety of different human diseases like Hutchinson–Gilford progeria syndrome (HGPS), Emery–Dreifuss muscular dystrophy, dilated cardiomyopathy and generalised lipodystrophy, collectively known as laminopathies, characterised by alteration of nuclear stiffness and misshapen/fragile nuclei [35–37]. Pathogenesis in muscular dystrophy and dilated cardiomyopathy caused by mutations in lamin A was found to be linked with altered heterochromatin distribution [38] and transcriptional landscape [39, 40].

Concurrently, nuclear shape is strongly associated with various pathologies, many stemming from *LMNA* mutations (laminopathies). Many cancers also exhibit abnormal nuclear shapes, often characterised by nuclear blebs—protrusions from the nuclear surface [41]. In progeria, blebbed regions are primarily lamin A, with little or no lamin B, and exhibit a larger lamin mesh size [42]. Atomic force microscopy (AFM) and micropipette aspiration have shown that the lamina is a major resistive element protecting nuclear shape [43–45]. However, experiments with MNase-treated nuclei demonstrated that lamins alone cannot maintain nuclear shape; the lamina buckles under stress without chromatin support, highlighting chromatin's role in nuclear shape maintenance [46, 47]. Hi-C contact maps revealed distinct chromatin organisation differences in wild-type, senescent and progeroid cells [48, 49]. Senescent cells showed increased long-range contacts, while progeroid cells exhibited reduced distal contacts [50]. Reduction of lamin B1 levels during senescence was associated with a loss of heterochromatin at the nuclear periphery. FRAP experiments indicated that lamins within the meshwork are relatively stable, suggesting a solid-like rather than liquid-like structure [51, 52]. The lamina was found to exhibit elastic properties, returning to its original shape after deformation [43]. Fast active microrheology has further revealed viscoelastic behaviour, with a time-dependent elastic response [53]. Importantly, *LMNA* mutations altered the lamina's viscoelastic properties, providing insights into how specific mutations affect nuclear mechanics [54].

To integrate these experimental studies into a theoretical understanding, a variety of simulation and computational models have been developed to systematically study the role of lamin proteins in maintaining large-scale genome organisation. Depending on the experimental context, and the relevant scales of the specific features, theoretical modelling has adopted multiple different approaches. In this review, we categorise the broad classes of modelling approaches used in the context of lamin–chromatin interactions and summarise some key results to illustrate the strength of each of these categories. While chromatin modelling in different contexts has a wide and rich literature, we focus in particular on modelling work that seeks to understand the role of lamins in regulating chromatin organisation and how disruptions in these interactions can help understand the pathologies of different disease conditions and lamin mutants.

## Coarse-Grained Polymer Modelling of Chromatin

Chromatin is modelled as a coarse-grained polymer chain (Fig. 9.1), where each bead represents a small genomic segment spanning several kilobase pairs (kbp) of chromatin [7, 55–62]. These beads are categorised into different types to capture the underlying chromatin states—primarily heterochromatin (HC) and euchromatin (EC) [50]. However, depending on the specific goals of the study and the resolution of the chromatin models, the number of constituent chromatin components can be expanded to reflect a more detailed chromatin organisation. For example, heterochromatin can be subdivided into facultative heterochromatin (which is reversible and associated with gene silencing) and constitutive heterochromatin (which is stably repressive and enriched at regions like centromeres and telomeres) [63–65]. Telomeres and centromeres themselves can be modelled as special bead types with distinct properties [63, 66]. Additionally, some beads can represent nucleolus-associated domains (NADs), which are genomic regions interacting with the nucleolus [67]. The classification of the chromatin beads into subtypes typically follows one of two approaches. Polymer physics-inspired models typically construct block-copolymer structures [58], with the different blocks representing different chromatin subtypes. The distribution of blocks within the copolymer structure can be either idealised or based on certain biological inputs [62]. The second approach adopts a data-driven classification scheme for the chromatin subtypes [68] (Fig. 9.1). For example, HC beads can be constructed based on the enrichment of histone modification H3K9me3 and lamin B1 ChIP-seq signals [50], which serve as reliable indicators of heterochromatin regions associated with the nuclear lamina. Incorporating these additional chromatin components significantly increases the complexity of the simulation. It becomes essential to define the nature of interactions between the various bead types, which are not always well characterised in the existing experimental data. The challenge lies not only in quantifying these interactions but also in filtering out the most relevant ones that drive the observed chromatin behaviour. Some interactions might be crucial for specific chromatin conformations, while others could be less significant or even redundant. Identifying the right set of governing interactions requires a careful balance between biological realism and computational feasibility, making this a challenging yet essential task in modelling chromatin architecture.

Another major computational challenge in modelling chromatin systems lies in the sheer size and complexity of the mammalian genome [2]. For polymer simulations of chromatin, simulating the entire genome at a high resolution, such as 10 kbp per bead or higher, becomes computationally prohibitive due to the immense number of beads required [50]. To overcome this computational challenge, there are two common strategies (Fig. 9.1). One practical and feasible approach is to focus on smaller genomic regions that contain a representative mix of heterochromatin (HC) and euchromatin (EC). This allows for simulations that mimic a localised section of the nucleus while maintaining high resolution and biological relevance [50, 69]. In these focused simulations, the chromatin segment is typically placed within a simulation box with periodic boundary conditions in the x- and y-directions, while the z-direction is confined to represent a specific region near the nuclear periphery [50]. This setup provides a realistic environment for studying chromatin organisation at the domain level, such as topologically associating domains (TADs), which are crucial for understanding local chromatin folding and regulatory interactions. At resolutions of <10 kbp per bead, the simulation can capture fine-scale features of chromatin architecture, including TAD boundaries, enhancer–promoter loops and lamina-associated domains (LADs). The second approach, in contrast, is to perform whole-genome simulations at a much coarser resolution, typically 100 kbp or even 1 Mb per bead [67, 70, 71]. While these models cannot capture the detailed organisation at the TAD scale, they are highly valuable for studying large-scale chromatin organisation, such as chromosome territories and the overall spatial positioning of chromosomes within the nucleus [72]. In such cases, chromatin is confined within a specified geometry to accurately reflect the experimentally measured density of intranuclear chromatin, ensuring that the model remains consistent with biological data.

An alternative approach to polymer-based simulations is to investigate large-scale features of chromatin organisation using continuum models of phase separation [73–77] (Fig. 9.1). These continuum models construct a phase-field representation of the nucleus, with field variables corresponding to individual chromosome territories and with a different set of field variables representing chromatin regions—broadly euchromatin and heterochromatin (with some models further subdividing HC into facultative and constitutive subtypes) within a Ginzburg–Landau formalism. This approach allows investigation of phase separation between euchromatin and the different heterochromatin compartments and systematic study of the different interaction parameters that reproduce different heterochromatin domains. However, such phase-field models are unable to capture finer details of chromatin organisation at TAD scales. Thus, it becomes crucial to carefully choose the scale and resolution of the model based on the specific goals of the study [78].

## Coarse-Grained Models of Nuclear Lamins

Structural changes in chromatin associated with different cell types such as rod cells in nocturnal mammals [34], diseases such as Hutchinson–Gilford progeria syndrome (HGPS) and other progeroid syndromes [79] and various laminopathies [80] have motivated the use of modelling to better understand the underlying drivers of these altered organisation. One of the key elements in the global organisation of chromatin is its interaction with the nuclear membrane through the nuclear lamina. Despite its importance, many polymer models of interphase chromosomes have focused predominantly on intra- and interchromosomal interactions, often neglecting the role of lamina-associated constraints in chromatin folding and organisation.

To address this gap, simulations that seek to explicitly understand the role of the lamins in chromatin organisation can provide valuable insights into the organisation principle of genome assembly and also help understand how changes in lamina-mediated interactions contribute to the abnormal chromatin organisation observed in disease states.

## Nuclear Lamina as a Static Continuum Surface

The simplest and most common approach to incorporate lamin–chromatin interactions is to model the nuclear lamina as a fixed surface with which chromatin interacts [55, 63, 67, 69–71, 81–83] (Fig. 9.2). The static surface model is based on the observation that the formation of the lamin meshwork in healthy cells requires a localisation of the lamin near the INM [42], a stable association between the lamins to form uniform polymer-network-like structures [12, 84] and a sufficient lamin concentration to occupy a large fraction of the inner nuclear surface [19]. This model allows quantification of the role of such surface-driven interactions in the spatial organisation of chromatin. Within the static surface model of the NL, there are three main approaches to understand chromatin organisation—mechanistic polymer models that focus on the role of the different interactions [85, 86], data-driven inverse models that aim to reconstruct polymer topologies from experimental data [82, 87, 88] and continuum field theoretic models that focus on large-scale genome organisation [73–75, 77, 89, 90].

Polymer modelling of lamin–chromatin interactions seeks to understand the role of entropy, interaction potentials and geometry in emergent organisation. Heteropolymer models with lamina-interacting and non-interacting beads showed that lamin–chromatin interaction strength influenced LAD length distribution. Simplified models with the appropriate choice of interaction strength and chromatin volume fraction could reproduce experimentally observed LAD length distributions, as well as chromosome territories due to peripheral confinement of heterochromatin beads [83]. Such coarse-grained systems also revealed that flexible chromatin made more stochastic NL contacts, and that tuning the lamina–polymer interaction strength could control polymer adsorption to the nuclear periphery [91]. Fine-grained polymer models at nucleosome resolution (200 bp), which focused on the different properties of non-interacting acetylated (active) and self-interacting non-acetylated (inactive) nucleosomes, revealed that LAD compaction by the NL increased the available volume for interior chromatin, promoting open conformations and active transcription. In the absence of the NL interaction, the lamina-associated TADs de-condensed into the interior, resulting in an overall increase in chromatin compaction, consistent with lamin knockdown experiments in *Drosophila* S2 cells [69]. In contrast, genome-scale simulations require coarse graining at larger scales (>40 kbp or TAD scale) [63, 70, 71]. These simulations can capture chromatin dynamics, complex interphase chromosome diffusion, radial organisation and role of the lamina in stabilising 3D architecture. LAD attachment to the periphery slowed chromatin motion, highlighting the role of lamina interactions in dynamics. While lamina interactions alone did not fully maintain territorial organisation, they significantly delayed chromosome mixing and reduced entanglement [70]. Multicomponent polymer models, incorporating euchromatin (E), facultative heterochromatin (F) and constitutive heterochromatin (C) based on Hi-C data, were able to identify a hierarchy of interaction strengths ($EE \approx EF < EC < FF < FC < CC$) that resulted in the emergence of an inverted nuclei-like structure matching Hi-C data [63] as had been observed in rod nuclei of nocturnal mammals which lack both lamin A/C and lamin B receptors [34, 92]. Introducing attractive interactions between constitutive heterochromatin and lamina restores the conventional nuclear structure, highlighting the role of lamina–heterochromatin interactions in driving radial organisation [63] (Fig. 9.3a).

In contrast to these mechanistic polymer models, which focus explicitly on the role of different interaction strengths and other physical parameters, an alternative way to understand chromatin conformations is provided by data-based inverse modelling [87, 88]. Chrom3D [93–95], a restraint-based whole-genome model with beads representing individual TADs, integrated Hi-C and lamin ChIP-seq data to incorporate distance constraints to generate an ensemble of structures. It preferentially positioned LADs (lamin-associated TADs as identified from ChIP-seq) at the periphery while minimising separation between interacting TADs as identified by Hi-C. In a lamin mutant (R388P), the model predicted an interior shift of LAD regions, potentially linked to altered gene expression [95]. The model's simplicity in principle allows scaling to finer resolutions, enabling detailed study of local chromatin properties [94]. Such inverse modelling, starting from Hi-C data, also allowed the reconstruction of genome organisation in normal and inverted nuclei, although no explicit lamin interactions were taken into account [96, 97]. These inverse models offer a way to understand genome organisation in different contexts; however, they do not offer tunable parameters to deconstruct the role of different physical mechanisms in the resultant structures.

A third approach to study genome organisation is through continuum field theoretic models [73–75, 77, 90]. The inverted chromosomal organisation in rod cells of nocturnal mammals was studied using a phase-field model of the whole nucleus [74, 75, 77]. A 2D two-component (euchromatin/heterochromatin) model showed that heterochromatin-to-euchromatin conversion rate, loss of heterochromatin–lamina affinity and filled nuclear volume were crucial for inverted nuclei formation, while decreased nuclear size and changes in nuclear shape from elliptical to circular were not significant factors [77]. A 2D three-component model [74], which partitions heterochromatin into constitutive and facultative regions, revealed that comparable constitutive and facultative heterochromatin–lamina interactions formed alternating droplet structures at the nuclear envelope and that total heterochromatin volume

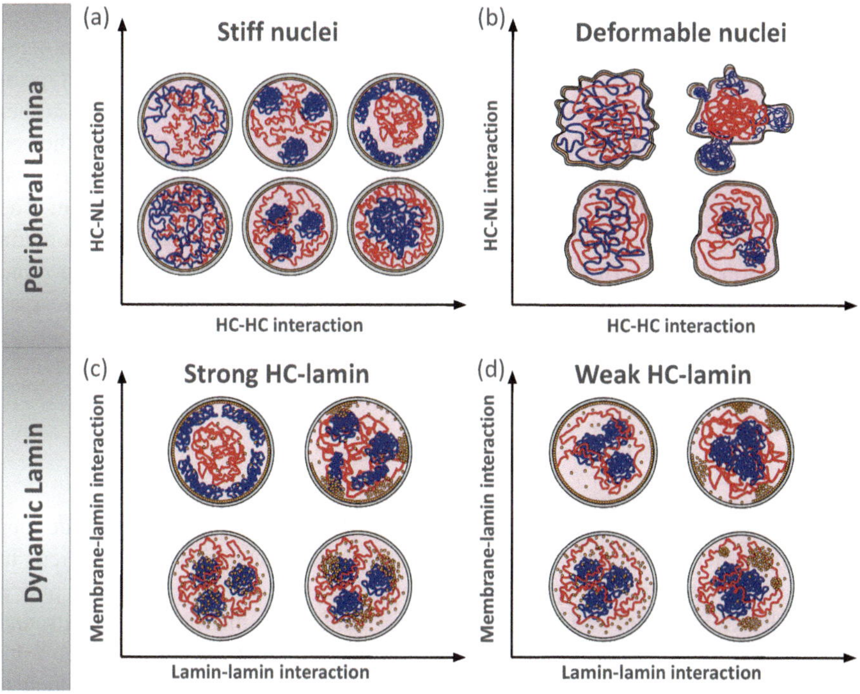


**Fig. 9.3** Representative lamin and chromatin conformations in the plane of different interaction strengths for a peripheral lamina layer (**a, b**) and dynamic lamin (**c, d**). Chromatin conformations in the heterochromatin–heterochromatin (HH) and heterochromatin–nuclear lamina (HL) plane for (**a**) stiff and (**b**) deformable nuclei. For stiff nuclei (**a**), at low HH and HL values, both heterochromatin (HC, blue) and euchromatin (EC, red) remain dispersed throughout the nucleus. As HH increases, compact heterochromatin domains begin to form. At high HL, for low HH, heterochromatin remains less compact and localised near the nuclear lamina, while moderate HH in the presence of HL results in the formation of heterochromatin clusters near the periphery. When both HH and HL are high, heterochromatin blobs adhere to the surface forming a peripheral HC layer. For low HL, or in the absence of HL, these heterochromatin domains remain dispersed through the nucleus for moderate HH values. At high HH interaction strengths (in the absence of HL), these separate clusters merge to form a heterochromatin aggregate in the nuclear core, resulting in an inverted nuclear configuration. For deformable nuclei (**b**), at low HH and HL, both heterochromatin and euchromatin remain dispersed. As HH increases, compact heterochromatin domains form. At low HL, small fluctuations are observed, whereas high HL induces nuclear deformations, resulting in bleb-like protrusions. Chromatin conformations in the lamin–lamin (LL) and nuclear membrane–lamin (ML) plane for (**c**) strong HC–lamin and (**d**) weak HC–lamin interactions. For strong HC–lamin interactions (**c**), at low LL and ML values, lamin remains dispersed throughout the nucleus with increased localisation near heterochromatin regions due to strong HC-lamin interactions. For low LL and high ML, lamin forms a layer on the nuclear membrane, which drives heterochromatin towards the periphery due to HC-lamin. High LL in the absence of ML leads to lamin clusters in the nucleoplasm. When both LL and ML are high, lamin aggregates may form near the periphery, depending on the relative strengths of LL and ML. Due to strong HC-lamin, heterochromatin tends to cluster around lamin aggregates. For weak HC–lamin interactions (**d**), at low LL and ML values, lamin remains dispersed in the nucleus. For low LL and high ML, lamin forms a membrane-associated layer but does not drive heterochromatin towards the periphery due to the absence of HC-lamin. High LL in the absence of ML results in lamin clusters in the nucleoplasm. When both LL and ML are high, lamin aggregates can form near the periphery, again depending on the relative strengths of LL and ML. With low HC-lamin, both heterochromatin and lamin clusters remain isolated

fraction influenced chromatin architecture and the connectedness of the lamin layer at the periphery. Turning on either facultative or constitutive heterochromatin–lamina interactions resulted in a layer of the corresponding heterochromatin subtype at the periphery, but interior droplets persisted due to strong interactions with the other heterochromatin subtype. A 3D whole-genome model [75] in *Drosophila* showed that heterochromatin–lamina (HL) interactions caused the dissolution of individual HC droplets inside chromosome territories and promoted a peripheral HC layer. The heterochromatin–lamina interaction also affected the kinetics of heterochromatin accumulation at the periphery with higher interaction strengths resulting in rapid quenching of heterochromatin dynamics and thus potentially in gene expression variability.

## Fluctuating Surface Models of Lamin

The static surface models [55, 63, 67, 69–71, 81–83] assume a rigid surface defines the nuclear shape and study lamin–chromatin interactions within this milieu. In contrast, the mechanical properties of the nuclear surface could potentially play a role in determining the surface properties and hence also the interaction with and organisation of chromatin. This is especially important in diseases like the Hutchinson–Gilford progeria syndrome (HGPS), which display altered nuclear morphology with bleb-like protrusions [18, 35], along with gross reorganisation of the nuclear lamina into distinct domains and increased mechanical stiffness. The bleb-like structures in progeria are rich in lamin A and largely devoid of lamin B [42]. Modelling nuclear bleb formation computationally requires relaxing the assumption of a rigid nuclear lamina/envelope (Fig. 9.2).

To capture the mechanical properties of the nuclear surface at mesoscopic scales, a continuum approach describes the lamin as a thin elastic shell with stretching and bending rigidities [98]. The lamin has two components, A-type and B-type, and the nuclear morphology is driven by two key parameters: the mesh size area ratio between A-type lamin and B-type lamin regions and the overall fraction of lamin A and lamin B within the nuclear lamina. When the mesh size of A-type lamin regions increases, these regions become less rigid and more prone to deformation. As a result, the A-rich surface becomes highly flexible, facilitating the formation of lobules—protrusions that arise due to the reduced mechanical resistance of the lamina. Interestingly, the number of blebs can be modulated by adjusting the proportion of lamin B. As the lamin B fraction increases, the nuclear surface stabilises, and the number of blebs progressively decreases. This balance between lamin A and lamin B plays a crucial role in determining the structural integrity and morphology of the nuclear envelope. By varying only these two model parameters, this continuum model can qualitatively reproduce experimentally observed nuclear membrane behaviour under different lamin mutation conditions. Specifically, it captures the emergence of nuclear blebs seen in various laminopathies, providing valuable insights into the physical mechanisms underlying these nuclear deformations.

Chromatin and lamins are two major structural components of the nucleus that maintain its shape and function, yet their relative roles in nuclear blebbing remained unclear. An important consideration is the role played by chromatin in influencing the surface properties of the lamina. In a polymeric shell model of the nuclear lamina, buckling of the empty shell was observed under tension, whereas buckling was relatively suppressed when the shell's interior was occupied by a cross-linked polymer tethered internally to the shell [46]. To better understand the role played by chromatin–lamina interactions in nuclear blebbing, different distribution and strengths of tether were studied [99]. Uniform distributions of chromatin tethers (mimicking NL proteins such as emerin, LBR, etc. [24] led to surface deformations but failed to form distinct blebs. Varying tethering stiffness and density resulted in creases and crumples rather than blebs. Two key factors to observe the formation of blebs were the formation of lamin domains [100], which were bounded by domain walls of low bending stiffness, and localisations of chromatin tethers at relatively less stiff lamin domain boundaries [99]. As these tethers pulled the boundaries inwards to create the necessary folds, simulations qualitatively recovered nuclear shapes with blebs similar to experimental observations [99]. Chromatin density has also been implicated in controlling shape fluctuations, with higher chromatin density leading to a stable lamina [66]. Interestingly, simulations pointed to the conclusion that nuclei with a conventional chromatin organisation consistently exhibit greater shape fluctuations compared to those with inverted chromatin arrangements [66]. Similarly, self-interactions between heterochromatin domains were found to significantly influence nuclear shape. When chromatin near the nuclear periphery became compact or condensed, it overpowered the elastic strength of the lamina, leading to disruptions in the morphology of softer nuclei, which may manifest as 'bleb-like' protrusions [66] (Fig. 9.3b), although those heterochromatin-filled blebs were structurally different from the experimentally observed euchromatin-rich blebs. Collectively, such studies involving model polymers along with a deformable lamina highlighted the intricate balance between various mesoscale interactions, chromatin–lamina interactions, distributions of lamin A/C versus B and chromatin density required to maintain a stable nuclear shape.

## Nuclear Lamina as an Assembly of Static Particles

Transient associations occur between lamins and the nuclear membrane due to interactions of lamin globular domains and their direct binding to the nuclear membrane (NM). Experimental evidence supports the idea that lamins are heterogeneous, contradicting the traditional view of the nuclear lamina as a uniform, homogeneous surface. A more realistic and effective model is to treat the lamina as a peripheral assembly composed of coarse-grained lamin particles positioned beneath the nuclear membrane (Fig. 9.2). This model allows for a more accurate modelling of the one-to-one interactions between the lamin and chromatin [50, 67, 101]. Additionally, this approach can incorporate the heterogeneity of lamins or polarity of lamina [102] by incorporating different types of coarse-grained particles and differing interactions with chromatin, although this remains unexplored.

Such a coarse-grained model of lamin was used to understand the aberrant chromosomal organisation in senescent and progeroid cell lines based on the mutual interaction between three primary components—euchromatin (EC), heterochromatin (HC) and nuclear lamina (NL) [50]. The behaviour of this model is primarily governed by two key interactions: HH, representing heterochromatin self-association strength, and HL, which defines the interaction strength between heterochromatin and the nuclear lamina (Fig. 9.3a). The phase space of these two parameters allows for the characterisation of chromatin organisation across different cellular states, including normal, senescent and diseased cells such as those observed in progeria. When both HL and HH are low, heterochromatin remains relatively open and loosely organised, with minimal attachment to the lamina. This configuration is reminiscent of the chromatin organisation observed in progeroid cells. In contrast, when HH is high while HL remains low, the heterochromatin collapses into dense clusters due to strong self-interactions, yet it remains largely detached from the lamina. This pattern resembles the chromatin organisation seen in senescent cells, where heterochromatin compaction increases, but peripheral anchoring is diminished, leading to structural abnormalities in the nucleus. When both HH and HL are sufficiently high, the heterochromatin forms dense clusters that are strongly tethered to the lamina, resulting in the characteristic peripheral organisation observed in healthy, growing cells. This state represents the normal chromatin configuration, where heterochromatin is compacted and anchored at the nuclear periphery, maintaining proper nuclear architecture and genome stability.

These results are broadly consistent with multi-chromosome, multi-component simulations [63] while enabling exploration of localised chromatin changes at smaller resolutions. The model also investigates chromatin organisation reversibility. While reverting interaction parameters to those of a growing nucleus restores chromatin organisation in simulations, complete wild-type recovery often requires higher interaction energies. This hysteresis, indicative of a first-order phase transition, suggests a metastable senescent state resisting full restoration, even under healthy conditions, offering a physical mechanism for senescent cell stability [50].

## Lamin as Dynamic Coarse-Grained Beads

As was evident from the previous sections, the key focus on modelling lamin–chromatin interactions had been the role of lamin in pinning heterochromatin domains to the nuclear envelope—and the combination of experimental studies and theoretical modelling demonstrated that these attractive interactions were a key driver of heterochromatin positioning and large-scale chromosome organisation. The initial assumption in these models has been the peripheral organisation of the lamin at the nuclear envelope, a hallmark of normal nuclei. However, experiments on lamin mutants have shown that in a wide variety of mutants [103–105], this peripheral organisation of lamin is itself disrupted, leading to changes in chromatin organisation [38, 106–109]. A recent work on cardiomyopathic mutations of lamin A, K97E and E161K [110–113] showed that there was a disruption in the peripheral organisation of lamin in these mutants, with large-scale lamin aggregates in the nuclear interior [114].

Particle-based coarse-grained models [50, 67, 101] of lamin offer a natural pathway to investigate the role of both peripheral and non-peripheral lamin in chromatin organisation. While many existing particle models constrained the lamin to form a peripheral layer at the nuclear envelope, this constraint could be relaxed to introduce lamin dynamics and study the organisation in terms of an effective interaction between lamin and the nuclear membrane (ML) along with a lamin–lamin (LL) interaction [114] (Fig. 9.2). Large ML interactions, together with weak LL and moderate heterochromatin–lamin interactions, reproduced the wild-type chromosomal organisation, with a uniform peripheral lamin layer and a radial organisation of heterochromatin preferentially located at the periphery. Disrupting the ML interaction with heightened LL interaction resulted in a disruption of the peripheral lamin organisation, with the simultaneous formation of lamin aggregates in the nucleoplasm (Fig. 9.3c, d). The HC-lamin interaction mediated the heterochromatin organisation around these lamin aggregates—for strong HC-lamin, HC domains colocalised with lamin aggregates (Fig. 9.3c), while, for weak HC-lamin, lamin aggregates and HC clusters remain isolated (Fig. 9.3d). This model was able to capture the altered heterochromatin positioning in two different *LMNA* mutants within the phase space of the lamin-lamin and HC-lamin interaction strengths, for weak ML (K97E) and moderate ML (E161K) strengths [114].

This new approach, which introduces the role of non-peripheral lamin, also offers a pathway to understand different possible structures in a variety of lamin mutants based on the relative interaction strengths of the three components—heterochromatin, lamin and the nuclear membrane. Tuning these interaction strengths can lead to a variety of structures, with different radial organisation of heterochromatin and lamin, which can result in different transcriptional profiles of the associated genes [114].

## Discussion and Outlook

This review classifies existing models of lamin–chromatin interactions into four distinct approaches. Most simulation studies investigating the role of lamin proteins utilise a static surface model of the nuclear lamina, motivated by the localisation of lamin near the inner nuclear membrane [55, 63, 69–71, 81–83]. These models have provided valuable insights into the role of lamin in maintaining radial genome organisation [55, 63, 71, 83] and how alterations in these lamin-mediated interactions can lead to altered pathologies [69]. However, by construction, these models overlook the flexible nature of the lamina itself [115], necessitating deformable surface models that can better capture surface fluctuations and their resultant impact on chromatin conformations [63, 95]. In addition, the peripheral lamina is not homogeneous [84], and such heterogeneities can influence surface fluctuations and, in turn, chromatin organisation. Further, experiments in lamin mutants point to the formation of multiple lamin aggregates in the nuclear interior, along with a general loss of lamina from the periphery [116]. Taken together, these observations have motivated coarse-grained particle models of lamin [50, 67, 114]. If the lamin beads are static, they allow for an alternative formulation of the static surface model [55, 63, 69–71, 81–83], better representing individual lamin–chromatin interactions. In contrast, dynamic lamin beads [114] allow a more flexible representation of lamin behaviour, capturing roles beyond structural integrity, including lamin aggregate formation in the nucleoplasm [116]. However, these coarse-grained particle-based models increase the computational cost and hence have been primarily used to study local features of chromatin structure rather than whole-genome organisation [114]. Selection of appropriate modelling frameworks is thus guided by the experimental system and the features of the genome organisation being studied.

From the context of chromatin, the two broad classes are bead-spring polymer models [50, 63, 69] and continuum field theoretic models [73–75, 77, 90]. Within bead-spring models, two primary approaches have been employed in the literature. The first and most common approach is mechanistic model [86], which involves starting with the biophysical properties of chromatin and chromatin-binding proteins, using 1D bioinformatics data as input, and then validating the model against available experimental data, including 3D-FISH, Hi-C and LAD distributions [50, 63]. The second approach is data-driven inverse modelling [96], which begins with Hi-C-like contact probability data [117] or other types of 3D experimental information [95] to predict chromatin organisation. Mechanistic models require an explicit description of the lamina, while inverse models can replicate lamina effects without prior assumptions. Hi-C, Micro-C and other conformation capture data-driven polymer modelling enable a detailed exploration of chromatin's 3D organisation, particularly in the context of disease, including changes of enhancer–promoter interactions, and TAD reorganisation in a mutation- and disease-specific manner [96]. However, despite their success in visualising chromatin structures and inferring interactions, inverse approaches have their own limitations. These do not allow tunable interactions, preventing systematic exploration of how gradual changes in certain parameters—such as interaction strength [50] or density [66]—lead to specific chromatin rearrangements or disease states. Since inverse models rely on experimental observations as inputs, these models often fall short of mechanistic models in terms of predictive power [86]. Ideally, a combination of both approaches can help build a comprehensive picture of the role of lamin in determining chromatin conformations in different scenarios. Phase-field models [73–75, 77, 90], which adopt a large-scale view of the whole nucleus, can potentially complement these results, although the differences in scales and assumptions between these models necessitate a careful comparison in order to ensure mutual consistency.

Despite growing research that seeks to understand the role of lamin–chromatin interactions, many facets of this organisation still need extensive investigation. In order to dissect the roles played by lamin proteins in chromatin organisation, further simulations that compare the strength of the lamin interaction with the nuclear membrane as compared to lamin–chromatin interactions can help in understanding heterochromatin domain formation and the adsorption/desorption properties of these domains to the nuclear membrane. While such studies have been carried out for various chromatin organising multivalent proteins [118, 119] and for loop extruding proteins such as cohesin and condensin [4], understanding the role of lamin proteins in these contexts still remains an open area of research. Beyond coarse-grained particle-based models of lamin, there is also a need to take into account the heterogeneous nature of the lamina, comprised of lamin A/C and B subtypes [12], as well as the filamentous nature of the laminar meshwork itself [120]. The laminar meshwork comprises extended filaments, with a high aspect ratio—approximately $14 \pm 2$ nm [19] in thickness and a mean fibre length of $380 \pm 122$ nm [19]—which is not captured within particle-based models. A related aspect is whether a homopolymer model is sufficient to describe the individual lamin filaments. Experimental data show that both lamins A and B share common structural features: a short N-terminal head domain, an $\alpha$-helical central rod domain and a globular tail domain [121]. While a detailed atomic-scale model of lamin, along with chromatin, would be computationally demanding, a coarse-grained approach could still capture important features of the filamentous meshwork. Such polymer models of the lamina itself, along with polymeric chromatin, can yield finer insights into the role of the lamina in chromatin organisation. The distinct roles of lamin A/C versus B in chromatin organisation and function also remain to be explored. While some of the fluctuating surface models of the nuclear lamina incorporate these heterogeneities [98, 99], these models are focused on shape fluctuations, and integrating these insights together with chromatin organisation remains an open problem. Chromatin polymer models have in some cases differentiated constitutive heterochromatin from facultative heterochromatin [63, 66], but the differential interaction of these subtypes with A- and B-type lamins [122, 123] needs to be integrated into polymer models. However, increasing the number of components in the model introduces many inter-component interactions which have not been determined experimentally, significantly complicating the phase space. Choosing the optimal coarse-graining scale, and determining the effective interaction space at this resolution remains a challenging task. Mechanistic polymer models can serve as a test bed for elucidating the different roles of these multiple interactions, potentially helping generate phase diagrams in the space of these possible interactions that can help classify different experimental morphologies [50, 63, 114]. Extensions of phase-field models, which incorporate lamin proteins as a distinct phase, can also play an important role in understanding how lamin self-interactions as well as mutual interactions between lamins and chromatin components can generate different organisations of lamin and consequently chromatin. These models can be particularly useful in the context of laminopathies, as they allow for insights into the system through a polymer physics perspective. Given the vast number of mutations reported in various laminopathies, present experimental studies are often mutation-specific, and such polymer physics-based approaches offer a pathway to understanding these different mutations within the lens of a few specific interactions.

Another major challenge is to integrate available experimental data on the mechanical response of the nuclear lamina into chromatin models. Experiments suggest that the mechanical response of the nucleus is controlled by three different regulating factors—the mechanical properties of the lamin meshwork and in particular the differences in this response for A- and B-type lamins [18], the overall chromatin and nuclear density [46] and the distribution of epigenetic marks along chromatin [124], with flexible and open chromatin preferentially occupying blebbed regions in mutant HGPS nuclei. Further simulations are required to tease apart the contributions of chromatin and the lamin meshwork and to disentangle the causes and effects of this nuclear mechanical response [115]. Both polymer-based mechanical models and potentially phase-field models can help understand different domains of this mechanical behaviour of the nuclei, with detailed polymer models elucidating local responses as compared to phase-field whole-genome models which can characterise large-length-scale behaviour. Similar open questions remain in the context of various laminopathies that display altered lamin distributions, with a loss of peripheral lamin and the formation of nucleoplasmic aggregates [116, 125]. While these have only recently been investigated with mechanistic polymer models, understanding the altered mechanical response in these mutant nuclei [54] and their role in the altered chromatin organisation remains to be understood.

Finally, data-driven inverse modelling has allowed for the reconstruction of polymer topologies in normal as compared to inverted nuclei [96, 97]. While these models do not explicitly take into account lamin interactions, the reconstruction procedure effectively incorporates the role of these lamin–chromatin contacts and reproduces structures consistent with experimental observations. Such data-driven inverse modelling then opens a potential new way to understand chromatin organisation in various laminopathies and could provide a broader framework, allowing chromatin structures to be classified based on standard observables in chromatin polymer modelling or polymer physics. Inverse models can also generate interaction matrices between different chromatin regions and identify altered interactions in mutant nuclei which can then be integrated with more mechanistic models of chromatin organisation [97]. Additionally, such data-driven modelling allows for reasonable estimates of chromatin locus sizes across different regions, such as euchromatin (EC), constitutive heterochromatin (cHC) and facultative heterochromatin (fHC), in a scale-dependent manner [59]. While present models mostly use uniform coarse-grained bead sizes and flexibility across these regions, a recent study highlighted that the heterogeneous flexibility of chromatin can lead to inverted nuclear configurations, where heterochromatin (HC) becomes phase-separated at the centre of the nucleus, even without explicitly assigning affinities between specific regions [126]. This alternative view challenges the conventional idea that heterochromatin interactions drive the establishment of inverted nuclei [63]. Hi-C-driven inverse modelling holds promise in addressing these debates, as it can derive stiffness values from population-averaged data [59], providing insights into chromatin's mechanical properties and their influence on genome organisation.

Despite extensive modelling of lamin-mediated chromatin organisation, the field remains open to improved modelling frameworks that can integrate different approaches and features towards building a comprehensive understanding of the crucial role of lamin proteins in genome organisation, chromosome dynamics and nuclear mechanics.

**Acknowledgement** SD acknowledges fellowship support from the PMRF, MoE, India. MKM acknowledges funding from Core Research Grant by Science and Engineering Research Board, India (Grant number: CRG/2022/008142).

**Competing Interest** The authors have no conflicts of interest to declare that are relevant to the content of this chapter.

## References

1. Cremer T, Cremer M (2010) Chromosome territories. Cold Spring Harb Perspect Biol 2:a003889
2. Misteli T (2020) The self-organizing genome: principles of genome architecture and function. Cell 183:28–45
3. Bonev B, Cavalli G (2016) Organization and function of the 3D genome. Nat Rev Genet 17:661–678
4. Fudenberg G, Imakaev M, Lu C et al (2016) Formation of chromosomal domains by loop extrusion. Cell Rep 15:2038–2049
5. Sanborn AL, Rao SS, Huang S-C et al (2015) Chromatin extrusion explains key features of loop and domain formation in wild-type and engineered genomes. Proc Natl Acad Sci 112:E6456–E6465
6. Rao SS, Huntley MH, Durand NC et al (2014) A 3D map of the human genome at kilobase resolution reveals principles of chromatin looping. Cell 159:1665–1680
7. Rosa A, Everaers R (2008) Structure and dynamics of interphase chromosomes. PLoS Comput Biol 4:e1000153
8. Burke B, Stewart CL (2013) The nuclear lamins: flexibility in function. Nat Rev Mol Cell Biol 14:13–24
9. Gruenbaum Y, Foisner R (2015) Lamins: nuclear intermediate filament proteins with fundamental functions in nuclear mechanics and genome regulation. Annu Rev Biochem 84:131–164
10. Aebi U, Cohn J, Buhle L, Gerace L (1986) The nuclear lamina is a meshwork of intermediate-type filaments. Nature 323:560–564
11. Stuurman N, Heins S, Aebi U (1998) Nuclear lamins: their structure, assembly, and interactions. J Struct Biol 122:42–66
12. Shimi T, Kittisopikul M, Tran J et al (2015) Structural organization of nuclear lamins a, c, B1, and B2 revealed by superresolution microscopy. Mol Biol Cell 26:4075–4086
13. Buxboim A, Kronenberg-Tenga R, Salajkova S et al (2023) Scaffold, mechanics and functions of nuclear lamins. FEBS Lett 597:2791–2805
14. Constantinescu D, Gray HL, Sammak PJ et al (2006) Lamin a/c expression is a marker of mouse and human embryonic stem cell differentiation. Stem Cells 24:177–185
15. Lenz-Böhme B, Wismar J, Fuchs S et al (1997) Insertional mutation of the drosophila nuclear Lamin Dm0 gene results in defective nuclear envelopes, clustering of nuclear pore complexes, and accumulation of annulate lamellae. J Cell Biol 137:1001–1016

16. Kittisopikul M, Shimi T, Tatli M et al (2021) Computational analyses reveal spatial relationships between nuclear pore complexes and specific lamins. J Cell Biol 220:e202007082
17. Turgay Y, Medalia O (2017) The structure of Lamin filaments in somatic cells as revealed by cryo-electron tomography. Nucleus 8:475–481
18. Shimi T, Pfleghaar K, Kojima S et al (2008) The a-and b-type nuclear Lamin networks: microdomains involved in chromatin organization and transcription. Genes Dev 22:3409–3421
19. Turgay Y, Eibauer M, Goldman AE et al (2017) The molecular architecture of lamins in somatic cells. Nature 543:261–264
20. Wilson KL, Foisner R (2010) Lamin-binding proteins. Cold Spring Harb Perspect Biol 2:a000554
21. Olins AL, Rhodes G, Welch DBM et al (2010) Lamin b receptor: multi-tasking at the nuclear envelope. Nucleus 1:53–70
22. Demmerle J, Koch AJ, Holaska JM (2012) The nuclear envelope protein emerin binds directly to histone deacetylase 3 (HDAC3) and activates HDAC3 activity. J Biol Chem 287:22080–22088
23. Holaska JM, Wilson KL (2007) An emerin "proteome": purification of distinct emerin-containing complexes from HeLa cells suggests molecular basis for diverse roles including gene regulation, mRNA splicing, signaling, mechanosensing, and nuclear architecture. Biochemistry 46:8897–8908
24. Manzo SG, Dauban L, van Steensel B (2022) Lamina-associated domains: tethers and looseners. Curr Opin Cell Biol 74:80–87
25. Ranade D, Pradhan R, Jayakrishnan M et al (2019) Lamin a/c and emerin depletion impacts chromatin organization and dynamics in the interphase nucleus. BMC Mol Cell Biol 20:1–20
26. Ye Q, Worman HJ (1996) Interaction between an integral protein of the nuclear envelope inner membrane and human chromodomain proteins homologous to drosophila HP1. J Biol Chem 271:14653–14656
27. Cook PR, Marenduzzo D (2009) Entropic organization of interphase chromosomes. J Cell Biol 186:825–834
28. Solovei I, Thanisch K, Feodorova Y (2016) How to rule the nucleus: divide et impera. Curr Opin Cell Biol 40:47–59
29. Van Steensel B, Belmont AS (2017) Lamina-associated domains: links with chromosome architecture, heterochromatin, and gene repression. Cell 169:780–791
30. Bizhanova A, Kaufman PD (2021) Close to the edge: heterochromatin at the nucleolar and nuclear peripheries. Biochimi et Biophys Acta (BBA)-Gene Regul Mech 1864:194666
31. Guelen L, Pagie L, Brasset E et al (2008) Domain organization of human chromosomes revealed by mapping of nuclear lamina interactions. Nature 453:948–951
32. Kind J, van Steensel B (2014) Stochastic genome-nuclear lamina interactions: modulating roles of Lamin a and BAF. Nucleus 5:124–130
33. Briand N, Collas P (2020) Lamina-associated domains: peripheral matters and internal affairs. Genome Biol 21:1–25
34. Solovei I, Wang AS, Thanisch K et al (2013) LBR and Lamin a/c sequentially tether peripheral heterochromatin and inversely regulate differentiation. Cell 152:584–598
35. Capell BC, Collins FS (2006) Human laminopathies: nuclei gone genetically awry. Nat Rev Genet 7:940–952
36. Maraldi NM, Lattanzi G, Capanni C et al (2006) Laminopathies: a chromatin affair. Adv Enzym Regul 46:33–49
37. Mounkes L, Kozlov S, Burke B, Stewart CL (2003) The laminopathies: nuclear structure meets disease. Curr Opin Genet Dev 13:223–230
38. Bianchi A, Mozzetta C, Pegoli G et al (2020) Dysfunctional polycomb transcriptional repression contributes to Lamin a/c–dependent muscular dystrophy. J Clin Invest 130:2408–2421
39. Bertero A, Rosa-Garrido M (2021) Three-dimensional chromatin organization in cardiac development and disease. J Mol Cell Cardiol 151:89–105

40. Tiwari V, Alam MJ, Bhatia M et al (2024) The structure and function of Lamin a/c: special focus on cardiomyopathy and therapeutic interventions. Life Sci 122489:122489
41. Zink D, Fischer AH, Nickerson JA (2004) Nuclear structure in cancer cells. Nat Rev Cancer 4:677–687
42. Goldman RD, Shumaker DK, Erdos MR et al (2004) Accumulation of mutant Lamin a causes progressive changes in nuclear architecture in Hutchinson–Gilford progeria syndrome. Proc Natl Acad Sci 101:8963–8968
43. Dahl KN, Kahn SM, Wilson KL, Discher DE (2004) The nuclear envelope lamina network has elasticity and a compressibility limit suggestive of a molecular shock absorber. J Cell Sci 117:4779–4786
44. Rowat AC, Foster LJ, Nielsen MM et al (2005) Characterization of the elastic properties of the nuclear envelope. J R Soc Interface 2:63–69
45. Sapra KT, Qin Z, Dubrovsky-Gaupp A et al (2020) Nonlinear mechanics of Lamin filaments and the meshwork topology build an emergent nuclear lamina. Nat Commun 11:6205
46. Banigan EJ, Stephens AD, Marko JF (2017) Mechanics and buckling of biopolymeric shells and cell nuclei. Biophys J 113:1654–1663
47. Stephens AD, Banigan EJ, Adam SA et al (2017) Chromatin and Lamin a determine two different mechanical response regimes of the cell nucleus. Mol Biol Cell 28:1984–1996
48. Chandra T, Ewels PA, Schoenfelder S et al (2015) Global reorganization of the nuclear landscape in senescent cells. Cell Rep 10:471–483
49. McCord RP, Nazario-Toole A, Zhang H et al (2013) Correlated alterations in genome organization, histone methylation, and DNA–Lamin a/c interactions in Hutchinson-Gilford progeria syndrome. Genome Res 23:260–269
50. Chiang M, Michieletto D, Brackley CA et al (2019) Polymer modeling predicts chromosome reorganization in senescence. Cell Rep 28:3212–3223
51. Broers JL, Machiels BM, van Eys GJ et al (1999) Dynamics of the nuclear lamina as monitored by GFP-tagged a-type lamins. J Cell Sci 112:3463–3475
52. Moir RD, Yoon M, Khuon S, Goldman RD (2000) Nuclear lamins a and B1: different pathways of assembly during nuclear envelope formation in living cells. J Cell Biol 151:1155–1168
53. Dahl KN, Engler AJ, Pajerowski JD, Discher DE (2005) Power-law rheology of isolated nuclei with deformation mapping of nuclear substructures. Biophys J 89:2855–2864
54. Mukherjee C, Kundu A, Dey R et al (2021) Active microrheology using pulsed optical tweezers to probe viscoelasticity of Lamin a. Soft Matter 17:6787–6796
55. Bajpai G, Pavlov DA, Lorber D et al (2021) Mesoscale phase separation of chromatin in the nucleus. elife 10
56. Barbieri M, Chotalia M, Fraser J et al (2012) Complexity of chromatin folding is captured by the strings and binders switch model. Proc Natl Acad Sci 109:16173–16178
57. Di Pierro M, Zhang B, Aiden EL et al (2016) Transferable model for chromosome architecture. Proc Natl Acad Sci 113:12168–12173
58. Jost D, Carrivain P, Cavalli G, Vaillant C (2014) Modeling epigenome folding: formation and dynamics of topologically associated chromatin domains. Nucleic Acids Res 42:9553–9561
59. Kadam S, Kumari K, Manivannan V et al (2023) Predicting scale-dependent chromatin polymer properties from systematic coarse-graining. Nat Commun 14:4108
60. Mirny LA (2011) The fractal globule as a model of chromatin architecture in the cell. Chromosom Res 19:37–51
61. Parmar JJ, Woringer M, Zimmer C (2019) How the genome folds: the biophysics of four-dimensional chromatin organization. Annu Rev Biophys 48:231–253
62. Shi G, Liu L, Hyeon C, Thirumalai D (2018) Interphase human chromosome exhibits out of equilibrium glassy dynamics. Nat Commun 9:3161
63. Falk M, Feodorova Y, Naumova N et al (2019) Heterochromatin drives compartmentalization of inverted and conventional nuclei. Nature 570:395–399
64. Saksouk N, Simboeck E, Déjardin J (2015) Constitutive heterochromatin formation and transcription in mammals. Epigenetics Chromatin 8:1–17

65. Trojer P, Reinberg D (2007) Facultative heterochromatin: is there a distinctive molecular signature? Mol Cell 28:1–13
66. Attar AG, Paturej J, Banigan EJ, Erbaş A (2024) Chromatin phase separation and nuclear shape fluctuations are correlated in a polymer model of the nucleus. Nucleus 15:2351957
67. Qi Y, Zhang B (2021) Chromatin network retards nucleoli coalescence. Nat Commun 12:6824
68. Chiang M, Brackley CA, Naughton C et al (2024) Genome-wide chromosome architecture prediction reveals biophysical principles underlying gene structure. Cell Genomics 4:100698
69. Ulianov SV, Khrameeva EE, Gavrilov AA et al (2016) Active chromatin and transcription play a key role in chromosome partitioning into topologically associating domains. Genome Res 26:70–84
70. Kinney NA, Sharakhov IV, Onufriev AV (2018) Chromosome–nuclear envelope attachments affect interphase chromosome territories and entanglement. Epigenetics Chromatin 11:1–18
71. Tolokh IS, Kinney NA, Sharakhov IV, Onufriev AV (2023) Strong interactions between highly dynamic lamina-associated domains and the nuclear envelope stabilize the 3D architecture of drosophila interphase chromatin. Epigenetics Chromatin 16:21
72. Fujishiro S, Sasai M (2024) Three-dimensional simulation of whole-genome structuring through the transition from anaphase to interphase. In: Computational methods for 3D genome analysis. Springer, pp 293–308
73. Cheng Q, Delafrouz P, Liang J et al (2022) Modeling and simulation of cell nuclear architecture reorganization process. J Comput Phys 449:110808
74. Laghmach R, Di Pierro M, Potoyan DA (2021) The interplay of chromatin phase separation and lamina interactions in nuclear organization. Biophys J 120:5005–5017
75. Laghmach R, Di Pierro M, Potoyan DA (2024) Four-dimensional mesoscale liquid model of nucleus resolves chromatin's radial organization. PRX life 2:013006
76. Laghmach R, Potoyan DA (2020) Liquid–liquid phase separation driven compartmentalization of reactive nucleoplasm. Phys Biol 18:015001
77. Lee SS, Tashiro S, Awazu A, Kobayashi R (2017) A new application of the phase-field method for understanding the mechanisms of nuclear architecture reorganization. J Math Biol 74:333–354
78. Chiariello AM, Abraham A, Bianco S et al (2024) Multiscale modelling of chromatin 4D organization in SARS-CoV-2 infected cells. Nat Commun 15:4014
79. Gordon LB, Rothman FG, López-Otín C, Misteli T (2014) Progeria: a paradigm for translational medicine. Cell 156:400–407
80. Worman HJ, Bonne G (2007) "Laminopathies": a wide spectrum of human diseases. Exp Cell Res 313:2121–2133
81. Annaël Brunet ND, Collas P (2021) Physical constraints in polymer modeling of chromatin associations with the nuclear periphery at kilobase scale. Nucleus 12:6–20
82. Korsak S, Banecki K, Plewczynski D (2024) Multiscale molecular modeling of chromatin with MultiMM: from nucleosomes to the whole genome. Comput Struct Biotechnol J 23:3537–3548
83. Maji A, Ahmed JA, Roy S et al (2020) A Lamin-associated chromatin model for chromosome organization. Biophys J 118:3041–3050
84. Nmezi B, Xu J, Fu R et al (2019) Concentric organization of a-and b-type lamins predicts their distinct roles in the spatial organization and stability of the nuclear lamina. Proc Natl Acad Sci 116:4307–4315
85. Brackey CA, Marenduzzo D, Gilbert N (2020) Mechanistic modeling of chromatin folding to understand function. Nat Methods 17:767–775
86. Chiang M, Brackley CA, Marenduzzo D, Gilbert N (2022) Predicting genome organisation and function with mechanistic modelling. Trends Genet 38:364–378
87. Imakaev MV, Fudenberg G, Mirny LA (2015) Modeling chromosomes: beyond pretty pictures. FEBS Lett 589:3031–3036

88. Rosa A, Zimmer C (2014) Computational models of large-scale genome architecture. Int Rev Cell Mol Biol 307:275–349
89. Laghmach R, Di Pierro M, Potoyan D (2022) A liquid state perspective on dynamics of chromatin compartments. Front Mol Biosci 8:781981
90. Laghmach R, Di Pierro M, Potoyan DA (2020) Mesoscale liquid model of chromatin recapitulates nuclear order of eukaryotes. Biophys J 118:2130–2140
91. Brunet A, Destainville N, Collas P (2021) Physical constraints in polymer modeling of chromatin associations with the nuclear periphery at kilobase scale. Nucleus 12:6–20
92. Feodorova Y, Falk M, Mirny LA, Solovei I (2020) Viewing nuclear architecture through the eyes of nocturnal mammals. Trends Cell Biol 30:276–289
93. Paulsen J, Collas P (2022) Modeling the 3D genome using hi-c and nuclear Lamin-genome contacts. In: Hi-C data analysis: methods and protocols, pp 337–352
94. Paulsen J, Liyakat Ali TM, Collas P (2018) Computational 3D genome modeling using Chrom3D. Nat Protoc 13:1137–1152
95. Paulsen J, Sekelja M, Oldenburg AR et al (2017) Chrom3D: three-dimensional genome modeling from hi-c and nuclear Lamin-genome contacts. Genome Biol 18:1–15
96. Shi C, Liu L, Hyeon C (2024) Hi-c-guided many-polymer model to decipher 3D genome organization. Biophys J 123:2574–2583
97. Shin S, Shi G, Thirumalai D (2023) From effective interactions extracted using hi-c data to chromosome structures in conventional and inverted nuclei. PRX Life 1:013010
98. Funkhouser CM, Sknepnek R, Shimi T et al (2013) Mechanical model of blebbing in nuclear Lamin meshworks. Proc Natl Acad Sci 110:3248–3253
99. Lionetti MC, Bonfanti S, Fumagalli MR et al (2020) Chromatin and cytoskeletal tethering determine nuclear morphology in progerin-expressing cells. Biophys J 118:2319–2332
100. Dahl KN, Scaffidi P, Islam MF et al (2006) Distinct structural and mechanical properties of the nuclear lamina in Hutchinson–Gilford progeria syndrome. Proc Natl Acad Sci 103:10271–10276
101. Kamat K, Lao Z, Qi Y et al (2023) Compartmentalization with nuclear landmarks yields random, yet precise, genome organization. Biophys J 122:1376–1389
102. Brahmachari S, Contessoto VG, Di Pierro M, Onuchic JN (2022) Shaping the genome via lengthwise compaction, phase separation, and lamina adhesion. Nucleic Acids Res 50:4258–4271
103. Evangelisti C, Bernasconi P, Cavalcante P et al (2015) Modulation of TGFbeta 2 levels by Lamin a in U2-OS osteoblast-like cells: understanding the osteolytic process triggered by altered lamins. Oncotarget 6:7424
104. Mukherjee C, Sengupta D, Maganti L et al (2022) Slower diffusion and anomalous association of R453W Lamin a protein alter nuclear architecture in AD-EDMD. RSC Adv 12:32129–32141
105. Sengupta D, Sengupta K (2024) Lamin a K97E leads to NF-b-mediated dysfunction of inflammatory responses in dilated cardiomyopathy. Biol Cell 116:e2300094
106. Briand N, Collas P (2018) Laminopathy-causing Lamin a mutations reconfigure lamina-associated domains and local spatial chromatin conformation. Nucleus 9:216–226
107. Dechat T, Pfleghaar K, Sengupta K et al (2008) Nuclear lamins: major factors in the structural organization and function of the nucleus and chromatin. Genes Dev 22:832–853
108. Puckelwartz MJ, Depreux FF, McNally EM (2011) Gene expression, chromosome position and Lamin a/c mutations. Nucleus 2:14342–14329
109. Taimen P, Pfleghaar K, Shimi T et al (2009) A progeria mutation reveals functions for Lamin a in nuclear assembly, architecture, and chromosome organization. Proc Natl Acad Sci 106:20788–20793
110. Arbustini E, Pilotto A, Grasso M et al (2009) Novel human pathological mutations. Gene symbol: LMNA. Disease: cardiomyopathy, dilated with conduction defects. Hum Genet 125:350–350

111. Ostlund C, Bonne G, Schwartz K, Worman HJ (2001) Properties of Lamin a mutants found in emery-dreifuss muscular dystrophy, cardiomyopathy and dunnigan-type partial lipodystrophy. J Cell Sci 114:4435–4445
112. Raharjo WH, Enarson P, Sullivan T et al (2001) Nuclear envelope defects associated with LMNA mutations cause dilated cardiomyopathy and emery-dreifuss muscular dystrophy. J Cell Sci 114:4447–4457
113. Song K, Dubé M-P, Lim J et al (2007) Lamin a/c mutations associated with familial and sporadic cases of dilated cardiomyopathy in koreans. Exp Mol Med 39:114–120
114. Nath S, Dutta S, Sarkar SD, Sengupta D, Mitra MK, Sengupta K (2025) Liquid–liquid phase separation of lamin drives altered chromatin organization in cardiomyopathic mutations of lamin A. Nucleic Acids Res 53(14):gkaf615
115. Stephens AD, Banigan EJ, Marko JF (2018a) Separate roles for chromatin and lamins in nuclear mechanics. Nucleus 9:119–124
116. Bhattacharjee P, Dasgupta D, Sengupta K (2017) DCM associated LMNA mutations cause distortions in lamina structure and assembly. Biochim et Biophys Acta (BBA) Gen Sub 1861:2598–2608
117. Lieberman-Aiden E, Van Berkum NL, Williams L et al (2009) Comprehensive mapping of long-range interactions reveals folding principles of the human genome. Science 326:289–293
118. Ancona M, Brackley CA (2022) Simulating the chromatin-mediated phase separation of model proteins with multiple domains. Biophys J 121:2600–2612
119. Guha S, Mitra MK (2023) Multivalent binding proteins can drive collapse and reswelling of chromatin in confinement. Soft Matter 19:153–163
120. de Leeuw R, Gruenbaum Y, Medalia O (2018) Nuclear lamins: thin filaments with major functions. Trends Cell Biol 28:34–45
121. Fisher DZ, Chaudhary N, Blobel G (1986) cDNA sequencing of nuclear lamins a and c reveals primary and secondary structural homology to intermediate filament proteins. Proc Natl Acad Sci 83:6450–6454
122. Forsberg F, Brunet A, Ali TML, Collas P (2019) Interplay of Lamin a and Lamin b LADs on the radial positioning of chromatin. Nucleus 10:7–20
123. Patil S, Sengupta K (2021) Role of a-and b-type lamins in nuclear structure–function relationships. Biol Cell 113:295–310
124. Stephens AD, Liu PZ, Banigan EJ et al (2018b) Chromatin histone modifications and rigidity affect nuclear morphology independent of lamins. Mol Biol Cell 29:220–233
125. Bhattacharjee P, Banerjee A, Banerjee A et al (2013) Structural alterations of Lamin a protein in dilated cardiomyopathy. Biochemistry 52:4229–4241
126. Girard M, De La Cruz MO, Marko JF, Erbaş A (2024) Heterogeneous flexibility can contribute to chromatin segregation in the cell nucleus. Phys Rev E 110:014403

# Chapter 10
# The Fall of the Armor: Lamin Dysregulation and a Wide Network of Laminopathies

**Hrishee Ray and Tapas K. Kundu**

**Abstract** The nuclear membrane of cells harbors a lamina as a critical structural component beneath itself which serves as a dynamic scaffold that maintains nuclear integrity and cellular homeostasis. The primary component of the nuclear lamina is lamins, a set of type V intermediate filament proteins, which form a meshwork to provide mechanical stability while modulating essential biological processes like DNA replication, chromatin organization, and gene expression via mechanotransduction. The clinical significance of lamins is exemplified by laminopathies, a heterogeneous group of genetic disorders arising from mutations in the genes encoding the proteins (*LMNA* for A-type lamins [A and C], *LMNB1* for lamin B1, and *LMNB2* for lamin B2) that manifest in the form of a variety of pathologies comprising muscular dystrophies, premature aging syndromes, cardiovascular abnormalities, and metabolic aberrations. Laminopathies, though rare, form the basis of many life-threatening conditions that lack potent therapeutic interventions. Lamins have also been shown to be dysregulated in a multitude of cancers, and research has uncovered a diabolical role of lamins in oncogenesis. The understanding of laminopathies and dysregulation of lamins resulting in disorders is critical in developing novel therapeutic strategies through drug repurposing and epigenetic modulation to curb the burden of the diseases.

**Keywords** Nuclear lamina · Lamin dysregulation · Laminopathies · LINC · Muscular dystrophy · EDMD · Cardiomyopathy · Progeria · Lipodystrophy · Charcot-Marie-Tooth disease · Lamins in cancer · Lonafarnib

H. Ray
Transcription and Disease Laboratory, Molecular Biology and Genetics Unit, Jawaharlal Nehru Centre for Advanced Scientific Research, Bengaluru, India

T. K. Kundu (✉)
Transcription and Disease Laboratory, Molecular Biology and Genetics Unit, Jawaharlal Nehru Centre for Advanced Scientific Research, Bengaluru, India

Department of Biochemistry, Tohoku University Graduate School of Medicine, Sendai, Japan
e-mail: tapas@jncasr.ac.in

© The Author(s), under exclusive license to Springer Nature Switzerland AG 2025 
K. Sengupta, A. Muchir (eds.), *Lamins in Genome Organization and Diseases*, Subcellular Biochemistry 115, https://doi.org/10.1007/978-3-032-00537-3_10

# Introduction

## *The Nucleus*

The cell theory formulated by Schleiden, Schwann, and Virchow mentions how cells serve as the structural and functional unit of all life forms [1]. The fundamental controls of cellular activity and behavior rest with the genetic material, i.e., the DNA. Prokaryotic cells do not bind their DNA with a subcellular organelle, while the necessity to protect the genetic material evolved in eukaryotic cells, giving rise to the nucleus. The genetic material of eukaryotes in association with various proteins and other biomolecules is bound by a double membrane that forms an organelle called the "nucleus." [2]

The nucleus is a large, membrane-bound organelle characteristic of eukaryotic cells, serving as the control center of the cell by housing the genetic material and orchestrating essential cellular activities such as expression of genes, replication of DNA, and RNA processing [3]. It occupies almost 10% of the cell volume and comprises the genome, organized into chromatin associated with histones and nonhistone proteins. The nucleus has several structural components that form key building blocks of its structure-function relationship. The *nuclear envelope* is a double membrane system that encloses the nucleus and separates its contents from the cytoplasm. Such a selective barrier is essential to allow for the administered transport of biomolecules such as RNA and proteins across the nuclear envelope through a system of nuclear pores in a bidirectional manner [4, 5]. *Nuclear pores* are complex protein-associated structures shaped like a basket with a sieve that allow passive diffusion of small molecules and ions while necessitating active transport for larger macromolecules, thus maintaining the nuclear environment crucial for processes like transcription and DNA replication [6]. Beneath the inner nuclear membrane (INM) lies the *nuclear lamina* which is a dense protein meshwork that provides mechanical support and aids in organizing the chromatin within the nucleus. The lamina is highly heterogeneous showcasing the involvement of multiple proteins that help in maintaining the structural integrity and architecture of the nucleus [7] (Fig. 10.1a). The *nucleoplasm* is a semifluid matrix filling the nuclear interior. The nucleoplasm supports the various nuclear components and serves as the medium for molecular interactions and critical cell processes [8]. The nucleus houses the chromatin, the complex of DNA and its associated proteins, including histones, nonhistone proteins, and transcription factors. Chromatin primarily exists in two forms: euchromatin, which is loosely packed and transcriptionally active, and heterochromatin, which is densely packed and generally transcriptionally repressed [9]. The organization of chromatin within the nucleus is nonrandom and leads to the formation of different domains depending on their associated functions and transcription. The organization of chromatin and regulation of chromatin structure is essential for gene regulation, as it defines the accessibility of different genes for the transcription machinery [10].

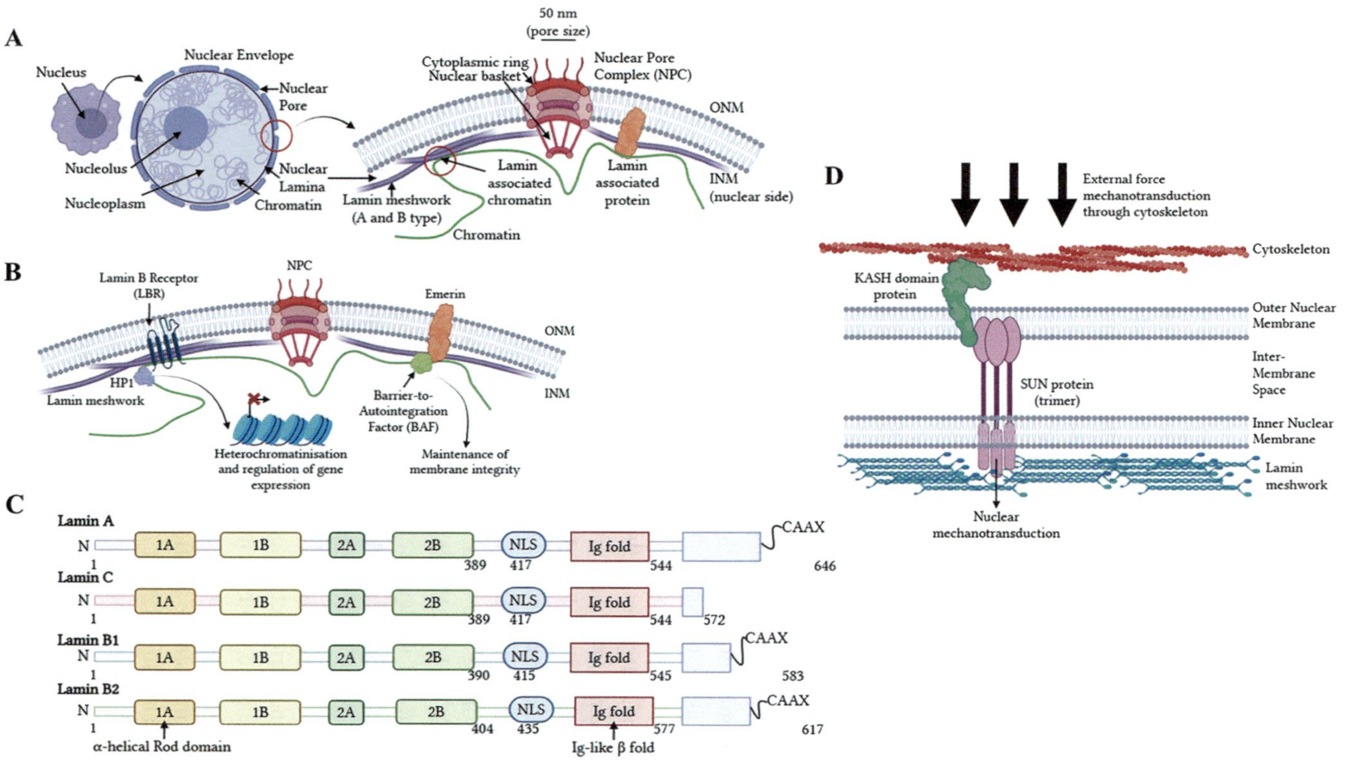


**Fig. 10.1** The nucleus and lamins. Lamins form the laminar meshwork just below the inner nuclear membrane (INM) and form an intricate structure characteristic of the nuclear lamina with associated partners (**a**). Lamins interact with different proteins such as heterochromatin protein 1 (HP1) to modulate gene expression in associated chromatin, while interactions with Emerin and barrier-to-autointegration factor (BAF) aid in maintaining membrane integrity and reforming the nuclear membrane post-mitosis (**b**). Lamins are coded by three genes, *LMNA* (has two isoforms: A and C), *LMNB1*, and *LMNB2*. Their domain structures reveal conserved alpha-helical rod domains and a characteristic Ig-fold besides the CAAX motif at the end of the sequence, essential for posttranslational modifications (**c**). Lamins interact with the cytoskeleton through mediating partners such as SUN proteins or KASH-domain-containing proteins to enable mechanotransduction to the nucleus, hence forming the linker of nucleoskeleton and cytoskeleton complex (LINC) (**d**)

## The Nuclear Lamina

Nuclear lamina is an important structural component of the nucleus, forming a dense fibrillar network beneath the inner nuclear membrane. This 30–100-nm-thick meshwork, composed mostly of intermediate filaments termed lamins and associated proteins, provides mechanical stability to the nucleus while regulating essential biological processes like DNA replication, cell division, and chromatin organization required for controlled transcription [7, 11]. Situated at the nuclear periphery, it anchors nuclear pore complexes (NPCs) that mediate transport between the nucleus and the cytosol, ensuring proper molecular trafficking and selective entry of substances into the nuclear milieu [12]. Beyond its structural role, the lamina acts as a dynamic scaffold influencing gene expression and cellular response to stress.

The nuclear lamina serves a repertoire of diverse functions pivotal to the maintenance of cellular homeostasis. It protects the nuclear shape and resists mechanical stress, preventing deformation during migration or physical strain [13]. During mitosis, lamins get phosphorylated that leads to their disassembly, enabling karyokinesis or the division of the nucleus [14]. The lamina also organizes chromatin by tethering specific genomic regions, therefore playing a part in transcriptional regulation as well. Further, it facilitates DNA repair mechanisms and apoptosis, ensuring genomic integrity and programmed cell death [15, 16]. Through their association with NPCs, the lamina ensures their spatial arrangement, which is critical for efficient nucleocytoplasmic transport.

The proteins primarily involved in building this reticulum are lamins. Lamins are also present in small quantities in the nucleoplasm. Nuclear lamins, which are type V intermediate filament proteins, compose an ornate and dynamic supramolecular scaffold fundamental to regular cell homeostasis [15]. These proteins orchestrate the spatial and functional organization of chromatin, regulate gene expression, maintain nuclear structural integrity, and serve as a key player in mechanotransduction. The lamin isoforms are classified into A-type (lamins A and C) and B-type (lamins B1 and B2) [17]. Each of these isoforms executes discrete and overlapping roles in maintenance of nuclear architecture, genome organization, and mechanotransduction [17, 18]. Besides lamins, proteins like lamina-associated proteins 1 and 2 (LAP1 and LAP2), Emerin, and lamin B receptor (LBR) are involved in forming the structural components of the nuclear lamina [19]. Structurally, the lamina comprises two concentric networks with distinct functions. Lamin B1 forms an outer meshwork near the inner nuclear membrane that is pretty flexible and harbors the ability to adapt to membrane curvature and stabilize nuclear morphology. In contrast, A-type lamins create a stiffer, inner lattice facing the nucleoplasm, contributing to mechanical rigidity [20]. Lamins assemble into dimers through coiled-coil interactions, which polymerize into filaments and further organize into a two-dimensional lattice [21]. This stratified arrangement balances flexibility and strength, allowing the nucleus to withstand mechanical stress while simultaneously maintaining functional plasticity.

The cross talk between lamins and chromatin is paramount to the establishment of nuclear domain organization and transcriptional regulation [22]. Lamins interact with lamina-associated domains (LADs), which are characterized by dense heterochromatic signatures enriched in repressive histone marks [7, 23, 24]. Both the isoforms of lamins contribute to the maintenance of chromosome topology (Fig. 10.1b). Their dysregulation relates to aberrations in chromatin compaction and epigenetic programming. The tethering of euchromatin and heterochromatin governed by lamins modulates gene mobility and polymerase accessibility during transcription. These LADs are large genomic regions (approximately 0.1–10 megabases) tethered to the nuclear periphery with lamins binding matrix attachment regions (MARs) and histones to enforce their repression [25]. Boundaries between LADs and euchromatin are often characterized by CTCF-binding sites (CCCTC-binding factor-binding sites), which ensure spatial segregation of the genetic material [26]. Disruption of the interactions between the lamins and chromatin alters gene expression contributing to senescence and diseases like cancer.

Dysregulation of the nuclear lamins is implicated in numerous pathologies. Mutations in lamin-encoding genes precipitate a heterogeneous spectrum of rare genetic disorders called laminopathies. Laminopathies commonly manifest in the form of musculoskeletal, metabolic, neurodevelopmental, and progeroid pathologies [27, 28].

This chapter illustrates the molecular architecture, cellular functions, pathological disturbances, and emergent therapeutic strategies pertaining to laminopathies. Premature aging disorders such as Hutchinson-Gilford progeria syndrome arise from *LMNA* gene mutations, causing abnormal lamin A processing and nuclear deformities [29]. Neurodegenerative diseases, including amyotrophic lateral sclerosis (ALS), correlate with TDP-43 aggregates that disrupt lamina integrity and nucleocytoplasmic transport [30]. Muscular dystrophies stemming from lamin A/C mutations weaken nuclear stiffness, compromising muscle cell viability [28]. In cancer, lamin dysfunction promotes genomic instability and metastasis by altering chromatin organization and DNA repair [31, 32]. These conditions highlight the role of lamins in maintaining cellular health and the consequences of its dysregulation. Therefore, nuclear lamina is a multifaceted structure integral to nuclear integrity and gene regulation. Its dynamic interplay with chromatin, membrane proteins, and NPCs underscores its importance in cellular function.

## *LINCing the Cytoplasm and Nucleus: Lamins*

Lamins are a subtype of type V intermediate filaments that constitute the primary structural component of the nuclear lamina, a dense meshwork underlying the inner nuclear membrane of metazoan cells. Lamins are broadly categorized into two main types: A and B [15]. This classification is based on sequence homology, biochemical characteristics, and expression patterns. *A-type* lamins, including lamin *A* and lamin *C*, are coded by the *LMNA* gene through alternative splicing mechanisms.

A-type lamin isoforms are usually expressed in differentiated cells and are associated with tissue-specific roles, with significant involvement in the regulation of gene expression. On the other hand, *B-type* lamins, comprising lamin *B1* and lamin *B2*, are encoded by the *LMNB1* and *LMNB2* genes, respectively. B-type lamins show ubiquitous expression in all somatic cells and are integral for cell viability [13, 33, 34]. The type of posttranslational modifications undergone by the two subtypes of lamins further underscores their difference [35]. The genes encoding lamins are highly conserved and exhibit distinct chromosomal localization. The *LMNA* gene is located on chromosome 1 (1q22), while the *LMNB1* gene is located on chromosome 5 (5q23.3) and the *LMNB2* gene in chromosome 19 (19p13.3) in humans [36–38]. These genes are annotated in gene ontology databases as being involved in the maintenance of nuclear structural integrity, chromatin organization, DNA replication, karyokinesis, and the regulation of gene expression. The *LMNA* gene produces multiple isoforms, including lamins A, C, AΔ10, and C2, through alternative splicing [39]. Some of these isoforms are gamete-specific and play a pivotal role in meiotic chromatin organization [40].

Lamins are known to assemble into higher-order structures through a highly regulated multistep process. Lamins get synthesized as soluble monomers in the cytosol and get imported into the nucleus by virtue of their nuclear localization signals (NLS) [41]. Each lamin monomer shows a tripartite structure comprising a central α-helical rod domain flanked by non-helical head and largely unstructured tail domains (Fig. 10.1d). Dimerization occurs through parallel coiled-coil interactions of the rod domains forming stable lamin dimers. These dimers further associate in a head-to-tail fashion to produce polar protofilaments [41, 42]. The formation of higher-order filamentous networks is regulated by lateral interactions between these protofilaments which are stabilized by interactions with nuclear envelope proteins and chromatin. The mature nuclear lamina is therefore a dynamic mesh-like structure with distinct and interconnected networks formed by A-type and B-type lamins [43]. The carboxy-terminal CaaX motif in most lamins (exception: lamin C) is prone to isoprenylation which aids in the proper targeting and assembly of lamins at the inner nuclear membrane [44]. Each of the lamins also possesses an Ig-fold, which is a β-fold sandwich comprising two β-sheets formed from nine β-strands joined by short loops [45].

Lamins play an important part in nucleocytoplasmic communication. This conversation is mediated by the *linker of nucleoskeleton and cytoskeleton complex (LINC)*. The LINC complex is a multiprotein assembly that spans the nucleolemma, and it physically links the lamina to the cytoskeleton. Core components of the LINC complex consist of SUN-domain proteins (SUN1 and SUN2) located in the inner nuclear membrane and KASH-domain proteins (Nesprins) in the outer nuclear membrane [46]. A-type lamins majorly interact directly with SUN proteins, anchoring the LINC complex to the nuclear lamina and the genome [47]. B-type lamins also contribute to the stability of the LINC complex, while A-type lamins are considered the principal nucleoskeletal partners in such contexts. The LINC complex acts as a conduit for mechanical signals between the cytoplasm and the nucleus, conferring upon the cell the ability to sense and respond to mechanical changes in the extracellular environment. It harbors important functions like nuclear positioning, migration, mechanotransduction, and the maintenance of nuclear rigidity. The LINC complex orchestrates changes in gene expression, cell differentiation, and tissue speciation by transmitting cytoskeletal signals to the nuclear lamina and chromatin [46, 48]. Disruption of LINC-lamin interactions results in defects in mechanosensation and aberrant responses to mechanical stress, besides aberrant nuclear shape, and loss of rigid integrity.

Lamins undergo a variety of posttranslational modifications (PTMs), which are crucial for the regulation of lamin function, assembly, and dynamics. The most prominent PTM of lamins is phosphorylation, which occurs at the conserved serine residues flanking the central rod domain [35, 49]. During mitosis, phosphorylation induces conformational changes in lamin that trigger disassembly and nuclear envelope breakdown essential for karyokinesis [50]. In contrast, dephosphorylation during telophase promotes lamin reassembly, thus furthering nuclear envelope reformation. Besides phosphorylation, lamins undergo isoprenylation (farnesylation) at the CAAX motif, carboxymethylation, and proteolytic cleavage. This proteolytic cleavage is notably involved in the maturation of prelamin A to lamin A for lamin maturation. These modifications are essential for proper nuclear envelope targeting and functional maturation of lamins from prelamins. Ubiquitination and SUMOylation of lamins have also been reported, influencing laminar interactions with other nuclear proteins and their turnover [51, 52].

Lamins are therefore multifaceted proteins involved in the regulation of nuclear health, and dysregulations of lamins manifest in the form of complex disorders termed as laminopathies.

## Laminopathies

The clinical significance of lamins is elucidated by a diverse group of genetic disorders collectively termed laminopathies. Laminopathies arise from mutations in lamin-encoding genes, mostly *LMNA*, and manifest as a spectrum of diseases affecting muscles, adipose tissue, peripheral nerves, and the cardiovascular system [53]. These conditions encompass myopathies such as Emery-Dreifuss muscular dystrophy (EDMD) and dilated cardiomyopathy, neurodevelopmental issues characterized by microcephaly, metabolic anomalies such as familial partial lipodystrophy, and premature aging disorders, notably Hutchinson-Gilford progeria syndrome [54]. A-type lamin mutations lead to nuclear fragility, transcriptional dysregulation, and mechanosensitive dysfunctions, engendering tissue-specific pathologies [55]. In contrast, mutations affecting B-type lamins are often embryonic lethal [56]. These observations corroborate the indispensable role of lamins in maintaining cellular viability and organismal development.

Lamin anomalies and spatial mislocalization are recurrently observed in neoplastic contexts. They perturb chromatin accessibility, cell cycle progression, and DNA repair mechanisms. Lamin mutations potentiate genomic instability, aid in epithelial-to-mesenchymal transition (EMT), and augment metastasis in cancers [57]. Although somatic mutations in lamin genes are rare in malignancies, their altered expression profiles exhibit a strong correlation with tumor progression and prognostic strategies [58]. Treatment modalities against laminopathies are under extensive research. While strategies targeting the mutant lamins directly are yet to be found, epigenetic modulators and symptom-controlling drugs are the most common ways to manage laminopathies.

The past two decades have seen researchers identify a plethora of laminopathies and characterize their genetic basis, clinical phenotypes, and molecular pathology. Cellular and animal models have been pivotal in delineating the roles of lamins and associated nuclear envelope proteins and their functions in tissue-specific manifestation of pathologies. Nevertheless, the mechanisms by which these disorders affect specific tissues remain poorly understood. LMNA encodes lamin A and lamin C proteins, which share homology in the primary sequence of their first 566 residues [59]. Lamin A further undergoes posttranslational modifications from its precursor protein prelamin A through a series of enzymatic reactions involving farnesylation (addition of a farnesyl [15-carbon isoprenoid] group to the -CAAX residue in the C-terminal domain of lamins), endoproteolytic cleavage, methylation, and removal of terminal amino acids [35, 59, 60].

The first evidence of a link between nuclear envelope dysfunction and disease was illustrated through EMD gene mutations resulting in X-linked EDMD [61]. Emerin, encoded by the *EMD* gene, is a type 2 integral membrane protein which localizes to the inner surface of the nuclear membrane [62]. Most *EMD* mutations result in absent or mislocalized Emerin, consequently resulting in their loss from the nuclear envelope, further destabilizing the integrity of the structure of the nucleus. Daniela Toniolo and colleagues identified mutations in a novel gene that manifested in the *X-linked Emery-Dreifuss muscular dystrophy (EDMD)* [61]. Two years after this, the transmembrane protein Emerin was found to be localized to the nuclear envelope, specifically the inner nucleolemma. Further, Ketty Schwartz and colleagues reported that mutations in the *LMNA* gene, encoding for both lamins A and C, resulted in autosomal dominant EDMD [63]. These findings expanded the scope of nuclear envelope-associated diseases, including Dunnigan-type familial partial lipodystrophy, mandibuloacral dysplasia type A, axonal periphery neuropathy, and Hutchinson-Gilford progeria syndrome (HGPS). All of the aforementioned conditions are referred to as laminopathies or nuclear envelopathies owing to their linkage to genes encoding nuclear envelope proteins. *LMNA* mutations are linked to at least six diseases affecting skeletal and cardiac muscles [58]. The diseases are autosomal dominant Emery-Dreifuss muscular dystrophy (AD-EDMD), limb girdle muscular dystrophy type 1B (LGMD-1B), dilated cardiomyopathy (DCM), Dunnigan's familial partial lipodystrophy (FPLD), mandibuloacral dysplasia (MAD), and Charcot-Marie-Tooth neuropathy type 2 B1 (CMT2B1) (Fig. 10.2a).

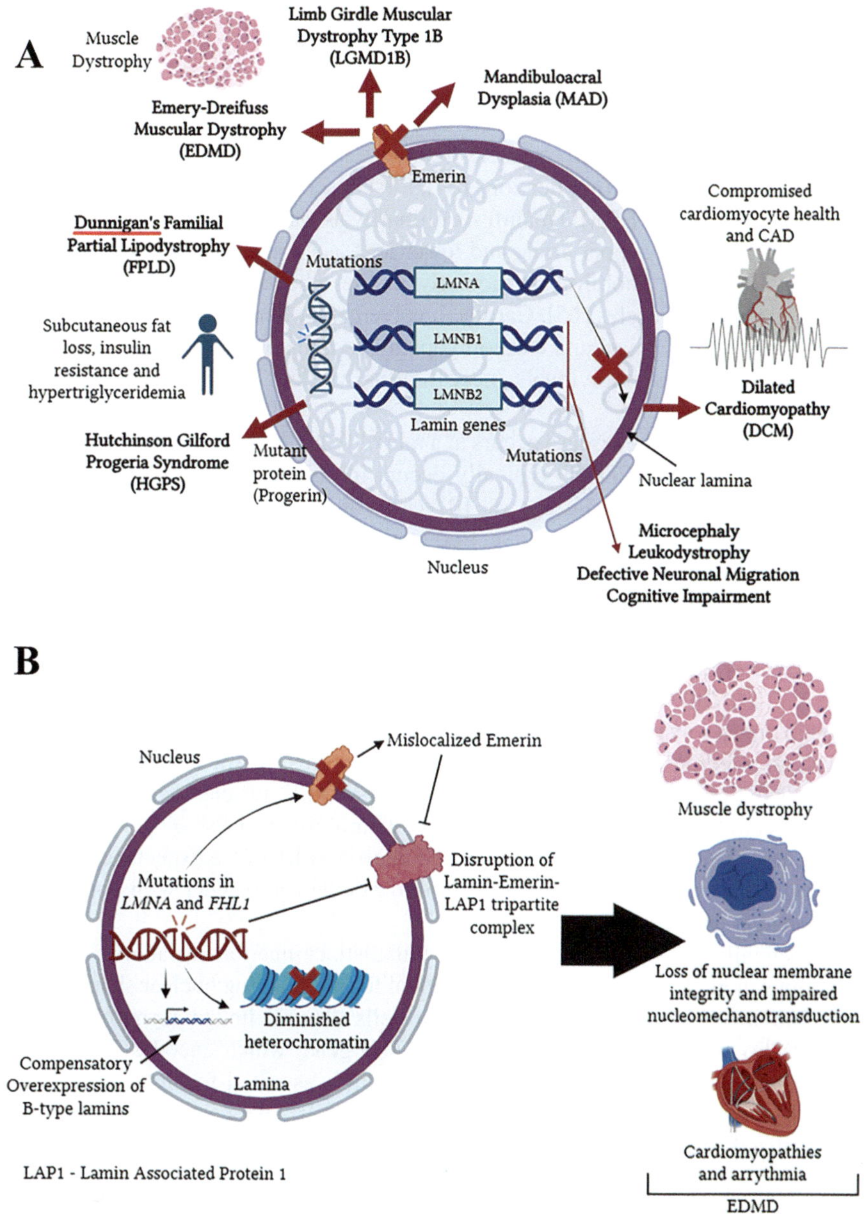


**Fig. 10.2** Laminopathies and their molecular mechanisms. Laminopathies are diseases that occur due to mutations in the lamin-encoding genes, which result in phenotypic abnormalities such as EDMD, FPLD, DCM, MAD, HGPS, LGMD1B, etc. (**a**) The causes of which can be attributed to dysregulated lamins that compromise nuclear membrane integrity and result in impaired nucleomechanotransduction. Emery-Dreifuss muscular dystrophy (EDMD) is the most common laminopathy that has a classical molecular pathology. Mutations in *LMNA* and *FHL1* result in multiple disrupted interactions and Emerin mislocalization that contribute to the disease phenotype (**b**)

## *Emery-Dreifuss Muscular Dystrophy (EDMD)*

Emery-Dreifuss muscular dystrophy (EDMD) is a rare, genetically heterogeneous neuromuscular disorder characterized by a distinctive clinical triad: early-onset joint contractures, progressive muscle weakness with a humero-peroneal distribution, and severe cardiac complications [64]. First described by Alan Emery and Fritz Dreifuss in the 1960s, EDMD manifests primarily as wasting and weakness of muscles in the shoulders, upper arms, and lower legs, accompanied by tender contractors that limit joint mobility, particularly in the elbows, Achilles tendons, and posterior cervical muscles [65]. Cardiac involvement, often manifesting as conduction defects, arrhythmias, and cardiomyopathy, is a major cause of morbidity and mortality in affected individuals, necessitating vigilant cardiac monitoring and intervention [66].

The incidence of the disease occurs at a rate of 0.39 per 100,000 individuals [67]. The clinical onset of EDMD typically occurs in childhood or adolescence, with symptoms appearing by the age of 10 years [68]. Muscle contractures develop unusually early in the disease course, frequently preceding muscle weakness. These contractors result in joint stiffness and deformities such as limited elbow extension and toe walking due to Achilles tendon tightness. Muscle weakness initially affects the humero-peroneal muscles, proximal upper limb muscles like biceps and triceps, and distal lower limb muscles such as the peroneal group while sparing the deltoids early on [69, 70]. Limb girdle muscle dystrophy type 1-B (LGMD 1-B) manifests through similar symptoms but is much milder in degree accompanied by a later onset and affecting only the pelvic and pectoral girdles. As the disease progresses, weakness extends to the pelvic girdle and other proximal muscles. Cardiac manifestations, including atrioventricular block, atrial paralysis, and risk of sudden cardiac death, often become evident by the second decade of life. Life expectancy may be reduced, although many patients survive into middle age or later with appropriate management and treatment [71, 72].

Genetically, EDMD is heterogeneous, with distinct inheritance patterns linked to mutations in nuclear envelope proteins critical for maintaining nuclear shape, integrity, and mechanotransduction in muscle cells. The X-linked form of EDMD (X-EDMD) stems from mutations in the *EMD* gene, which encodes Emerin, an integral inner nuclear membrane protein. Emerin is essential for nuclear envelope stability and chromatin organization. Mutations in *EMD* typically lead to absent or mislocalized Emerin, disrupting nuclear architecture and impairing gene regulation [73]. Autosomal dominant EDMD (AD-EDMD) arises from mutations in the *LMNA* gene, encoding A-type lamins (lamins A and C), which form the nuclear lamina's intermediate filament network [74]. These lamins are crucial for stabilizing the nucleus, organizing the chromatin, and regulating gene expression. Autosomal recessive forms, though rare, have also been described.

At the molecular level, the pathology of EDMD illustrates a disruption of nuclear envelope integrity and mechanotransduction cascades. Loss of Emerin in X-EDMD results in a destabilized nuclear envelope and altered chromatin organization, impairing transcriptional regulation [73]. In AD-EDMD, mutations in *LMNA* lead to defective nucleolemma assembly, nuclear deformation, and aberrant signaling [63, 75]. Studies utilizing cellular models have demonstrated that mutant lamins are expressed ubiquitously but disproportionately affect striated muscle cells due to their heightened sensitivity to mechanical stress. These mutant proteins interfere with nucleo-cytoskeletal coupling and compromise the ability of the nucleus to hold up against mechanical forces, manifesting in progressive muscle degeneration [76].

Single-gene testing facilitates diagnosis, with *EMD* mutations most commonly being implicated in X-linked EDMD, followed by mutations in *FHL1* [77]. *LMNA* mutations predominate in the cases of autosomal dominant or recessive inheritance. Dominant *LMNA* mutations comprise amino acid substitutions, deletions, splicing defects, and haploinsufficiency, all of which disrupt the function of A-type lamins. Experimental models, involving *LMNA*-null mice, exhibit nuclear deformation, impaired mechanotransduction, and defective nuclear positioning [78]. Cryoelectron tomography of fibroblasts harboring *LMNA* mutations reveals increased nuclear surface area, diminished heterochromatin organization, and compensatory upregulation of B-type lamins [79, 80]. Loss of A-type lamins also leads to mislocalized Emerin from the inner nuclear membrane to the endoplasmic reticulum, though some mutant lamin A variants only cause partial mislocalization [81]. Both A-type lamins and Emerin interact with LAP1, another nuclear membrane protein implicated in muscular dystrophies and cardiomyopathies, suggesting a multiprotein complex essential for striated muscle integrity maintenance [82]. The disruption of this lamin-Emerin-LAP1 tripartite complex compromises nuclear rigidity and mechanotransduction cascades. This leads to aberrant muscle function and progressive muscle wasting characteristic of EDMD. The altered nuclear mechanics impair the ability of the muscle cells to respond ably to mechanical stress, ultimately resulting in cellular dysfunction and tissue degeneration [78] (Fig. 10.2b). Such insights into the molecular pathology of EDMD have shed light on the potential therapeutic targets aimed at restoring nuclear envelope stability and mechanosensing.

Clinically, the diagnosis of EDMD is supported by characteristic symptoms and confirmed by genetic testing [83]. Muscle biopsy and imaging may reveal dystrophic changes, but genetic analysis of *EMD* and *LMNA* remains the gold standard [69]. Cardiac evaluation is critical due to the high risk of conduction defects and arrhythmias; a pacemaker or implantable cardioverter-defibrillator (ICD) might be necessary, as noted in [84, 85]. Management of the disease includes physiotherapy to maintain muscle strength and flexibility, surgical interventions to release contractures, and cardiac monitoring to prevent life-threatening complications. Although there are no cures, early diagnosis and multidisciplinary care improve the quality of life and survival of EDMD patients [69].

## *Dilated Cardiomyopathy (DCM)*

Dilated cardiomyopathy (DCM) is a primary myocardial disorder characterized by dilation and aberrant contraction of the left or both ventricles in the absence of abnormal loading conditions or significant coronary artery disease (CAD). The pathological expansion of the ventricular chambers results in systolic dysfunction, ultimately compromising cardiac output. DCM arises from a variety of etiologies, but a significant subset is attributable to genetic mutations, including those affecting the nuclear envelope proteins lamins A and C (A-type lamins), classifying these cases as laminopathies [86–88].

The incidence of DCM is approximately 7 per 100,000 people annually, with a prevalence in around 1 in 2500 individuals, making it the most frequent form of nonischemic cardiomyopathy worldwide [89]. DCM accounts for a substantial proportion of heart failure cases and is a leading cause of heart transplantation. Familial forms constitute up to 50% of cases, and among these mutations in the *LMNA* gene are among the most penetrant and of pathological significance, often associated with early onset of the disease and an elevated risk of arrhythmias and sudden cardiac death [90]. The first clinical identifications of DCM date to the mid-twentieth century, when it was recognized as a distinct entity from ischemic and hypertensive heart disorders. Over subsequent decades, advances in molecular genetics and cardiac imaging facilitated the identification of familial clusters and the recognition of DCM as a genetically heterogeneous disease. The discovery of *LMNA* mutations in the late 1990s and early 2000s established a direct link between nuclear envelope defects and DCM, providing a mechanistic framework for laminopathies [88, 91].

Clinically, DCM manifests with symptoms of progressive heart failure, including exertional dyspnea, fatigue, orthopnea, and peripheral edema. Palpitations and syncope may also occur as a result of ventricular arrhythmias, which are highly prevalent in laminopathy-associated DCM. Physical examination may also reveal displaced cardiac apex, S3 gallop, and signs of volume overload. In advanced stages, patients are at increased risk of thromboembolic events and sudden cardiac death, often prior to overt heart failure symptoms becoming apparent [92].

At the molecular level, laminopathy-associated DCM is caused by mutations in the *LMNA* genes [89–91]. Pathogenic *LMNA* variants typically lead to structural destabilization of the nuclear lamina, resulting in nuclear shape aberrations, impaired gene expression, and increased cellular susceptibility to mechanical stress. In cardiomyocytes, these nuclear defects disrupt the transcriptional regulation of genes involved in contractile function, mitochondrial metabolism, and cell survival [93]. Proteomic and transcriptomic analyses reveal that DCM is associated with altered expression of proteins involved in the tricarboxylic acid cycle (TCA cycle), oxidative phosphorylation, and cardiac muscle contraction [94]. Moreover, *LMNA* mutations are linked to increased nuclear fragility, aberrant mechanotransduction, and enhanced apoptosis, culminating in progressive myocardial dysfunction and fibrosis [86, 95].

## *Dunnigan's Familial Partial Lipodystrophy (FPLD)*

Dunnigan's familial partial lipodystrophy (FPLD), also known as FLPD type 2 or the Dunnigan variety, is a rare autosomal dominant laminopathy defined by a distinctive pattern of fat redistribution and metabolic abnormalities [58]. The hallmark of FPLD is the progressive loss of subcutaneous adipose tissue from the limbs, trunk, and buttocks, accompanied by abnormal fat accumulation in the face, neck, and upper back regions. This redistribution often results in a characteristic "buffalo hump" due to fat deposition around the neck and supraclavicular area [96]. The onset of these changes typically occurs around puberty and disproportionately affects females, who tend to exhibit more severe phenotypes [97]. The incidence of FPLD is estimated at around 1.3 to 4.7 cases per million, which is a probable underestimation due to low diagnostic rates and phenotypic heterogeneity [98]. FPLD is the most common form of familial lipodystrophy and is genetically linked to heterozygous missense mutations in the *LMNA* gene [99]. Mandibuloacral dysplasia (MAD) exhibits a similar fat redistribution phenotype besides craniofacial skeletal anomalies such as jaw abnormalities and osteolysis in fingertips [100]. Both of these conditions are associated with LMNA mutations that are clustered at the carboxy-terminal globular domain [99].

Clinically, FPLD manifests with selective, progressive lipoatrophy of the extremities and trunk, leading to a muscular appearance of the limbs and prominent veins due to subcutaneous fat loss [101]. Concurrently, fat accumulates in the face, neck, axillary, and pelvic regions, producing a cushingoid or pseudo-athletic phenotype [102]. Affected women often show additional endocrine abnormalities such as hirsutism, oligomenorrhea, polycystic ovaries, and infertility, reflecting the syndrome's systemic metabolic impact [97, 103]. Metabolic complications are prominent and comprise of insulin resistance, type 2 diabetes mellitus, hypertriglyceridemia, hepatosteatosis, acanthosis nigricans, hypertension, and premature atherosclerosis, significantly increasing the risk of cardiovascular conditions. Some patients have also been reported to have developed pancreatitis and liver cirrhosis as sequels of metabolic dysregulation.

The molecular pathology of FPLD centers on mutations in the *LMNA* gene that disrupt the nuclear lamina's stability and regulatory functions within adipocytes. These mutations impair nuclear envelope stability and chromatin organization, resulting in defective adipocyte differentiation, increased apoptosis, and abnormal lipid storage. The selective vulnerability of adipose tissue to nuclear envelope dysfunction explains the regional fat loss, while systemic insulin resistance and dyslipidemia arise from impaired adipocyte function and altered endocrine signaling [104–106]. Studies have elucidated that mutant A-type lamins compromise nuclear mechanics and gene regulation networks critical for adipogenesis and metabolic homeostasis.

## *Hutchinson-Gilford Progeria Syndrome (HGPS)*

Hutchinson-Gilford progeria syndrome (HGPS) is a rare and devastating genetic disorder classified as a laminopathy, defined by features of accelerated aging manifesting in early childhood. HGPS is caused by specific mutations in the *LMNA* gene [107, 108]. This disorder is most commonly associated with a de novo single-nucleotide substitution (c.1824C > T, p.G608G) in exon 11 of *LMNA*, which activates a cryptic splice site and leads to the translation of a truncated, aberrant protein termed progerin. Progerin lacks 50 amino acids near its C-terminal domain, rendering it permanently farnesylated and unable to integrate properly into the nuclear lamina [109]. The incidence of HGPS is estimated at approximately 1 in 20 million live births, and only 350–400 cases have been reported worldwide. The disorder affects both sexes and all ethnic backgrounds equally [108]. HGPS was first described by Jonathan Hutchinson in 1886 and further characterized by Hastings Gilford in 1897 [110]. The molecular basis of the disease was elucidated only in 2003 with the identification of the causative *LMNA* mutation [107].

Clinically, children with HGPS appear normal at birth but begin to exhibit symptoms within the first 2 years of life. Hallmark features include profound growth failure, loss of subcutaneous fat, alopecia, sclerodermatous skin alterations, joint contractures, and distinctive craniofacial features such as micrognathia, prominent eyes, and a beaked nose [111]. Despite their appearance, cognitive development is typically normal. The most life-threatening complications are progressive cardiovascular disease, particularly severe premature atherosclerosis, which results in myocardial infarction or stroke and leads to a median life expectancy limited to only 13–15 years [108, 112].

At the molecular level, the accumulation of progerin in the nuclear lamina causes profound abnormalities in the nuclear morphology, including nuclear blebbing, thickening of the lamina, loss of peripheral heterochromatin, and clustering of nuclear pores [113]. Progerin disrupts nuclear architecture, dysregulates gene expression, impairs DNA repair and mitosis, accelerates telomere shortening, and induces genomic instability [113, 114]. These cellular defects limit proliferative capacity and drive the premature senescence observed characteristically in HGPS. Notably, progerin also accumulates in vascular cells, directly contributing to severe cardiovascular pathology defining the syndrome [115].

**Charcot-Marie-Tooth neuropathy type 2 B1 (CMT2B1)** stems from mutations in *LMNA* that are homozygous recessive. These mutants affect axon density and lead to demyelination of axons manifesting in the form of peripheral muscle wasting. This autosomal recessive axonal peripheral sensorimotor polyneuropathy is characterized by progressive distal muscle weakness and atrophy, primarily affecting the lower limbs and later involving proximal muscles [115–117]. Sensory impairment may also be present, but foot deformities are typically mild or absent. The disease predominantly occurs in families from Northwestern Africa, with an onset detected in the second decade of life. CMT2B1 is caused by a single amino acid substitution p.R644C in *LMNA* [118]. The incidence of CMT2B1 is rare and geographically restricted, reflecting its founder mutation status [119]. Symptoms include muscle weakness, sensory loss, and gradual progression to proximal muscle atrophy, significantly impairing motor function [116].

Further research is essential to elucidate the effect on nuclear architecture and chromatin organization, which may provide insights into broader molecular mechanisms underlying neurological diseases associated with LMNA mutations.

B-type lamins, encoded by the separate genes LMNB1 and LMNB2, are expressed ubiquitously from early developmental stages and play essential roles in cephalogenesis [120]. Deficiencies in lamin B1 or B2 in mice disrupts neuronal migration and cortical organization, illustrating their importance in neurodevelopment [121]. Mutations in LMNB1 are linked to primary microcephaly, while mutations in both LMNB1 and LMNB2 contribute to syndromic microcephaly. Further, LMNB1 duplications cause autosomal dominant leukodystrophy, a disorder that results in demyelinated neurons [122]. Overexpression of lamin B1 in mice leads to myelin defects, axonal degeneration, and impairment of cognitive function [123]. Despite their ubiquitous expression, B-type lamins are particularly crucial for central nervous system integrity and function. Understanding their precise roles in neurodevelopment and CNS maintenance may inform therapeutic strategies for related neurodevelopmental and neurodegenerative disorders. Further research is essential to elucidate their interactions with nuclear architecture and chromatin organization, which may provide insights into broader molecular mechanisms underlying neurological diseases.

## Lamins in Cancer

Lamin dysregulation has keen associations with malignancies [57]. The *LMNA* gene encoding A-type lamins (A and C) is implicated in diverse cancer-related processes [32, 57]. *LMNA*'s role, as a structural component of the nuclear lamina and a mechanotransducive regulator, positions it at the nexus of tumorigenesis, metastasis, and chemoresistance. Recent studies have explicated how the dysregulation of *LMNA* perturbs oncogenic pathways, modulates protein-protein interactions, and further influences cellular phenotypes across varieties of cancers, offering novel insights into its potential as a therapeutic target and molecular biomarker.

Most malignancies have an altered expression profile of mostly A-type lamins governed by epigenetic mechanisms. Colorectal cancer has been shown to possess overexpression of *LMNA* that activates the epithelial-to-mesenchymal transition (EMT) pathway and enhances tumor motility and metastatic potential [124, 125]. Dysregulation of the *LMNA*-V6 promoter by G-quadruplexes has resulted in a characteristic upregulation in the expression of *LMNA* [126]. miRNAs such as miR-9 are also involved in modifying the presence of the protein in the cells, by selectively targeting LMNA mRNA [127]. Interestingly, polyphenols of *Artemisia annua* L. (pKAL) origin exhibit antitumor effects in CRC by upregulation of *LMNA* through p53-dependent mechanisms, which in turn induce apoptosis and cell cycle arrest [128]. The mutant *LMNA* expression also enables cancer to enhance its invasive potential through stabilized nuclear architecture and modulation of the PI3K/AKT/PTEN pathway [129, 130]. But besides the overexpression, the recurrence of higher stages of CRC is also correlated with decreased expression of *LMNA* [125]. Prostate cancer also reveals a heterogeneous expression of *LMNA*, with elevated levels corresponding to the invasive tumor margins compared to the inner core of cells [131]. This spatial variability corroborated enhanced invasive potential [130]. B-type lamins also showcase an overexpression pathology in prostate carcinoma [132]. Furthermore, *LMNA*'s symbiotic behavior with Emerin in breast cancer enhances nuclear stability, potentially substantiating resistance to mechanical stress during metastasis as seen in MCF7 cells [133]. These observations underscore the utility of *LMNA* as a biomarker for tumor staging and a candidate for targeted therapies aimed at curbing metastasis of neoplasms.

Interestingly, *LMNA* exhibits a context-dependent role in onco-progression, acting as both an oncogenic facilitator and suppressor. In breast cancer, reduced *LMNA* expression induces genomic instability via the dysfunction of the p53/p21 pathway, thereby promoting aneuploidy and polyploidy, while miR-129 overexpression in triple negative breast cancer (TNBC) downregulates *LMNA*, suggesting its therapeutic potential in curbing the invasiveness of TNBCs [128, 134–136]. *LMNA* also downregulates YAP (Yes-associated protein)/TAZ nuclear translocation in bone-invasive cancers, limiting oncogenicity [136]. Hepatocellular carcinoma (HCC) studies correlate *LMNA* aberration to metabolic reprogramming through SPAG4 (sperm-associated antigen 4) and LINC complex interactions, upregulating sterol regulatory element-binding protein 1 (SREBP1) that regulates lipid metabolism and drives tumorigenesis, with hypoxia exacerbating its role via lysine posttranslational modifications (crotonylation) on residues K265/270 to accentuate proliferation and retard senescence [137–140]. Ovarian cancer shows altered lamin phosphorylation due to disrupted LMNA-TPX2 interactions which destabilizes nuclear architecture stemming from negative regulation of *LMNA* by TPX2, while *LMNA* downregulation disrupts heat shock protein 90 (Hsp90) distribution, potentially derailing oncogenic signaling [141, 142]. Microenvironmental growth differentiation factor 15 (GDF15) has been shown to disrupt *LMNA*-c-Fos binding, liberating c-Fos to activate EMT-related genes and metastasis in colon cancer [143]. Beyond mere structural roles, *LMNA* stabilizes tumor suppressors like *pRb* to safeguard cell cycle checkpoints yet paradoxically enhances oncoproteins like SREBP1 in HCC, highlighting its diabolical role in response to dysregulation [137, 138, 144].

Therapeutic exploitation of *LMNA*-related mechanisms is gaining a fresh impetus. For instance, targeting the miR-129/*LMNA* axis in TNBC or leveraging pKAL's p53-mediated *LMNA* induction in the pretext of colorectal cancer exemplifies precision approaches. Similarly, inhibition of *LMNA*'s interaction with TPX2 or Hsp90 could also disrupt ovarian cancer progression. Nevertheless, challenges persist in reconciling *LMNA*'s dichotomous roles across cancers, where its overexpression or suppression may yield contrasting outcomes depending on tissue type and genetic proforma.

## Therapeutic Regimes Against Laminopathies

Modern treatment regimens focus on mitigating aberrant protein processing, regulating dysregulated signaling cascades, and employing gene therapy techniques to restore normal lamin A function.

Therapeutic strategies for DCM are primarily aimed at mitigating heart failure symptoms and preventing arrhythmia. Standard pharmacological treatments include angiotensin-converting enzyme inhibitors (ACEi), angiotensin receptor blockers, beta-blockers, and aldosterone antagonists, which collectively reduce morbidity and mortality [145, 146]. In patients with advanced disease, device-based therapies such as implantable cardioverter-defibrillators (ICDs) and cardiac resynchronization therapy are indicated to prevent sudden cardiac attacks and improve cardiac function [147]. For end-stage cases refractory to medical therapy, left ventricular assist devices (LVADs) and orthotopic heart transplantation (OHT) remain definitive options [148, 149]. Therapeutic interventions for FPLD focus on managing metabolic complications and preventing cardiovascular disease. Lifestyle modifications, including dietary optimization and regular physical activity, are foundational [150, 151]. Pharmacologic treatments typically involve insulin sensitizers such as metformin, lipid-lowering agents like fibrates or statins, and, in selected cases, recombinant leptin therapy (e.g., metreleptin) to ameliorate leptin deficiency and improve glycemic and lipid control [152–154]. Regular monitoring for diabetes, hepatosteatosis, and cardiovascular disease is essential [152].

Therapeutic strategies for HGPS have focused on targeting the aberrant farnesylation of progerin. The first FDA-approved treatment, *lonafarnib*, is a farnesyltransferase inhibitor that reduces progerin accumulation at the nuclear envelope and has been shown to improve cardiovascular outcomes and survival in affected children [155–157]. Additional approaches under investigation include antisense oligonucleotides to correct splicing, gene editing, and agents that enhance progerin clearance or ameliorate downstream cellular defects [158–161]. However, as FTIs do not completely eliminate progerin, combination therapies with agents such as statins are being utilized to enhance treatment efficacy.

Gene therapy also offers promising avenues for correcting laminopathies resulting from genetic defects. Adeno-associated virus (AAV)-mediated gene transfer has been of much importance in delivering functional *LMNA* genes to muscle tissues, improving muscle integrity in Emery-Dreifuss muscular dystrophy (EDMD) models [162, 163]. RNA interference (RNAi) approaches are also being investigated to selectively degrade mutant *LMNA* transcripts which further reduce the production of pathogenic or truncated proteins [164]. Furthermore, CRISPR-Cas9 technology is a revolutionary strategy for precise genome editing, though concerns related to off-target effects and delivery efficiency must be addressed prior to clinical implementation.

Downstream signaling pathways can also be targeted to provide additional therapeutic opportunities. Histone deacetylase (HDAC) inhibitors have demonstrated much potential in restoring nuclear architecture and chromatin organization, ameliorating the cellular aberrations observed in laminopathies [165]. Moreover, modulation of the TGF-β pathway by aberrant lamins, involved in fibrosis and inflammation, is under investigation as a therapeutic modality [166]. Supportive therapies, comprising physical therapy and cardiac interventions, also remain critical for managing symptoms in muscular laminopathies such as EDMD and LGMD. Other epidrugs, such as histone lysine acetyltransferase (KAT) activators, have also been utilized to override laminopathies. Examples include small molecule inhibitors of NAT10 (a KAT) that rescue nuclear morphology defects in *LMNA*-deficient cells by increasing the expression of *LMNA* through more acetylation in the promoter regions [167]. Epigenetic modulators such as KAT activators or inhibitors, histone lysine methyltransferase (KMT) activators or inhibitors, etc., altering the expression of the lamin genes in systems where they are dysregulated can offer important insight into deriving new-age therapeutics against laminopathies. A multidisciplinary approach integrating molecular and symptomatic interventions using both clinical and medical tools is essential for improving patient outcomes, with ongoing research focusing on combination therapies and biomarker discoveries to optimize treatment strategies.

## Conclusion

The nucleus, as the defining organelle of eukaryotic cells, exemplifies the evolutionary leap in cellular compartmentalization and genetic regulation. Its complex architecture, comprising the nuclear envelope, nuclear pores, nucleoplasm, chromatin, and nuclear lamina, enables a highly orchestrated environment for genome maintenance, gene expression, and cellular homeostasis. Nuclear lamina, a sophisticated meshwork of type V intermediate filament proteins known as lamins, is pivotal in maintaining nuclear shape, mechanical resilience, and chromatin organization. The discovery and subsequent molecular characterization of lamins, particularly the distinction between A-type and B-type isoforms, have illuminated their discrete and overlapping roles in nuclear structure, gene regulation, and mechanotransduction. The dynamic assembly of lamins into higher-order networks, their regulated post-translational modifications, and their interactions with associated proteins, such as LAP1, LAP2, Emerin, and the lamin B receptor, underscore the lamina's centrality in both structural integrity and nuclear function.

Advances in imaging and molecular biology have revealed the lamina's essential role in tethering chromatin domains, particularly lamina-associated domains (LADs), thereby influencing the spatial organization of the genome and the epigenetic landscape. The interplay between lamins and chromatin, mediated through direct protein-DNA and protein-protein interactions, governs transcriptional accessibility and gene silencing, contributing to the precise regulation of cellular identity and differentiation. Furthermore, the LINC complex, anchored by lamins, physically couples the nuclear interior to the cytoskeleton, enabling the transmission of mechanical signals and facilitating processes such as nuclear positioning, migration, and cellular response to mechanical stress.

Crucially, the dysregulation of lamins resulting in laminopathies has profound pathological consequences, as evidenced by the spectrum of laminopathies. Mutations in lamin-encoding genes disrupt nuclear mechanics, chromatin organization, and signal transduction, manifesting as diverse clinical phenotypes ranging from premature aging syndromes and muscular dystrophies to metabolic and neurodegenerative disorders. Common laminopathies are summarized in Table 10.1,

**Table 10.1** Common laminopathies and a summary of their causative mutations, symptoms, and treatment modalities

| Disease | Abbreviation | Mutation | Symptoms | Treatments |
|---|---|---|---|---|
| Emery-Dreifuss muscular dystrophy | EDMD | *LMNA* (autosomal dominant AD-EDMD), *EMD* (X-linked EDMD) | Joint contractures, progressive muscle weakness, cardiac complications | Adeno-associated virus (AAV)-mediated gene therapy, symptom control |
| Dilated cardiomyopathy | DCM | *LMNA* | Dilation and aberrant contractions of ventricles, CAD | ACEi, angiotensin receptor blockers, beta-blockers, ICDs |
| Familial partial lipodystrophy (Dunnigan type) | FPLD | Codon 482 of *LMNA* [168] | Abnormal fat accumulation, "buffalo hump," cushingoid appearance | Insulin sensitizers (metformin), statins, leptin therapy (metreleptin), diet, and lifestyle control |
| Limb-girdle muscular dystrophy 1B (LGMD1B) | LGMD1B | *LMNA* | Progressive muscle weakness in limbs and girdles, motor difficulties, calf muscle hypertrophy [169] | Symptom management, cardiac complication management |
| Mandibuloacral dysplasia | MAD | *LMNA* (p.R527L, p. M540T, etc.) [170, 171] | Mandibular hypoplasia, stubby digits (acro-osteolysis) lipodystrophy | Symptomatic care |
| Charcot-Marie-tooth disease type 2 B1 | CMT2B1 | p.R644C in *LMNA* | Distal muscle weakness and atrophy Demyelination of axons, sensory impairment | Symptomatic care, physical therapy |
| Hutchinson-Gilford progeria syndrome | HGPS | p.G608G in *LMNA* (progerin) | Growth failure, loss of subcutaneous fat, micrognathia, beaked nose | FTi (lonafarnib), gene therapy |
| Pelger-Huet anomaly | PHA | Lamin B receptor [172] | Abnormally shaped neutrophil nuclei | Benign condition no treatment |

corroborating with the discussion above. These insights not only highlight the multifaceted contributions of lamins to nuclear and cellular physiology but also position it as a focal point for understanding disease mechanisms and developing targeted therapeutic strategies. Besides laminopathies, dysregulation of lamins also plays a diabolical role in carcinogenesis and can serve as a possible therapeutic target. The ongoing elucidation of lamin dynamics, its molecular interactions, and its integration with cellular signaling networks continues to expand our understanding of the nucleus as both a guardian of genomic integrity and a responsive hub of cellular function. Research on drugs and small molecules that can act as therapeutics against these rare disorders is essential to tackle laminopathies, and one of the major developments in the field is the use of epidrugs in modulating the expression of mutant or dysregulated lamin genes.

**Acknowledgments** TKK is a recipient of the J C Bose National Fellowship. HR acknowledges the M. Sc. Interdisciplinary Biosciences Program of JNCASR, Bangalore, India.

## References

1. Ribatti D (2018) An historical note on the cell theory. Exp Cell Res 364(1):1–4
2. Gabaldón T, Pittis AA (2015) Origin and evolution of metabolic sub-cellular compartmentalization in eukaryotes. Biochimie 119:262–268
3. Cooper GM. The cell: a molecular approach. 2nd edition. Sunderland: Sinauer Associates; 2000. Chapter 8, The Nucleus
4. Pederson T (2011) The nucleus introduced. Cold Spring Harb Perspect Biol 3(5):a000521. Published 2011 May 1
5. Hetzer MW (2010) The nuclear envelope. Cold Spring Harb Perspect Biol 2(3):a000539
6. Adam SA (2001) The nuclear pore complex. Genome Biol 2(9):REVIEWS0007
7. Wong X, Melendez-Perez AJ, Reddy KL (2022) The nuclear lamina. Cold Spring Harb Perspect Biol 14(2):a040113. Published 2022 Feb 1
8. Caragine CM, Haley SC, Zidovska A (2019) Nucleolar dynamics and interactions with nucleoplasm in living cells. Elife. 8:e47533Published 2019 Nov 26
9. Murakami Y (2013) Heterochromatin and euchromatin. In: Dubitzky W, Wolkenhauer O, Cho KH, Yokota H (eds) Encyclopedia of systems biology. Springer, New York, NY
10. Van Bortle K, Corces VG (2012) Nuclear organization and genome function. Annu Rev Cell Dev Biol 28:163–187
11. Aebi U, Cohn J, Buhle L, Gerace L (1986) The nuclear lamina is a meshwork of intermediate-type filaments. Nature 323(6088):560–564
12. Dobrzynska A, Gonzalo S, Shanahan C, Askjaer P (2016) The nuclear lamina in health and disease. Nucleus 7(3):233–248
13. Gruenbaum Y, Goldman RD, Meyuhas R et al (2003) The nuclear lamina and its functions in the nucleus. Int Rev Cytol 226:1–62
14. Ottaviano Y, Gerace L (1985) Phosphorylation of the nuclear lamins during interphase and mitosis. J Biol Chem 260(1):624–632
15. Gerace L, Huber MD (2012) Nuclear lamina at the crossroads of the cytoplasm and nucleus. J Struct Biol 177(1):24–31
16. Shimi T, Butin-Israeli V, Adam SA, Goldman RD (2010) Nuclear lamins in cell regulation and disease. Cold Spring Harb Symp Quant Biol 75:525–531

17. Parry DA, Conway JF, Steinert PM (1986) Structural studies on lamin. Similarities and differences between lamin and intermediate-filament proteins. Biochem J 238(1):305–308
18. Butin-Israeli V, Adam SA, Goldman AE, Goldman RD (2012) Nuclear lamin functions and disease. Trends Genet 28(9):464–471
19. Shaw DJ, Eggleton P, Young PJ (2008) Joining the dots: production, processing and targeting of U snRNP to nuclear bodies. Biochim Biophys Acta 1783(11):2137–2144
20. Nmezi B, Xu J, Fu R et al (2019) Concentric organization of A- and B-type lamins predicts their distinct roles in the spatial organization and stability of the nuclear lamina. Proc Natl Acad Sci USA 116(10):4307–4315
21. Ahn J, Jeong S, Kang SM, Jo I, Park BJ, Ha NC (2020) Separation of coiled-coil structures in lamin A/C is required for the elongation of the filament. Cells 10(1):55. Published 2020 Dec 31
22. Cremer T, Cremer M (2010) Chromosome territories. Cold Spring Harb Perspect Biol 2(3):a003889
23. Robson MI, de Las Heras JI, Czapiewski R, Sivakumar A, Kerr ARW, Schirmer EC (2017) Constrained release of lamina-associated enhancers and genes from the nuclear envelope during T-cell activation facilitates their association in chromosome compartments. Genome Res 27(7):1126–1138
24. Dixon JR, Selvaraj S, Yue F et al (2012) Topological domains in mammalian genomes identified by analysis of chromatin interactions. Nature 485(7398):376–380. Published 2012 Apr 11
25. Yáñez-Cuna JO, van Steensel B (2017) Genome-nuclear lamina interactions: from cell populations to single cells. Curr Opin Genet Dev 43:67–72
26. Lazniewski M, Dawson WK, Rusek AM, Plewczynski D (2019) One protein to rule them all: the role of CCCTC-binding factor in shaping human genome in health and disease. Semin Cell Dev Biol 90:114–127
27. Worman HJ, Bonne G (2007) "Laminopathies": a wide spectrum of human diseases. Exp Cell Res 313(10):2121–2133
28. Broers JL, Ramaekers FC, Bonne G, Yaou RB, Hutchison CJ (2006) Nuclear lamins: laminopathies and their role in premature ageing. Physiol Rev 86(3):967–1008
29. Eriksson M, Brown WT, Gordon LB et al (2003) Recurrent de novo point mutations in lamin A cause Hutchinson-Gilford progeria syndrome. Nature 423(6937):293–298
30. Yabata H, Riku Y, Miyahara H, Akagi A, Sone J, Urushitani M, Yoshida M, Iwasaki Y (2023) Nuclear expression of TDP-43 is linked with morphology and ubiquitylation of cytoplasmic aggregates in amyotrophic lateral sclerosis. Int J Mol Sci 24(15):12176
31. Broers JL, Ramaekers FC (2014) The role of the nuclear lamina in cancer and apoptosis. Adv Exp Med Biol 773:27–48
32. Sakthivel KM, Sehgal P (2016) A novel role of lamins from genetic disease to cancer biomarkers. Oncol Rev 10(2):309. Published 2016 Nov 14
33. Moir RD, Yoon M, Khuon S, Goldman RD (2000) Nuclear lamins A and B1: different pathways of assembly during nuclear envelope formation in living cells. J Cell Biol 151(6):1155–1168
34. Goldman RD, Gruenbaum Y, Moir RD, Shumaker DK, Spann TP (2002) Nuclear lamins: building blocks of nuclear architecture. Genes Dev 16(5):533–547
35. Simon DN, Wilson KL (2013) Partners and post-translational modifications of nuclear lamins. Chromosoma 122(1–2):13–31
36. Genschel J, Schmidt HH (2000) Mutations in the LMNA gene encoding lamin A/C. Hum Mutat 16(6):451–459
37. Lin F, Worman HJ (1995) Structural organization of the human gene (LMNB1) encoding nuclear lamin B1. Genomics 27(2):230–236
38. Damiano JA, Afawi Z, Bahlo M et al (2015) Mutation of the nuclear lamin gene LMNB2 in progressive myoclonus epilepsy with early ataxia. Hum Mol Genet 24(16):4483–4490
39. Sweet ME, Taylor MRG, Mestroni L (2017) Chapter 12 – Diseases of the nuclear membrane. In: Jefferies JL, Blaxall BC, Robbins J, Towbin JA (eds) Cardioskeletal myopathies in children and young adults. Academic, London; San Diego, pp 233–248. ISBN 9780128000403

40. DeBoy E, Puttaraju M, Jailwala P, Kasoji M, Cam M, Misteli T (2017) Identification of novel RNA isoforms of LMNA. Nucleus 8(5):573–582
41. Stuurman N, Heins S, Aebi U (1998) Nuclear lamins: their structure, assembly, and interactions. J Struct Biol 122(1–2):42–66
42. Herrmann H, Aebi U (2016) Intermediate filaments: structure and assembly. Cold Spring Harb Perspect Biol 8(11):a018242. Published 2016 Nov 1
43. Gruenbaum Y, Margalit A, Goldman R et al (2005) The nuclear lamina comes of age. Nat Rev Mol Cell Biol 6:21–31
44. Young SG, Meta M, Yang SH, Fong LG (2006) Prelamin A farnesylation and progeroid syndromes. J Biol Chem 281(52):39741–39745
45. Dhe-Paganon S, Werner ED, Chi YI, Shoelson SE (2002) Structure of the globular tail of nuclear lamin. J Biol Chem 277(20):17381–17384
46. Stewart CL et al (2007) Blurring the boundary: the nuclear envelope extends its reach. Science 318:1408–1412
47. Fong LG, Ng JK, Lammerding J et al (2006) Prelamin A and lamin A appear to be dispensable in the nuclear lamina. J Clin Invest 116(3):743–752
48. King MC (2023) Dynamic regulation of LINC complex composition and function across tissues and contexts. FEBS Lett 597(22):2823–2832
49. Machowska M, Piekarowicz K, Rzepecki R (2015) Regulation of lamin properties and functions: does phosphorylation do it all? Open Biol 5(11):150094
50. Heald R, McKeon F (1990) Mutations of phosphorylation sites in lamin A that prevent nuclear lamina disassembly in mitosis. Cell 61(4):579–589
51. Gareau JR, Lima CD (2010) The SUMO pathway: emerging mechanisms that shape specificity, conjugation, and recognition. Nat Rev Mol Cell Biol 11(12):861–871
52. Khanna R, Krishnamoorthy V, Parnaik VK (2018) E3 ubiquitin ligase RNF123 targets lamin B1 and lamin-binding proteins. FEBS J 285(12):2243–2262
53. Rankin J, Ellard S (2006) The laminopathies: a clinical review [published correction appears in Clin Genet. 2007 Mar;71(3):293]. Clin Genet 70(4):261–274
54. Lattanzi G, Maggi L, Araujo-Vilar D (2018) Laminopathies. Nucleus 9(1):543–544
55. Al-Saaidi RA, Rasmussen TB, Birkler RID, Palmfeldt J, Beqqali A, Pinto YM, Nissen PH, Baandrup U, Mølgaard H, Hey TM, Eiskjær H, Bross P, Mogensen J (2018) The clinical outcome of *LMNA* missense mutations can be associated with the amount of mutated protein in the nuclear envelope. Eur J Heart Fail 20:1404–1412
56. Gruenbaum Y, Foisner R (2015) Lamins: nuclear intermediate filament proteins with fundamental functions in nuclear mechanics and genome regulation. Annu Rev Biochem 84:131–164
57. Irianto J, Pfeifer CR, Ivanovska IL, Swift J, Discher DE (2016) Nuclear lamins in cancer. Cell Mol Bioeng 9(2):258–267
58. Schreiber KH, Kennedy BK (2013) When lamins go bad: nuclear structure and disease. Cell 152(6):1365–1375
59. Dittmer TA, Misteli T (2011) The lamin protein family. Genome Biol 12(5):222
60. Goldberg MW, Huttenlauch I, Hutchison CJ, Stick R (2008) Filaments made from A- and B-type lamins differ in structure and organization. J Cell Sci 121(Pt 2):215–225
61. Bione S, Maestrini E, Rivella S et al (1994) Identification of a novel X-linked gene responsible for Emery-Dreifuss muscular dystrophy. Nat Genet 8(4):323–327
62. Manilal S, Nguyen TM, Sewry CA, Morris GE (1996) The Emery-Dreifuss muscular dystrophy protein, emerin, is a nuclear membrane protein. Hum Mol Genet 5(6):801–808
63. Bonne G, Barletta M, Varnous S et al (1999) Mutations in the gene encoding lamin A/C cause autosomal dominant Emery-Dreifuss muscular dystrophy. Nat Genet 21:285–288
64. Bonne G, Leturcq F, Ben YR (1993–2025) Emery-Dreifuss muscular dystrophy. 2004 Sep 29 [Updated 2019 Aug 15]. In: Adam MP, Feldman J, Mirzaa GM et al (eds) GeneReviews® [Internet]. University of Washington, Seattle
65. Dreifuss FE, Hogan GR (1961) Survival in x-chromosomal muscular dystrophy. Neurology 11:734–737

66. Bialer MG, McDaniel NL, Kelly TE (1991) Progression of cardiac disease in Emery-Dreifuss muscular dystrophy. Clin Cardiol 14(5):411–416
67. Mah JK, Korngut L, Fiest KM et al (2016) A systematic review and Meta-analysis on the epidemiology of the muscular dystrophies. Can J Neurol Sci 43(1):163–177
68. Bonne G, Capeau J, De Visser M et al (2002) 82nd ENMC international workshop, 5th international Emery-Dreifuss muscular dystrophy (EDMD) workshop, 1st workshop of the MYO-CLUSTER project EUROMEN (European muscle envelope nucleopathies), 15-16 September 2000, Naarden, The Netherlands. Neuromuscul Disord 12(2):187–194
69. Heller SA, Shih R, Kalra R, Kang PB (2020) Emery-Dreifuss muscular dystrophy. Muscle Nerve 61(4):436–448
70. Bonne G, Mercuri E, Muchir A et al (2000) Clinical and molecular genetic spectrum of autosomal dominant Emery-Dreifuss muscular dystrophy due to mutations of the lamin A/C gene. Ann Neurol 48(2):170–180
71. Faiella W, Bessoudo R (2018) Cardiac manifestations in Emery-Dreifuss muscular dystrophy. CMAJ 190(48):E1414–E1417
72. Boriani G, Gallina M, Merlini L et al (2003) Clinical relevance of atrial fibrillation/flutter, stroke, pacemaker implant, and heart failure in Emery-Dreifuss muscular dystrophy: a long-term longitudinal study. Stroke 34(4):901–908
73. Koch AJ, Holaska JM (2014) Emerin in health and disease. Semin Cell Dev Biol 29:95–106
74. Raffaele Di Barletta M, Ricci E, Galluzzi G et al (2000) Different mutations in the LMNA gene cause autosomal dominant and autosomal recessive Emery-Dreifuss muscular dystrophy. Am J Hum Genet 66(4):1407–1412
75. Vaughan A, Alvarez-Reyes M, Bridger JM et al (2001) Both emerin and lamin C depend on lamin A for localization at the nuclear envelope. J Cell Sci 114(Pt 14):2577–2590
76. Nicolas HA, Akimenko MA, Tesson F (2019) Cellular and animal models of striated muscle laminopathies. Cells 8(4):291. Published 2019 Mar 29
77. Gueneau L, Bertrand AT, Jais JP et al (2009) Mutations of the FHL1 gene cause Emery-Dreifuss muscular dystrophy. Am J Hum Genet 85(3):338–353
78. Lammerding J, Schulze PC, Takahashi T et al (2004) Lamin A/C deficiency causes defective nuclear mechanics and mechanotransduction. J Clin Invest 113(3):370–378
79. Muchir A, Medioni J, Laluc M et al (2004) Nuclear envelope alterations in fibroblasts from patients with muscular dystrophy, cardiomyopathy, and partial lipodystrophy carrying lamin A/C gene mutations. Muscle Nerve 30(4):444–450
80. Widyastuti HP, Norden-Krichmar TM, Grosberg A et al (2020) Gene expression profiling of fibroblasts in a family with *LMNA*-related cardiomyopathy reveals molecular pathways implicated in disease pathogenesis. BMC Med Genet 21:152
81. Muchir A, Massart C, van Engelen BG, Lammens M, Bonne G, Worman HJ (2006) Proteasome-mediated degradation of integral inner nuclear membrane protein emerin in fibroblasts lacking A-type lamins. Biochem Biophys Res Commun 351(4):1011–1017
82. Manilal S, Sewry CA, Pereboev A et al (1999) Distribution of emerin and lamins in the heart and implications for Emery-Dreifuss muscular dystrophy. Hum Mol Genet 8(2):353–359
83. Menezes MP, Waddell LB, Evesson FJ et al (2012) Importance and challenge of making an early diagnosis in LMNA-related muscular dystrophy. Neurology 78(16):1258–1263
84. van Berlo JH, de Voogt WG, van der Kooi AJ et al (2005) Meta-analysis of clinical characteristics of 299 carriers of LMNA gene mutations: do lamin A/C mutations portend a high risk of sudden death? J Mol Med (Berl) 83(1):79–83
85. Anselme F, Moubarak G, Savouré A et al (2013) Implantable cardioverter-defibrillators in lamin A/C mutation carriers with cardiac conduction disorders. Heart Rhythm 10(10):1492–1498
86. Schultheiss HP, Fairweather D, Caforio ALP et al (2019) Dilated cardiomyopathy. Nat Rev Dis Primers 5:32
87. Japp A, Gulati A, Cook S et al (2016) The diagnosis and evaluation of dilated cardiomyopathy. JACC 67(25):2996–3010

88. Jakobs PM, Hanson EL, Crispell KA et al (2001) Novel lamin A/C mutations in two families with dilated cardiomyopathy and conduction system disease. J Card Fail 7(3):249–256
89. Mestroni L, Brun F, Spezzacatene A, Sinagra G, Taylor MR (2014) Genetic causes of dilated cardiomyopathy. Prog Pediatr Cardiol 37(1–2):13–18
90. Brodsky GL, Muntoni F, Miocic S, Sinagra G, Sewry C, Mestroni L (2000) Lamin A/C gene mutation associated with dilated cardiomyopathy with variable skeletal muscle involvement. Circulation 101(5):473–476
91. Hershberger RE, Hanson EL, Jakobs PM et al (2002) A novel lamin A/C mutation in a family with dilated cardiomyopathy, prominent conduction system disease, and need for permanent pacemaker implantation. Am Heart J 144(6):1081–1086
92. Spector ZZ, Meliones C, Idriss SF (2019) 27 – Arrhythmias and pacing. In: Ungerleider RM, Meliones JN, McMillan KN, Cooper DS, Jacobs JP (eds) Critical heart disease in infants and children, 3rd edn. Elsevier, pp 326–350.e3., ISBN 9781455707607
93. Wallace M, Zahr H, Perati S et al (2023) Nuclear damage in *LMNA* mutant iPSC-derived cardiomyocytes is associated with impaired lamin localization to the nuclear envelope. Mol Biol Cell. Published online August, 16, 2023
94. Santos JL, Miranda J, Lagos C, Cortés V (2024) Case report: concurrent de novo pathogenic variants in the LMNA gene as a cause of sporadic partial lipodystrophy. Front Genet 15
95. Fontana M, Barison A, Botto N et al (2013) CMR-verified interstitial myocardial fibrosis as a marker of subclinical cardiac involvement in LMNA mutation carriers. J Am Coll Cardiol Img 6(1):124–126
96. Hussain I, Garg A (2016) Lipodystrophy syndromes. Endocrinol Metab Clin N Am 45(4):783–797
97. Garg A (2000) Gender differences in the prevalence of metabolic complications in familial partial lipodystrophy (Dunnigan variety). J Clin Endocrinol Metab 85(5):1776–1782
98. Chiquette E, Oral EA, Garg A, Araújo-Vilar D, Dhankhar P (2017) Estimating the prevalence of generalized and partial lipodystrophy: findings and challenges. Diabetes Metab Syndr Obes 10:375–383. Published 2017 Sep 13
99. Belo SP, Magalhães ÂC, Freitas P, Carvalho DM (2015) Familial partial lipodystrophy, Dunnigan variety – challenges for patient care during pregnancy: a case report. BMC Res Notes 8:140. Published 2015 Apr 11
100. Jéru I, Nabil A, El-Makkawy G, Lascols O, Vigouroux C, Abdalla E (2021) Two decades after Mandibuloacral dysplasia discovery: additional cases and comprehensive view of disease characteristics. Genes 12(10):1508
101. Garg A (2011) Clinical review#: lipodystrophies: genetic and acquired body fat disorders. J Clin Endocrinol Metab 96(11):3313–3325
102. Drac H, Madej-Pilarczyk A, Gospodarczyk-Szot K, Gaweł M, Kwieciński H, Hausmanowa-Petrusewicz I (2010) Familial partial lipodystrophy associated with the heterozygous LMNA mutation 1445G>A (Arg482Gln) in a polish family. Neurol Neurochir Pol 44(3):291–296
103. Shackleton S, Lloyd DJ, Jackson SN et al (2000) LMNA, encoding lamin A/C, is mutated in partial lipodystrophy. Nat Genet 24(2):153–156
104. Fernández-Pombo A, Sánchez-Iglesias S, Cobelo-Gómez S, Hermida-Ameijeiras Á, Araújo-Vilar D (2021) Familial partial lipodystrophy syndromes. Presse Med 50(3):104071
105. Burn J, Baraitser M (1986) Partial lipoatrophy with insulin resistant diabetes and hyperlipidaemia (Dunnigan syndrome). J Med Genet 23(2):128–130
106. Fernandez-Pombo A, Diaz-Lopez EJ, Castro AI et al (2023) Clinical Spectrum of *LMNA*-associated type 2 familial partial lipodystrophy: a systematic review. Cells 12(5):725. Published 2023 Feb 24
107. Brune T, Bonne G, Denecke J et al (2004) Progeria: a new kind of Laminopathy-- report of the First European Symposium on Progeria and creation of EURO-Progeria, a European Consortium on Progeria and related disorders. Pediatr Endocrinol Rev 2(1):39–45
108. Lamis A, Siddiqui SW, Ashok T, Patni N, Fatima M, Aneef AN (2022) Hutchinson-Gilford progeria syndrome: a literature review. Cureus 14(8):e28629. Published 2022 Aug 31

109. Piekarowicz K, Machowska M, Dzianisava V, Rzepecki R (2019) Hutchinson-Gilford progeria syndrome-ta status and prospects for gene therapy treatment. Cells 8(2):88. Published 2019 Jan 25
110. "Genetic: Hutchinson-Gilford Syndrome." UXL Encyclopedia of diseases and disorders. Encyclopedia.com 2025. https://www.encyclopedia.com
111. Hamczyk MR, del Campo L, Andrés V (2018) Aging in the cardiovascular system: lessons from Hutchinson-Gilford progeria syndrome. Annu Rev Physiol 80:27–48
112. Dorado B, Andrés V (2017) A-type lamins and cardiovascular disease in premature aging syndromes. Curr Opin Cell Biol 46:17–25
113. Wu D, Flannery AR, Cai H, Ko E, Cao K (2014) Nuclear localization signal deletion mutants of lamin A and progerin reveal insights into lamin A processing and emerin targeting. Nucleus 5(1):66–74
114. Cao K, Capell BC, Erdos MR, Djabali K, Collins FS (2007) A lamin A protein isoform overexpressed in Hutchinson-Gilford progeria syndrome interferes with mitosis in progeria and normal cells. Proc Natl Acad Sci USA 104(12):4949–4954
115. Coll-Bonfill N, Mahajan U, Shashkova EV, Lin CJ, Mecham RP, Gonzalo S (2023) Progerin induces a phenotypic switch in vascular smooth muscle cells and triggers replication stress and an aging-associated secretory signature. Geroscience 45(2):965–982
116. Liang JJ, Grogan M, Ackerman MJ, Goodsell K (2016) LMNA-mediated Arrhythmogenic right ventricular cardiomyopathy and Charcot-Marie-Tooth Type 2B1: a patient-discovered unifying diagnosis. J Cardiovasc Electrophysiol 27(7):868–871
117. Goizet C, Yaou RB, Demay L et al (2004) A new mutation of the lamin A/C gene leading to autosomal dominant axonal neuropathy, muscular dystrophy, cardiac disease, and leuconychia. J Med Genet 41(3):e29
118. Kang SM, Yoon MH, Park BJ (2018) Laminopathies; mutations on single gene and various human genetic diseases. BMB Rep 51(7):327–337
119. Hamadouche T, Poitelon Y, Genin E et al (2008) Founder effect and estimation of the age of the c.892C>T (p.Arg298Cys) mutation in LMNA associated to Charcot-Marie-Tooth subtype CMT2B1 in families from North Western Africa. Ann Hum Genet 72(Pt 5):590–597
120. Lee JM, Jung HJ, Fong LG, Young SG (2014) Do lamin B1 and lamin B2 have redundant functions? Nucleus 5(4):287–292
121. Brady GF, Kwan R, Bragazzi Cunha J, Elenbaas JS, Omary MB (2018) Lamins and lamin-associated proteins in gastrointestinal health and disease. Gastroenterology 154(6):1602–1619.e1
122. Zhang Y, Li J, Bai R et al (2019) LMNB1-related adult-onset autosomal dominant leukodystrophy presenting as movement disorder: a case report and review of the literature. Front Neurosci 13:1030Published 2019 Oct 21
123. Koufi FD, Neri I, Ramazzotti G et al (2023) Lamin B1 as a key modulator of the developing and aging brain. Front Cell Neurosci 17:1263310Published 2023 Aug 31
124. Willis ND, Cox TR, Rahman-Casañs SF et al (2008) Lamin A/C is a risk biomarker in colorectal cancer. PLoS One 3(8):e2988. Published 2008 Aug 20
125. Foster CR, Robson JL, Simon WJ et al (2011) The role of lamin A in cytoskeleton organization in colorectal cancer cells: a proteomic investigation. Nucleus 2(5):434–443
126. Nishikawa T, Kuwano Y, Nakata M, Rokutan K, Nishida K (2021) Multiple G-quadruplexes in the LMNA promoter regulate LMNA variant 6 transcription and promote colon cancer cell growth. Biochim Biophys Acta Gene Regul Mech 1864(10):194746
127. Zhao J, Zhang H, Pan C, He Q, Zheng K, Tang Y (2024) Advances in research on the relationship between the LMNA gene and human diseases (review). Mol Med Rep 30(6):236
128. Thompson SL, Compton DA (2010) Proliferation of aneuploid human cells is limited by a p53-dependent mechanism. J Cell Biol 188(3):369–381
129. Carnero A, Paramio JM (2014) The PTEN/PI3K/AKT pathway in vivo, cancer mouse models. Front Oncol 4:252. Published 2014 Sep 23

130. Kong L, Schäfer G, Huajie B, Zhang Y, Zhang Y, Klocker H (2012) Lamin A/C protein is overexpressed in tissue-invading prostate cancer and promotes prostate cancer cell growth, migration and invasion through the PI3K/AKT/PTEN pathway. Carcinogenesis 33(4):751–759
131. Hong JH, Liang ST, Wang AS et al (2022) LMNB1, a potential marker for early prostate cancer progression. Am J Cancer Res 12(7):3390–3404. Published 2022 Jul 15
132. Coradeghini R, Barboro P, Rubagotti A et al (2006) Differential expression of nuclear lamins in normal and cancerous prostate tissues. Oncol Rep 15(3):609–613
133. Hansen E, Rolling C, Wang M et al (2024) Emerin deficiency drives MCF7 cells to an invasive phenotype. Sci Rep 14:19998
134. Capo-chichi CD, Cai KQ, Smedberg J, Ganjei-Azar P, Godwin AK, Xu XX (2011) Loss of A-type lamin expression compromises nuclear envelope integrity in breast cancer. Chin J Cancer 30(6):415–425
135. Setijono SR, Park M, Kim G, Kim Y, Cho KW, Song SJ (2018) miR-218 and miR-129 regulate breast cancer progression by targeting lamins. Biochem Biophys Res Commun 496(3):826–833
136. Chiarini F, Paganelli F, Balestra T et al (2022) Lamin A and the LINC complex act as potential tumor suppressors in Ewing Sarcoma. Cell Death Dis 13:346
137. Liu T, Yu J, Ge C et al (2022) Sperm associated antigen 4 promotes SREBP1-mediated de novo lipogenesis via interaction with lamin A/C and contributes to tumor progression in hepatocellular carcinoma. Cancer Lett 536:215642
138. Su F, Koeberle A (2024) Regulation and targeting of SREBP-1 in hepatocellular carcinoma. Cancer Metastasis Rev 43(2):673–708
139. Park S, Hall MN (2025) Metabolic reprogramming in hepatocellular carcinoma: mechanisms and therapeutic implications. Exp Mol Med 57(3):515–523
140. Zhang D, Tang J, Xu Y et al (2022) Global crotonylome reveals hypoxia-mediated lamin A crotonylation regulated by HDAC6 in liver cancer. Cell Death Dis 13(8):717. Published 2022 Aug 17
141. Meng X, Cao J, Zheng H et al (2023) TPX2 promotes ovarian tumorigenesis by interacting with lamin A/C and affecting its stability. Cancer Med 12(8):9738–9748
142. Wang Y, Chen Q, Wu D et al (2021) Lamin-A interacting protein Hsp90 is required for DNA damage repair and chemoresistance of ovarian cancer cells. Cell Death Dis 12(8):786. Published 2021 Aug 12
143. Ding Y, Hao K, Li Z et al (2020) c-Fos separation from lamin A/C by GDF15 promotes colon cancer invasion and metastasis in inflammatory microenvironment. J Cell Physiol 235(5):4407–4421
144. Nitta RT, Jameson SA, Kudlow BA, Conlan LA, Kennedy BK (2006) Stabilization of the retinoblastoma protein by A-type nuclear lamins is required for INK4A-mediated cell cycle arrest. Mol Cell Biol 26(14):5360–5372
145. Altinier A, Paldino A, Gigli M et al (2019) Current management and treatment. In: Sinagra G, Merlo M, Pinamonti B (eds) Dilated cardiomyopathy: from genetics to clinical management [internet]. Springer. Chapter 13. Available from: https://www.ncbi.nlm.nih.gov/books/NBK553842/, Cham
146. Verdonschot JAJ, Hazebroek MR, Ware JS, Prasad SK, Heymans SRB (2019) Role of targeted therapy in dilated cardiomyopathy: the challenging road toward a personalized approach. J Am Heart Assoc 8(11):e012514
147. Notaristefano F, Ambrosio G (2019) Defibrillator and non-ischaemic dilated cardiomyopathy: a never ending story. Eur Heart J Supp 21., Issue Supplement_B:B5–B6
148. McNamara N, Narroway H, Williams M et al (2021) Contemporary outcomes of continuous-flow left ventricular assist devices-a systematic review. Ann Cardiothorac Surg 10(2):186–208
149. George TJ, Arnaoutakis GJ, Shah AS (2011) Surgical treatment of advanced heart failure: alternatives to heart transplantation and mechanical circulatory assist devices. Prog Cardiovasc Dis 54(2):115–131

150. Brown RJ, Araujo-Vilar D, Cheung PT et al (2016) The diagnosis and Management of Lipodystrophy Syndromes: a multi-society practice guideline. J Clin Endocrinol Metab 101(12):4500–4511
151. Robbins DC, Danforth E Jr, Horton ES, Burse RL, Goldman RF, Sims EA (1979) The effect of diet on thermogenesis in acquired lipodystrophy. Metabolism 28(9):908–916
152. Bagias C, Xiarchou A, Bargiota A, Tigas S (2020) Familial partial lipodystrophy (FPLD): recent insights. Diabetes Metab Syndr Obes 13:1531–1544. Published 2020 May 6
153. Araújo-Vilar D, Santini F (2019) Diagnosis and treatment of lipodystrophy: a step-by-step approach. J Endocrinol Investig 42(1):61–73
154. Kountouri A, Korakas E, Maratou E, Ikonomidis I, Balampanis K, Liatis S, Tentolouris N, Toulas P, Kousathana F, Giatzakis C et al (2023) Familial partial lipodystrophy: clinical features, genetics and treatment in a Greek referral center. Int J Mol Sci 24(15):12045
155. Capell BC, Erdos MR, Madigan JP et al (2005) Inhibiting farnesylation of progerin prevents the characteristic nuclear blebbing of Hutchinson-Gilford progeria syndrome. Proc Natl Acad Sci USA 102(36):12879–12884
156. Dhillon S (2021) Lonafarnib: first approval [published correction appears in Drugs. 2021 Apr;81(5):619. doi:10.1007/s40265-021-01500-6.]. Drugs 81(2):283–289
157. Murtada SI, Mikush N, Wang M et al (2023) Lonafarnib improves cardiovascular function and survival in a mouse model of Hutchinson-Gilford progeria syndrome. Elife 12:e82728Published 2023 Mar 17
158. Arun A, Nath AR, Thankachan B, Unnikrishnan MK (2024) Hutchinson–Gilford progeria syndrome: unraveling the genetic basis, symptoms, and advancements in therapeutic approaches. Ther Adv Rare Dis:5
159. Kreienkamp R, Gonzalo S (2020) Metabolic dysfunction in Hutchinson-Gilford progeria syndrome. Cells 9(2):395. Published 2020 Feb 8
160. Harhouri K, Navarro C, Depetris D et al (2017) MG132-induced progerin clearance is mediated by autophagy activation and splicing regulation. EMBO Mol Med 9(9):1294–1313
161. Lai WF, Wong WT (2020) Progress and trends in the development of therapies for Hutchinson-Gilford progeria syndrome. Aging Cell 19(7):e13175
162. Azibani F, Brull A, Arandel L et al (2018) Gene therapy via trans-splicing for LMNA-related congenital muscular dystrophy. Mol Ther Nucleic Acids 10:376–386
163. Kieserman JM, Myers VD, Dubey P, Cheung JY, Feldman AM (2019) Current landscape of heart failure gene therapy. J Am Heart Assoc 8:e012239
164. Huang S, Chen L, Libina N et al (2005) Correction of cellular phenotypes of Hutchinson-Gilford progeria cells by RNA interference. Hum Genet 118(3–4):444–450
165. Mattioli E, Andrenacci D, Mai A et al (2019) Statins and histone deacetylase inhibitors affect lamin A/C – histone deacetylase 2 interaction in human cells. Front Cell Dev Biol 7:6. Published 2019 Jan 31
166. Evangelisti C, Bernasconi P, Cavalcante P et al (2015) Modulation of TGFbeta 2 levels by lamin A in U2-OS osteoblast-like cells: understanding the osteolytic process triggered by altered lamins. Oncotarget 6(10):7424–7437
167. Larrieu D, Britton S, Demir M, Rodriguez R, Jackson SP (2014) Chemical inhibition of NAT10 corrects defects of laminopathic cells. Science 344(6183):527–532
168. Cao H, Hegele RA (2000) Nuclear lamin A/C R482Q mutation in canadian kindreds with Dunnigan-type familial partial lipodystrophy. Hum Mol Genet 9(1):109–112
169. D'Este G, Spagna M, Federico S et al (2025) Limb-girdle muscular dystrophies: a scoping review and overview of currently available rehabilitation strategies. Muscle Nerve 71(2):138–146.
170. Al-Haggar M, Madej-Pilarczyk A, Kozlowski L et al (2012) A novel homozygous p.Arg527Leu *LMNA* mutation in two unrelated Egyptian families causes overlapping mandibuloacral dysplasia and progeria syndrome. Eur J Hum Genet 20:1134–1140

171. Bai S, Lozada A, Jones MC, Dietz HC, Dempsey M, Das S (2014) Mandibuloacral dysplasia caused by LMNA mutations and uniparental disomy. Case Rep Genet 2014:508231
172. Best S, Salvati F, Kallo J et al (2003) Lamin B-receptor mutations in Pelger-Huët anomaly. Br J Haematol 123(3):542–544
173. Aaronson RP, Blobel G (1974). On the attachment of the nuclear pore complex. J Cell Biol. 62(3):746–54. PMID: 4853439; PMCID: PMC2109221

# Chapter 11
# Dynamic Post-translational Modifications of Nuclear Lamins in Cellular Function and Pathophysiology

**Anjali Sharma, Nabanita Das, Rohini Bhatt, Supriya Varsha Bhagat, and Tapas K. Kundu**

**Abstract** Nuclear lamins, a crucial type of intermediate filament protein, primarily form the inner nuclear membrane and are essential for maintaining nuclear integrity throughout various cell cycle stages. However, recent research has uncovered their broader functions as key scaffolds in nuclear sub-compartmentalization, 3D genome organization, and gene regulation. These functions are dynamically regulated by several post-translational modifications (PTMs), including phosphorylation, acetylation, SUMOylation, methylation, ubiquitination, farnesylation, and O-GlcNAcylation. Lamin PTMs influence chromatin stability, nuclear organization, stress responses, cellular differentiation, metabolism, and ageing. The pathological implications of lamin dysfunction are profound. Altered PTM patterns have been associated with multiple disorders, including laminopathies, metabolic syndromes, premature ageing diseases like Hutchinson–Gilford progeria, and even cancer.

This chapter discusses how dysregulated lamin PTMs lead to nuclear instability and chromatin disorganization and contribute to disease progression. Understanding how PTMs affect lamin function opens avenues for therapeutic strategies targeting lamin-related disorders. This research is critical for developing innovative treatments aimed at restoring nuclear integrity and normal cellular function, ultimately improving disease outcomes.

---

Anjali Sharma, Nabanita Das, Rohini Bhatt and Supriya Varsha Bhagat contributed equally with all other contributors.

---

A. Sharma · N. Das · R. Bhatt · S. V. Bhagat
Molecular Biology and Genetics Unit, Jawaharlal Nehru Centre for Advanced Scientific Research, Bangalore, India

T. K. Kundu (✉)
Molecular Biology and Genetics Unit, Jawaharlal Nehru Centre for Advanced Scientific Research, Bangalore, India

Department of Biochemistry, Tohoku University Graduate School of Medicine, Sendai, Japan
e-mail: tapas@jncasr.ac.in

© The Author(s), under exclusive license to Springer Nature Switzerland AG 2025

K. Sengupta, A. Muchir (eds.), *Lamins in Genome Organization and Diseases*, Subcellular Biochemistry 115, https://doi.org/10.1007/978-3-032-00537-3_11

**Keywords** Nuclear lamins · PTMs · Laminopathies · Chromatin organization · Cell cycle · Stress · Cancer · Ageing

## Introduction

Lamins, classified as type V intermediate filament (IF) proteins, are essential components of the nuclear lamina in all metazoans. They form a dense 30–100-nm-thick filamentous network that upholds nuclear structure, regulates chromatin architecture, ensures genome stability, and maintains the cytoskeletal organization. Lamins are classified into types A and B, each encoded by distinct genes that share significant sequence homology, structural characteristics, and biochemical properties. The *LMNA* gene produces lamins A and C, including the primary A-type lamins lamin A delta 10 and lamin C2 as well as the toxic isoform progerin/LAΔ50 [138]. *LMNB1* and *LMNB2* encode the major B-type lamins, lamin B1 and B2, with LMNB2 also giving rise to an additional isoform, lamin B3 [79, 138]. Structurally, all lamin proteins consist of three domains: N-terminal head domain, central coiled-coil forming rod domain, and a largely unstructured C-terminal tail domain. The alpha helical rod domain contains four subdomains 1A, 1B, 2A, and 2B connected by linkers, while the C-terminal tail includes a nuclear localization signal (NLS), an immunoglobulin-fold (IgG-fold) domain, and a conserved CAAX (C, cysteine; a, aliphatic residue; X, any residue) motif (except in lamin C) at the tail domain. Interestingly, a small fraction of soluble lamin A is also found in nucleoplasm [29]. The C-terminal tail domain mediates interactions with the DNA [123]. Lamins undergo a sequential assembly process: first, they form a coiled-coil dimer through head-to-head association of two α-helical coiled-coil rod domains. These dimers then polymerize in a head-to-tail fashion, ultimately arranging themselves into 3.5 A° tetrameric filaments through an antiparallel organization [30, 116, 127].

Altered expression of lamins results in abnormal nuclear morphology [140]. LMNA mutations, as well as alterations in lamin A/C expression and protein modifications, have been implicated in various pathological processes and are the primary causes of laminopathies which include muscular dystrophies, cardiomyopathies, lipodystrophy, neuropathy, and progeroid syndromes [141, 9, 87, 106]. LMNB1 duplications have been observed in autosomal dominant leukodystrophy (ADLD), leading to a degenerative phenotype characterized by progressive loss of white matter in the brain[146, 144, 145, 146]. Disruptions caused by the dysregulated expression of lamin proteins contribute to fatal diseases like amyotrophic lateral sclerosis (ALS)[148]. While *LMNA* mutations account for most laminopathies, fewer reports have also linked B-type (*LMNB1* and *LMNB2*) lamin, prelamin A processing protein (*ZMPSTE24, ZMPSTE54*), and lamin-binding protein (*EMD, TMPO, LBR*, and *LEMD3*) mutations to human diseases due to their essential role during embryonic and perinatal development [64, 131]. To date, over

500 disease-causing mutations have been identified in LMNA, with most being autosomal dominant [27]. Laminopathies are categorized into four groups based on the tissues they impact (Table 11.1):

1. *Muscle-related diseases* such as Emery–Dreifuss muscular dystrophy (EDMD), limb–girdle muscular dystrophy type 1B (LGMD1B), LMNA-related congenital muscular dystrophy (LCMD), and dilated cardiomyopathy with conduction disease (DCM-CD).
2. *Metabolic disorders*, including Dunnigan-type familial partial lipodystrophy (FPLD) and leukodystrophy (linked to LMNB1 mutations), as well as metabolic syndrome (MS) linked to LMNA.
3. *Peripheral neuronal diseases*, like Charcot–Marie–Tooth disease (AD-CMT).
4. *Early ageing disorders*, such as Hutchinson–Gilford progeria syndrome (HGPS), atypical HGPS, atypical Werner syndrome, and mandibuloacral dysplasia.

Most disease-causing mutations are missense mutations, scattered across the LMNA gene with different mutations in the same residue leading to distinct diseases, with varying severity among individuals [131]. Laminopathy-causing mutations in human lamin A/C are often conserved in *Caenorhabditis elegans* lamin (Ce-lamin), allowing researchers to study their effects on filament assembly. The E145K mutation in human lamin A, linked to progeroid syndrome, has a homologous Q159K mutation in Ce-lamin. This mutation results in filaments with 17 nm repeats, containing only two protofilaments instead of the three or four observed in wild-type Ce-lamin [7]. Similarly, mutations such as ΔK32 (ΔK46 in Ce-lamin), T150P (T164P in Ce-lamin), and L530P (L535P in Ce-lamin) which cause Emery–Dreifuss muscular dystrophy (EDMD) in humans lead to structural changes in Ce-lamin filaments [4, 5, 27]. Consequently, laminopathy-linked mutations may directly impact filament structure and stability, providing key insights into disease mechanisms. Lamin-associated abnormalities likely result from a combination of mechanisms ranging from weakening of nuclear structure, disruption of gene expression, and cell death, thus affecting various tissues differently. Although B-type lamin mutations are rare, they have been implicated in heart, brain, and nervous system disorders [26]. Ongoing research continues to uncover how lamin defects disrupt nuclear structure and cellular function, offering new insights into potential therapeutic approaches.

The functionality of lamins is modulated by various post-translational modifications, including phosphorylation, acetylation, SUMOylation, methylation, ubiquitination, farnesylation, and O-GlcNAcylation (Fig. 11.1).

**Farnesylation** is a crucial post-translational modification in lamin biogenesis, affecting prelamin A (lamin A precursor), lamin B1, and lamin B2. This process involves the addition of a farnesyl group to the C-terminal cysteine of proteins containing a CAAX motif [35]. Lamin C does not undergo farnesylation, as it lacks a CAAX motif, despite being encoded by the same gene as lamin A. While lamins B1 and B2 remain permanently farnesylated at their C-termini, prelamin A undergoes four modifications to mature into lamin A. Farnesylation of the CAAX motif initi-

**Table 11.1** A summary of laminopathy-causing mutations in the *LMNA*, *LMNB1*, and prelamin A processing protein (*ZMPSTE54*) with the associated clinical phenotypes [24, 91]

| Disease | Mutation gene | Mutation residue | Clinical phenotypes |
|---|---|---|---|
| *Muscular dystrophy (diseases of striated muscles)* | | | |
| Muscular dystrophy, congenital (MDC) | *LMNA* | N39S, R50P, R249W, L302P, E358K, L380S, R543P, R455P, N456D | Dropped head syndrome due to weak neck extensors, diffuse myopathy during the first 24 months of age, followed by cardiomyopathy |
| Emery–Dreifuss muscular dystrophy (autosomal dominant [AD-EDMD], autosomal recessive [AR-EDMD], and X-linked [XL-EDMD]) | *LMNA*, *EMD*, *FHL1* | Q6X, E33G, L35V, N39S, A43T, Y45C, R50S, I63S, E66G, R89C, R133P, L140P, T150P, R190Q, R196S, H222P/Y, G232E, L248P, R249Q/W, S268P, L271P, S295P, R336Q, E358K, M371K, R377L, R401C, V442A, D446V, G449D, R453W, L454P, D461Y, W467R, R527P, L530P, R541H/S/P, G602S, R624H | Rigidity of spine and neck, progressive muscle weakness, contractures of elbows, heel, and spine, cardiomyopathy, cardio arrhythmias |
| Malouf syndrome (MLF) | *LMNA* | A57P, L59R | Intellectual disability, congestive cardiomyopathy, broad nasal base, bone abnormalities, micrognathia |
| Dilated cardiomyopathy A) | *LMNA* | Q6X, L59R, R60G, E82K, L85R, R89L, K97E, S143P, E161K, R190W, D192G, N195K, E203K/G, L215P, R225X, Y267C, E317K, A347K, R349L, R399C, R435C, R541C/S, R644C | The heart is weakened and elongated, with conduction problems |
| Limb–girdle muscular dystrophy (LGMD1 B) | *LMNA* | R25G, Y259X, E358K, R377H/L, R399C, Y481H, | Classic limb–girdle pattern of muscle atrophy, with lower limb muscular weakness and wasting |

(continued)

**Table 11.1** (continued)

| Disease | Mutation gene | Mutation residue | Clinical phenotypes |
|---|---|---|---|
| *Lipodystrophy syndromes* | | | |
| Autosomal dominant Dunnigan familial partial lipodystrophy (FPLD2) | *LMNA* | R25W, R60G, R62G, △K208, D230N, G465D, R482W/Q/L/R/N, S573L, R582H, R584H | Accumulation of fat in the face and neck with progressive loss of fat from limbs; associated with diabetes mellitus |
| Autosomal recessive mandibuloacral dysplasia (MADA/B) | *LMNA* | V440M, R471C, R527C/H, A529V | Partial lipodystrophy, insulin resistance, skeletal abnormalities, hypertriglyceridemia, progeria syndrome similarity |
| Barraquer–Simons syndrome (heterozygous mutation) | *LMNB2* | | Similar phenotype as observed in FPLD2 |
| Autosomal dominant lipoatrophy with diabetes, hepatic steatosis, hypertrophic cardiomyopathy, and leukomelanodermic papules (LDHPC) | *LMNA* | | Hepatic steatosis, hypertrophic cardiomyopathy, leukomelanodermic papules |
| *Neuropathy (accelerated ageing disorders)* | | | |
| Charcot–Marie–tooth disorder type 1B (CMT2B1) | *LMNA* | R298C | Sensory loss, axonal peripheral neuropathy, distal muscle weakness, and atrophy |
| Autosomal dominant atypical Werner syndrome (AWS) | *LMNA* | A57P, R133L, L140R, D300N, Q656Q | Premature ageing |
| Hutchinson–Gilford progeria syndrome (HGPS) | *LMNA* | T10I, S143F, E145K, R471C, R527C, T528M, M540T, K542N, E578V, V607V, G608S/G, T623S, R644C | Alopecia, loss of subcutaneous fat, progressive joint contractures, nail dystrophy, hearing loss, dental crowding, and children die early due to premature ageing |
| Autosomal dominant restrictive dermopathy lethal (RD) | *ZMPSTE54* | | Hyperkeratosis of epidermis, polyhydramnios, pulmonary hypoplasia, foetal akinesia |
| *Leukodystrophies* | | | |
| Autosomal dominant leukodystrophy (ADLD) | *LMNB1* | | The white matter of the CNS is affected, particularly the autonomic, cerebellar, and pyramidal systems |

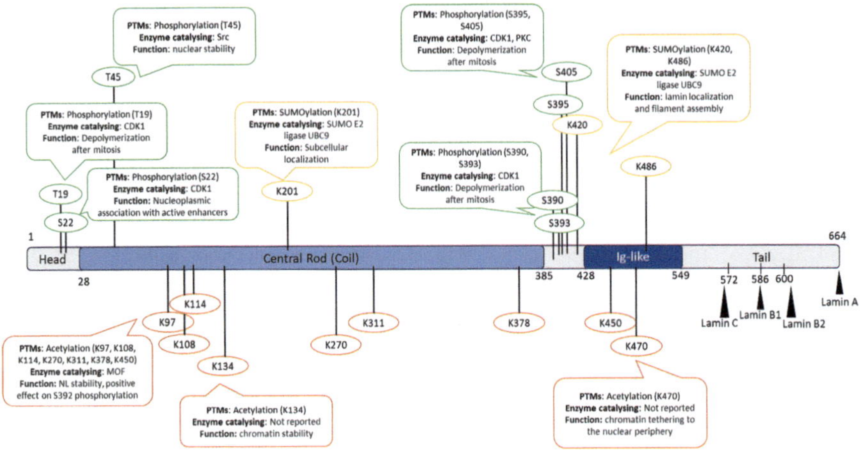


**Fig. 11.1** Lamin PTMs, their mediators, and physiological implications

ates the process, followed by cleavage of the terminal tripeptide by Ras-converting enzyme 1 (RCE1) or ZMPSTE24. Isoprenylcysteine carboxyl methyltransferase (ICMT) then methylates the exposed farnesylated cysteine, and ZMPSTE24 removes the final 15 tail domain residues, including the C-terminal farnesylcysteine methyl ester.

**Phosphorylation** is one of the most extensively studied post-translational modifications (PTMs) in lamins, playing a crucial role in their assembly, solubility, and interaction with chromatin. While phosphorylation occurs across the head, rod, and tail domains, it is most frequent in the head and tail regions [83, 118]. Lamin A has over 70 identified phosphorylation sites, which play key roles in structural integrity and function. Several enzymes phosphorylate lamins, including Cdk1 (targets Ser22 and Ser392 in lamin A/C and analogous residues in B-type lamins), PKC (phosphorylates S22/S392 and S395/S405), AKT (targets S404), Scr (specifically phosphorylates Tyr45), and EGFR that phosphorylates several tyrosine residues, including Y45, Y81, Y211, Y359, Y376, Y481, and Y646 [20, 22, 43, 126]. Phosphorylation regulates lamin polymerization and depolymerization, where lamin A phosphorylation enables the dynamic reorganization of nuclear envelope and chromatin interactions and B-type lamin phosphorylation impacts assembly state and nuclear envelope localization [43, 45, 49, 52, 75]. Overall, lamin phosphorylation is essential for nuclear assembly, chromatin organization, and multiple nuclear functions, serving as a key regulatory mechanism in cellular processes.

**Acetylation** is a key post-translational modification (PTM) of lamins, contributing to nuclear stability, cell cycle regulation, DNA repair, and genome integrity. Most lamin A/C acetylation sites are found within the coiled-coil rod domain (K97, K108, K114, K270, K311, and K378). In addition, K417 in NLS and K450/K470 in IgG-fold domains are also important acetylation sites. Lamin B1 has six acetylation sites (K33, K123, K157, K181, K271, and K483). Lamin B2 has six acetylation sites, which include K47, K81, K393, and K520. Karoutas et al. [61] identified that loss of the lysine acetyltransferase MOF or its associated NSL-complex members (KANSL2, KANSL3) leads to nuclear abnormalities, as abrogation of lamin A/C acetylation leads to an increase in S392 phosphorylation, resulting in higher lamin solubility, defective phosphorylation dynamics, and gross nuclear instability. Overall, lamin acetylation is essential for maintaining proper nuclear architecture.

**SUMOylation** is a conserved post-translational modification (PTM) where small ubiquitin-like modifiers (SUMOs) attach to lysine residues on target proteins, including lamins. SUMOylation involves three key enzymes: SUMO E1 (SAE1/SAE2 heterodimer), SUMO E2-conjugating enzyme (UBE2I/UBC9), and SUMO E3 ligases, which facilitate substrate specificity. Lamin A is a known target for SUMOylation [93, 139], interacting with UBC9. It contains a SUMOylation consensus sequence (CKXE/MKEE) near K201 in the rod domain, making it susceptible to SUMO2 modification, while SUMO1 and SUMO3 may modify its tail domain. SUMOylation is crucial for lamin function, influencing nuclear stability, transcription, apoptosis, and cell cycle regulation. SUMO-interacting motif 3 (SIM3) of lamin A interacts with SUMO2-modified proteins, contributing to nuclear lamina reorganization during mitosis [93]. Laminopathy mutations can disrupt SUMOylation. Loss of SUMOylation at K221 of lamin A contributes to familial cardiomyopathies [136].

**Carboxymethylation and isoprenylation** enhance lamin hydrophobicity [19, 66]. Evidence suggests that lamin A/C contains an arginine methylation site which is modified by PRMT and depletion of intracellular methyl group donor s-adenosylmethionine (SAM) disrupts nuclear lamina maturation [54, 63]. These modifications are crucial for cell stability and development, highlighting their significance in nuclear function and disease mechanisms.

**Ubiquitination** plays a significant role in protein stability and disease progression in laminopathies. RNF123, an E3 ubiquitin ligase, is upregulated in cells with rod domain lamin A mutations and mediates the ubiquitination and proteasomal degradation of pRb, LAP2a, and lamin B1 [62]. HECW2, another E3 ubiquitin ligase, regulates the lamin A-binding protein PCNA via PCNA-interacting protein (PIP) and lamin B1, and this interplay contributes to maintaining the nuclear lamina [70]. Disruptions in this balance may contribute to disease pathogenesis. These findings suggest that ubiquitin-mediated degradation of lamins and lamin-associated proteins could be a key factor in laminopathy progression, affecting nuclear integrity and cell cycle regulation.

**O-GlcNAcylation** is a significant post-translational modification (PTM) affecting lamin A [119]. This modification involves the enzymatic addition of O-GlcNAc to serine/threonine residues by O-GlcNAc transferase (OGT). Lamin A shows robust O-GlcNAcylation at specific tail domain residues (385–646), while lamin B1, lamin C, and progerin (Δ50) remain unmodified. Hart et al. [47] identified 11 O-GlcNAc sites in a region (601–645) unique to lamin A, with up to seven sugars per peptide. The deletion of Δ50 and Δ35 (596–664) reduces O-GlcNAcylation, indicating that these residues—absent in progeria—are essential for OGT substrate recognition and modification.

More broadly, lamin PTMs play a crucial role in cellular processes and disease mechanisms. In this chapter, we have discussed the implications of lamin PTMs in chromatin maintenance and stability, nuclear organization, cell cycle, metabolism, nucleophagy, stress, ageing and senescence, cellular differentiation, and finally diseases and disorders.

## Implications of Lamins PTMs in Normal Physiology

### *Chromatin Localization*

The eukaryotic genome is organized into well-defined territories which include both whole chromosomes and small domains [73]. Lamins act as a nucleoplasmic scaffold that directly or indirectly determines chromosome positioning throughout the nucleus and helps maintain genome integrity. Lamin–chromatin interactions involve direct and indirect interaction with N- and C-terminal tail domains of core histones [53] and DNA [6, 117, 123]. These multifaceted interactions are regulated by site-specific post-translational modifications of lamins which contribute to chromatin organization during cell cycle as well as chromatin stability. Acetylation in the coiled-coil rod domain (K97, K108, K114, K270, K311, and K378) of lamin A/C and the immunoglobulin-fold domain (K450 and K470) influence the chromatin tethering to the nuclear periphery [118]. Another acetylation at K134 lamin B1 acts as toggle controlling the chromatin stability by promoting error-free HR over NHEJ [98]. Acetylated K134 inhibits chromatin anchoring of DSBs to the lamina, which is a crucial step for productive cNHEJ, thus biasing towards homologous recombination, where chromatin mobility is essential. SUMOylation at K201, K420, and K486 has been reported to be important for lamin localization and filament assembly. Phosphorylation of lamin B1 at S395, S405, and S408 by PKCα leads to disassembly of the fibrous meshwork. This lamin B disassembly results in nuclear envelope breakdown which is needed for the formation of extracellular chromatin and protein trap structures called NETs (neutrophil extracellular traps) [77]. During inflammatory response, NETs pack the decondensed chromatin and are important for the propagation of inflammation by induction of proinflammatory cytokines. The potential mechanism includes the translocation of cytosolic PKCα to nucleus where it gets phosphorylated and activated by diacylglycerol (DAG). pPKCα phosphorylates lamin B followed by the lamin disassembly and nuclear envelope rupture. pPKCα-mediated lamin B phosphorylation impairments result in attenuated nuclear envelope rupture and NET formation. Similarly, NET formation is also regulated by CDK4/6-mediated phosphorylation of lamin A/C [3]. Additionally, methylation of lamin B increases resistance to proteolysis by caspases, preventing chromatin condensation and DNA fragmentation [68]. Farnesylation of lamin B1 is necessary to maintain chromatin positioning within the nuclear lamina during neuronal migration [59]. Deficiency of lamin B1 or B2 farnesylation is linked to neuronal defects. In addition, lamina might regulate global genomic organization through phase-separated compartments.

## *Nuclear Organization*

For the proper maintenance of nuclear shape and its integrity, PTM-mediated regulation of lamin and its cellular localization plays a pivotal role. In addition, mutations and altered expression of lamins can lead to aberrant nuclear stiffness and its loss of integrity. Phosphorylation of lamin A/C contributes majorly to the nuclear size maintenance and its subnuclear localization [100]. A detailed phosphoproteomic study has revealed that during interphase lamin A gets phosphorylated at 20 sites distributed over three regions, namely, N-terminal containing seven sites, between rod domains and Ig-fold having ten sites, and C-terminal domain containing three sites. Interestingly, eight of these phosphorylation sites, including S22 and S392, have been characterized as high turnover sites. Mutational-based studies have evidently shown that these phosphorylation sites, with S22 imparting the dominant effect, guide the ratio of nuclear periphery versus nucleoplasmic subsets of lamin A [67, 98]. Lamin phosphorylation also controls its nuclear localization. The nuclear import of lamin A is facilitated via PKC-mediated phosphorylation of lamin A at S403 and S404 near its nuclear localization signal (NLS) [75]. Interestingly, upon phosphorylation of lamin A at S268, a conserved site across the species, nuclear size decreases. This phenomenon was first observed during the early development of *Xenopus* where cPKC-mediated lamin B3 phosphorylation results in nuclear size reduction [33]. Another study reveals that AKT-mediated phosphorylation of mature lamin A S404 is essential for nuclear stability, and its mutation leads to nuclear abnormalities, a phenomenon similar to Emery–Dreifuss muscular dystrophy [15, 90]. Lamins also contribute to cellular mechanosensing as they bridge between the nuclear envelope and the cytoskeleton. The dynamic phosphorylation of lamin A/C at S22 contributes to substratum stiffness. Increase in the stiffness of the substratum results in increased cell spreading and elevated cytoskeleton tension which causes decreased S22 phosphorylation coupled with nuclear stiffness and increased myosin IIA expression [14]. During initiation and completion of mitosis in higher eukaryotic cells, the dynamic assembly and disassembly of nuclear lamina plays a pivotal role. A recent study shows that CDK1-mediated phosphorylation of endogenous lamin A/C is a key factor in the regulation of nuclear lamina. Both N-terminal Thr19/Ser22 and C-terminal Ser390/Ser392 get phosphorylated during mitosis, and 4D phospho-mimicking mutant T19D/S22D/S390D/S392D shows complete loss of lamin A filamentous structure in the interphase cells. Moreover, the non-phosphorylatable 4A mutant T19A/S22A/S390A/S392A has been reported to protect depolymerization of lamin A during mitosis. This engineered non-phosphorylatable mutant T19A/S22A in the LMNA gene also resulted in the elicited micronucleus formation and nuclear disruption during telophase. Thus, orchestration of both phosphorylation and dephosphorylation of lamin A is crucial for proper envelope dynamicity, nuclear envelope plasticity, and nuclear integrity during the cell cycle [80]. Studies have also illustrated that lamin PTMs, including acetylation, methylation, and SUMOylation, also play a crucial role in maintaining nuclear integrity. A study by Karoutas et al. has demonstrated that lamin A/C gets acetylated by MOF/MYST1/KAT8 at K311 and is necessary for protecting the nucleus from nuclear blebbing and micronuclei formation. Mechanistically, loss of lamin A/C acetylation leads to increase in lamin solubility and its defective phosphorylation which in turn results in compromised nuclear mechanic stability [61]. Additionally, lamin B1 (LMNB1) gets methylated at K417 by euchromatic histone methyltransferase (EHMT 1 and 2) which leads to its stability and nuclear peripheral localization. This methylation of LMNB1 is crucial for heterochromatin anchorage at the nuclear periphery as studies in ageing human iBlasts reveal that depletion in the level of EHMTs correlates with a decrease in LMNB1 methylation and peripheral heterochromatin domain organization [112]. Another PTM known as farnesylation has been recognized as a fundamental role player in lamin function. Both lamin B and prelamin A harbour a C-terminal CaaX motif in which cysteine residue gets farnesylated. This PTM acts as a lipid anchor contributing to the close association of lamin B1 and the nuclear membrane. Studies show that non-farnesylated lamin B1 in developing mice causes death after birth due to excessive neurodevelopmental defects and abnormal nuclear organization in which chromatin gets detached from the nuclear periphery [59].

## *Cell Cycle*

The eukaryotic nucleus is a highly organized yet dynamic structure that changes its organization as the cell progresses through the cell cycle. During interphase, nuclear lamins are primarily located at the nuclear lamina, thereby interacting with inner nuclear membrane proteins and heterochromatin. However, a portion of lamin A also resides in the nuclear interior [12]. During early mitosis, as the nuclear envelope disappears, the lamins become diffusible in the cytoplasm [92]. As mitosis concludes, lamins reassemble to form the nuclear lamina in daughter cells. This disassembly and reassembly process is regulated by phosphorylation [110] by cyclin-dependent kinase 1 (Cdk1) and protein kinase C (PKC) that phosphorylate lamin A to trigger lamina disassembly [31, 48], while protein phosphatase 1a dephosphorylates it during telophase to allow reformation [125]. Different phosphorylation patterns on lamin A significantly affect its assembly, turnover, and mobility. A- and B-type lamins follow distinct pathways during nuclear envelope disassembly and reassembly processes. Lamin B1 is observed to accumulate at the surface of chromosomes during the transition from anaphase to telophase, while lamin A only begins associating with the reforming nucleus after key nuclear envelope components, such as nuclear pore complexes, are in place. Following early G1, lamin A gradually relocates from the nuclear interior to the periphery over several hours [102]. Although lamin A and lamin C are often studied together due to their shared origin via alternative splicing and assumed functional similarity, emerging research suggests they may differ in biological roles and localization [2]. Notably, during mitotic exit, lamin A becomes part of the nuclear envelope, while lamin C remains in the nuclear interior, associated with LAD (lamina-associated domain) clusters. Phosphorylation at different residues of lamins, predominantly by Cdk1, is important throughout the cell cycle process. Specific phosphorylation sites, often referred to as 'mitotic sites', include Ser22 and Ser392 on lamin A/C, Ser23 and Ser393 on lamin B1, and several residues (Thr34, Ser37, Ser405) on lamin B2. Recent studies suggest that phosphorylation of lamins during interphase also has functional consequences. Modification of lamin A/C at Ser22 has been linked to cardiac conduction issues, and lamin C appears more susceptible to phosphorylation than lamin A, possibly due to its positioning within the nucleus. Other modifications, such as phosphorylation at Tyr45 by Src and several additional sites, can influence nuclear stability outside mitosis. During interphase, lamin A is also reported to be phosphorylated at Ser390, Ser404, Thr424, and Ser652, which is linked with maintaining nuclear structural integrity [138]. In eukaryotic cells, nuclear pore complexes (NPCs) are extensive structures that allow the selective transportation of macromolecules [25]. Nuclear pore distribution and density change throughout the cell cycle in human cells. Researchers have identified distinct regions in the nuclear envelope of HeLa S3 cells, known as 'pore-free islands' that lack nuclear pores, and they appear during the telophase and early G1 phase. As cells progress toward the S phase, these regions gradually disappear, and pore density increases. Interestingly, these pore-free areas are rich in inner nuclear membrane proteins such as emerin and lamin A/C but not lamin B. Similar patterns were observed in other human cell types, including U-2 OS and IMR90 cells. Reducing lamin A/C levels through siRNA led to the loss of these islands, while increasing lamin A/C levels promoted their formation or stability, implicating the structural and regulatory role of lamin A/C in maintaining pore distribution and determining nuclear subdomains [84].

Loss or mutation of A-type lamins has been linked to numerous problems in cell proliferation, including either an abnormal increase in cell division, cell cycle arrest, or early onset of cellular ageing [129]. Research indicates that lamins A and C play an important role in controlling the pRb/E2F pathway, which is essential for the transition from the G1 phase to the S phase of the cell cycle. A widely accepted model suggests that the active, hypophosphorylated form of pRb forms a complex with lamins A/C and LAP2α (most important nuclear envelope transmembrane protein). When A-type lamins are missing, this complex becomes unstable, leading to the breakdown of pRb [57]. Beyond pRb, lamins A/C also physically interact with cyclin D3, another key player in regulating progression through G1 [88].

## *Metabolism*

Lamins, which form a vital part of the nuclear lamina, are increasingly recognized for their involvement in regulating cellular metabolism and providing structural support to the nucleus. Specifically, A-type lamins (lamins A and C) influence several metabolic processes by modulating gene expression and cellular signalling pathways. They play a key role in organizing chromatin and interact with metabolic regulators such as SREBP1, which controls lipid metabolism, and PPARγ, which is important for fat storage and differentiation. Lamins also regulate crucial metabolic signalling pathways like mTOR and AMPK, essential for nutrient sensing and maintaining energy balance within cells [17]. Mutations in the LMNA gene can disturb these pathways, leading to metabolic abnormalities. Furthermore, lamins directly impact mitochondrial function, influencing mitochondrial biogenesis and dynamics (Ref). Lamin defects are often associated with reduced mitochondrial activity and elevated oxidative stress. The CaaX is critical as it undergoes post-translational modifications essential for lipid metabolism [71]. The correlation between lamins and lipid metabolism has been well studied. Lamin B1 overexpression suppresses lipid synthesis genes and reduces myelin-enriched lipids, which increases the risk of developing autosomal dominant leukodystrophy (ADLD) (Ref). Lamin A/C also activates NF-κB, which subsequently upregulates proinflammatory genes, leading to obesity-induced insulin resistance in adipose tissue macrophages (ATMs) (Ref).

## *Nucleophagy*

Nucleophagy is a selective form of autophagy that mediates the degradation of nuclear constituents, including the nuclear lamina. Lamin A/C-LC3 interactions have been reported to be influenced by lamin SUMOylation [76]. The authors also reported nucleophagy in human MDA-MB-231 and MCF-7 breast cancer cell lines, upon induction of DNA damage. DNA damage induces the nuclear accumulation of UBC9 ligase, resulting in SUMOylation of lamin A/C, which is required for their interaction with LC3.

# Lamin PTMs in Ageing and Senescence

Alterations in nuclear shape and compromised nuclear lamina are interlinked to altered chromatin functions and genomic stability. At the organismal level, nuclear localization defects and chromatin deregulations are associated with ageing and are strongly influenced by lamins. Here we discuss the role of post-translational modifications of lamins in ageing and senescence-like phenotypes. Specific downregulation of lamin B1 and secretion of proinflammatory cytokines known as senescence-associated secretory phenotype (SASP) have been studied as universal marker of cellular senescence [38, 121]. One of the lamin B PTMs, carboxyl methylation, resists its proteolysis by interleukin converting enzyme (ICE)-like proteases (e.g. caspases), thereby preventing the chromatin condensation and DNA fragmentation, causing apoptotic cell death [68]. Similarly, mutations in lamin A have been observed in premature ageing diseases HGPS and RD. The processing of mature lamin A requires farnesylation and carboxymethylation of its C-terminal CAXX motif. C-terminal isoprenylation and carboxyl methylation of lamins increase their hydrophobicity, thus promoting their interaction with nuclear membrane proteins [19, 66]. Mutated prelamin A fails to clip off its farnesylated tail, and accumulation of permanently farnesylated progerin leads to the HGPS and NGPS (Hutchinson–Gilford and the Néstor–Guillermo progeria syndromes, respectively). Farnesyltransferase inhibitors (FTIs), antisense oligonucleotides, etc., are being applied for reducing the accumulation of prelamin A or progerin to delay the onset of progeroid features in HGPS cells [128, 132]. A study by Glynn W and Glover W [41] showed that the G608G mutation leads to altered pre-mRNA splicing, producing progerin (LAΔ50), an irreversibly farnesylated truncated prelamin A responsible for premature ageing. Similarly, restrictive dermopathy (RD) arises from mutations in ZMPSTE24 or homozygous LMNA mutations, causing an accumulation of farnesylated and methylated prelamin A [147] [120, 133]. ZMPSTE24 deficiency can lead to DNA damage, delayed repair, and genomic instability. Mouse models, including ZMPSTE24 knockouts and LMNAG609G knock-ins, replicate many premature ageing phenotypes observed in human laminopathies.

## The Interplay Between Stress and PTMs of Nuclear Lamins

For ensuring the proper progression of events like chromatin organisation, DNA transcription, DNA replication, and DNA repair, maintaining the integrity of the nuclear envelope is crucial. The integrity and structural stability of the nucleus are supported by an underlying meshwork structure of rigid, gel-like intermediate filaments, namely, lamina. Nuclear lamins are physically connected to the cytoskeleton via the LINC complex. Mechanistically, activation of ATR kinase (DNA damage response kinase) due to cellular stress has been shown to rupture NE via lamin A/C phosphorylation at the S282 residue. In the absence of this phosphorylation, lamin A/C forms aggregates, which further leads to aberrant lamin organization. Moreover, loss of this phosphorylation perturbed the association of lamin A/C with the LINC complex and macro-H2A.2. This also results in the detachment of lamina from both cytoskeleton and chromatin, restricting the DNA damage-induced NE rupture [69]. Acetylation at K134 of lamin B1 at the nuclear periphery plays a major role in regulating DNA repair, cell cycle progression, and nuclear periphery maintenance. During infectious stress like herpesvirus type 1 (HSV-1), LMNB1 acetylation suppresses virus production by blocking lamina disruption. Further, during cellular stress like DNA damage, this K134 acetylation inhibits canonical nonhomologous end joining by impeding the recruitment of 53BP1 at the DNA damage foci. This results in a delayed DNA damage resolution and stable activation of G1/S checkpoint [98]. During human cytomegalovirus infection, which is prevalent in immunocompromised individuals, LMNB1 acetylation acts as a host response and functionally stabilizes the nuclear periphery and its integrity, thus impeding viral capsid nuclear egress and suppression of virus production [100]. Among all the organs, the brain is considered to be the most vulnerable to oxidative stress due to its high oxygen dependency, enriched polyunsaturated fatty acid environment, and the presence of excitotoxic amino acids. Also, proper persistence of neural stem cells/progenitor cells (NSPCs) is important throughout the mammalian lifespan for maintaining balanced tissue homeostasis. Interestingly, it has been reported that forkhead box O3 (FOXO3)-mediated induction of cGAS/STRING and type I interferon (IFN-I) responses modulates the neurogenic differentiation of NSPCs during the elevated cellular oxidative stress. Under this cellular stress, both FOXO3 and its transcriptional target glycine-N-methyltransferase (GNMT) get activated, and upregulation of GNMT results in the depletion of SAM levels. Further, a decrease in the intracellular pool of SAM negatively regulates the carboxymethylation and maturation of nuclear lamin, which leads to cytosolic leakage of DNA/chromatin fragments and activation of cGAS/STING-IFN-I cascade to deplete neurogenic differentiation [54]. The lamin A/C levels are also known to increase with matrix stiffness and elicit nuclear mechanical responses in the differentiation of mesenchymal stem cells. Low lamin A/C, on the other hand, correlates with less nucleus spread exhibited by MSCs on soft matrix. Studies show that matrix elasticity coregulates lamin A/C level and myosin IIA. Lamin A/C is rapidly phosphorylated in response to decreased cytoskeleton tension, leading to nuclear softening and lamin A/C turnover [14]. Androgen receptor serves as a connecting bridge between lamin A/C and the protein phosphatase PPP1, and loss of AR leads to nucleoplasmic redistribution of lamin A/C. Interestingly, this leads to depletion in lamin-PPP1 association and results in the elevated phosphorylation of lamin A/C at Ser 301, which binds to the transcriptional regulatory region of CAF effector genes, transforming them to active tumour-promoting CAFs in the cancer microenvironment [40]. Another group has performed a TMT-labelled quantitative lysine crotonylome analysis, including 12 pairs of hepatocellular carcinoma tissue along with the adjacent liver tissue, and has reported that lamin A undergoes crotonylation at K265/270. This study also reveals that hypoxic stress in liver cancer leads to downregulation of HDAC6-mediated decrotonylation of lamin A and crotonylation at K265/27, which promotes its subcellular position and promotes cell proliferation in liver cancer, along with preventing cellular senescence [137]. Mitochondrial dysregulation and oxidative stress are two of the most prevalent key features in the case of age-related neurodegenerative disorders like Parkinson's disease, Alzheimer's, and Huntington's disorder. A recent study revealed that mitochondrial stress triggers the neuronal cell death process during which lamin B1 gets phosphorylated at the T575 site via proteolytic activation of PKC δ and leads to nuclear membrane disassembly. Interestingly, their data revealed that nigral DAergic neurons collected from the postmortem brains of PD patients show higher activation of PKC δ, lamin B1 phosphorylation, and lamin B1 loss when compared to healthy control brains from a parallel age group [18].

## Role of Lamins in Cellular Differentiation

A-type lamins form a highly dynamic pool in the nucleoplasm and are found to exist as 3.5-nm-thick lamin filament at the peripheral lamina as resolved by cryo-electron tomography [127]. B-type lamins are expressed ubiquitously in embryonic and adult cells and form stable structures in the nuclear interior, whereas A-type lamins are particularly upregulated during differentiation [101]. A-type lamins are critical regulators of gene expression, cellular differentiation, DNA replication, and chromatin organization [1, 27, 134]. Lamin A influences key cellular signalling pathways Rb/E2F, Wnt/β-catenin, TGF-β, and Notch pathway through interactions with specific transcription factors [86]. Lamins interact with chromatin called lamina-associated domains (LADs), which are important regions for the regulation of gene expression. These domains house a variety of genes related to cell differentiation, and their expression status—whether active or silent—is influenced by their interaction with the lamina. LADs are classified as constitutive (cLADs), which are conserved across cell types, and facultative (fLADs), which vary by cell type and developmental stage. LADs usually contain few genes and show low gene activity. These regions tend to replicate DNA later and carry chemical markers like H3K9me2/3, which are linked to gene silencing. Genes tend to be expressed when LADs are separated from the lamina, while attachment to the lamina is typically linked to gene repression [10]. LADs can be quite large—up to ten million base pairs—and make up about one-third of the human genome.

Lamin A/C is crucial for regulating essential cell functions like migration, growth, differentiation, and homeostasis. Cell differentiation is influenced by mechanical forces, with A-type lamins playing a key role in mechanosensing [56] and responding to these cues through interactions with the cytoskeleton and extracellular matrix (ECM). These signals lead to changes in chromatin organization and nuclear proteins, which can activate or repress differentiation-related genes [124]. The lamin A/C-to-lamin B ratio differs depending on the cell type. Tissue and nuclear stiffness affect lamin A/C expression:

- Soft matrices (like fat tissue) reduce lamin A/C levels, promoting mesenchymal stem cells (MSCs) to adipogenic differentiation and contributing to fat phenotype development [50]. Embryonic stem cells have very low levels of lamin A/C, displaying deformable nuclei [74].
- Intermediate stiffness favours muscle (myocytes) cell differentiation with moderate lamin A/C expression [34].
- Stiff matrices (like bone) increase lamin A/C, stabilizing the nucleus against mechanical stress, hampering the cells with their migration potential, and promoting differentiation of stiff matrices towards a bone phenotype [124]. During

osteogenic differentiation, MSCs migrate to the bone where they adhere to the rigid ECMs and the mechanical cues like matrix stiffness and altered 3D geometry trigger an upregulation of lamin A/C. This promotes transition from a migratory to a differentiated osteogenic state, facilitating osteogenic differentiation.

Thus, lamin A/C levels increase during osteogenesis and decrease during adipogenesis of MSCs. Lamin A/C overexpression supports gene activation for differentiation, enhances stress responses, and activates regulators like YAP1, which is involved in cell growth and survival. Mechanical input from the environment helps guide tissue-specific gene expression through these processes. Hence, lamin A/C is key to connecting external mechanical cues with cellular function and fate.

- *Adipogenic differentiation*: Mutations in the *LMNA* gene, which codes for lamin A/C proteins in the nucleus, can lead to laminopathies. Some of these include forms of partial lipodystrophy tied to metabolic disorders. Recent studies have discovered that the p.R482W mutation disrupts the epigenetic regulation of long-range enhancers controlling the anti-adipogenic microRNA gene MIR335 in human adipocyte precursor cells. Specifically, the mutation compromises the normal, differentiation-dependent binding of lamin A to the MIR335 locus, thereby altering the chromatin landscape. Instead of promoting repressive H3K27 methylation, the mutation favours H3K27 acetylation at MIR335 enhancer regions, leading to transcriptional activation. Additionally, the mutation facilitates the spatial reorganization and clustering of MIR335 enhancers and promoter elements, contributing to the overexpression of MIR335 following adipogenic induction [105]. Lamin A/C may regulate MSC differentiation through interactions with Notch pathway components through emerin-dependent suppression of Notch. One study by Perepelina et al. demonstrated that this mutation (R482L) suppresses Notch activation and reduces adipogenic differentiation in MSCs, leading to nuclear abnormalities such as blebbing and altered morphology in fibroblasts [107].
- *Osteogenic differentiation*: The Wnt/β-catenin signalling pathway plays a pivotal role in the differentiation in osteogenic differentiation. β-catenin enhances osteoblast formation and matrix production by directly interacting with lamin A/C, promoting its nuclear translocation and nuclear activity. It suppresses adipogenic and chondrogenic differentiation, making this pathway vital for bone regeneration. The Notch signalling pathway also influences MSC differentiation. Notch activation promotes osteogenesis in a dose- and tissue-dependent manner, while its inhibition favours adipogenesis. Lamin A/C modulates Notch signalling through chromatin regulation and direct protein. Studies have shown that various LMNA mutations affect Notch activity and MSC differentiation. Mutations such as R527C and R471C influence the expression of Notch-related and osteogenic genes in a cell-type-specific manner [108].
- *Cardiogenic differentiation*: Recent studies using induced pluripotent stem cells (iPSCs) with tissue-specific LMNA mutations have shown that lamin–chromatin interactions are disrupted in genomic regions with distinct characteristics. iPSCs are good model to study laminopathies since they contain endogenous levels of lamin A/C [122]. Differentiation of iPSCs carrying cardiac-specific LMNA

mutations (T10I and R541C) into cardiomyocytes (iPS-CMs), adipocytes (iPS-adips), and hepatocytes (iPS-heps) revealed that lamina-associated domains (LADs) exhibit cell-type-specific organization. Notably, these LMNA mutations had a more pronounced detrimental effect on iPS-CMs than on adipocytes or hepatocytes, highlighting the tissue-specific vulnerability of cardiac cells [114].

- *Myogenic differentiation*: Differentiation in myoblasts is accompanied by a decrease in the soluble, nucleoplasmic pool of lamin A/C which relies on both pRb and cyclin D3 [89, 97]. Autosomal dominant Emery–Dreifuss muscular dystrophy (AD-EDMD), where lamin A/C is either reduced or mutated, is linked to disruptions in the Eb-MyoD pathway. It is associated with decreased pRb levels and a failure to properly hyperphosphorylate the remaining pRb, blocking myoblast differentiation [11, 60].

## Modified Lamins in Diseases and Disorders

Mutations and post-translational modifications in lamins contribute to diverse diseases, including premature ageing syndromes, muscular dystrophies, cardiomyopathies, cancer progression, immune evasion, and neurodegeneration, to name a few. Targeting lamin-associated pathways offers potential therapeutic strategies in these conditions.

### *Viral Infection*

Breaching the nuclear envelope is essential during the two phases of viral infections, i.e. during the initial phase and the nuclear egress of the viral progeny to initiate the next cycle of infection. Viruses employ diverse mechanisms to break the nuclear envelope barrier. The mechanisms include increasing the mobility of the lamin B receptor to destabilize the nuclear envelope [104], by changing the lamin A/C protein levels [13], and aberrant post-translational modification of lamins [110] [111]. Lamin A/C phosphorylation by cyclin-dependent kinase 1 (CDK1) is essential for mitotic disassembly and further for nuclear lamina breakdown. Intriguingly, the phosphorylation of lamin A/C to break the nuclear lamina is copied by the herpesvirus to facilitate their escape from the nucleus. The herpesvirus must traverse through the nuclear membrane to enter the cytoplasm of the infected cell, where it is packaged; however, the viral progeny cannot access the nuclear membrane because of the intact nuclear lamina. Viral kinase, UL97, phosphorylates lamin A at serine 22 to facilitate the lamina disassembly and escape of the viral progeny (Fig. 11.2) [46, 72, 94]. Baculoviruses use a nuclear egress mechanism similar to herpesviruses. Altered lamin B phosphorylation has also been implicated in baculovirus nuclear escape [135].

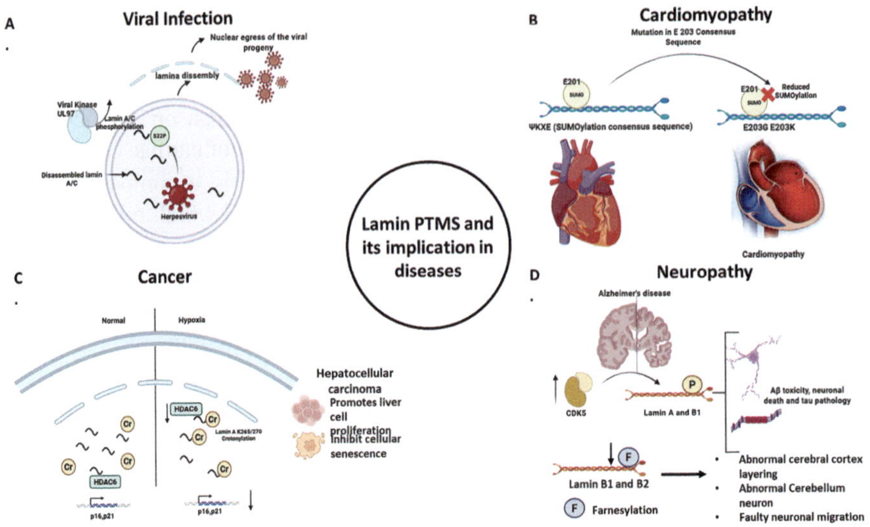


**Fig. 11.2** Post-translational modifications of lamin protein and its implication in various diseases. (**a**) Viral infection: Lamin A/C phosphorylation at serine 22 by viral kinase UL97 in herpesvirus causes nuclear egress of viral progeny. (**b**) Mutation at lamin A E203 to E203G/E203K reduces the SUMOylation at lamin A E201, resulting in familial cardiomyopathies. (**c**) In hypoxia conditions, decreased levels of HDAC6 result in increased levels of lamin crotonylation, causing liver cancer progression. (**d**) Phosphorylation of lamins A and B1 causes Alzheimer's pathology; lack of lamin B1 and B2 farnesylation results in abnormal brain health

## *Cardiomyopathy*

Small ubiquitin-like modifiers (SUMOs) catalyse the SUMOylation modification of lamins. SUMO family consists of four members, SUMO1, SUMO2, and SUMO3, which are ubiquitously expressed in all the eukaryotic cells, and the newly discovered member SUMO4, which is uniquely present in placenta, renal, pancreatic, and immune cells. Three critical enzymes SUMO E1, SUMO E2-conjugating enzyme, and E3 ligases catalyse the covalent and reversible SUMOylation of lysine residues of the target proteins [32, 58]. SUMOylation is indispensable for regulating localization, function, interactions, and stability of target proteins and influences nuclear import/export, transcription, apoptosis, and cell cycle regulation [39]. Dilated cardiomyopathy (DCM), the most common type of cardiomyopathy, is characterized by ventricular dilation, arrhythmia, and fibrosis. Over 200 mutations in LMNA have been identified to be responsible for DCM [8, 51, 81, 103]. Lamin mutations account for 6–8% of DCM cases [109]. Lamin SUMOylation at lysine at position 201 has been implicated in familial cardiomyopathies and conduction-system disease [36]. The conserved glutamic acid at the position 203 sits in the SUMOylation consensus sequence ΨKXE and is essential for the lysine SUMOylation at the 201st residue [113]. In line with this, both E203G and E203K mutant forms of lamin A show reduced SUMOylation compared to the wild-type protein (Fig. 11.2). A similar reduction is observed in skin fibroblasts from a patient carrying the E203K lamin A mutation. Additionally, the subcellular localization of the E203G and E203K mutant lamin A proteins is altered in a way that closely resembles the localization pattern seen in the lamin A K201R SUMO site mutant [136]. These mutations appear to be the first known human disease-associated changes that affect a critical residue in a SUMOylation consensus motif, leading to reduced SUMOylation of the resulting mutant protein.

## *Cancer*

Cancer cells have dysregulated gene expression, genomic instability, and abnormal nuclear shape. As lamin A/C plays many roles in the cell, including regulating gene expression and maintaining proper nuclear shape [115], it is likely lamin A/C also plays a role in the development and/or maintenance and propagation of cancer cells. Lamin A/C interacts with many promoters and is associated with gene expression and signalling pathways. Positioning of lamins with respect to promoters affects whether the gene is repressed or activated. The positioning of lamins upstream, proximal to the transcriptional start site, suppresses gene expression, while upstream distal causes increased expression. Lamin A/C can also regulate the histone post-translational modifications, thus altering the gene expression even in genes not directly associated with lamins [82]. Phosphorylated lamins reside in the nucleus, interacting with the enhancers and regulating gene expression [55]. Lamins can be overexpressed or underexpressed in many different cancer types. Since lamins play such diverse roles and their expression is highly variable in different tissue types, their effect in various cancer types is highly variable with no overall expression patterns [23, 37, 42, 95, 130]. Lamin A/C and lamin B levels change in various cancers compared to the normal tissue, which suggests that lamins either play a role in cancer progression or alter in response to it. Lamins constrain the cancer cell migration potential and modulate tumour growth. The lower levels of lamin, in general, confer a growth advantage. In addition, the lamina might indirectly affect cancer progression by altering gene expression via either transcription factors or changes in chromosomal conformation through LADs [149]. The effects of lamins in some cancers are further complicated by the fact that lamin A and lamin C expressions are not always affected equally [42]. Tumours employ various strategies to evade the immune system and flourish. One of the strategies is to mimic the anti-inflammatory mechanism of CNS by overexpressing NAT8L (N-acetyltransferase 8-like) and its metabolite NAA (N-acetyl aspartate). NAA suppresses natural killer cells and CD8+ T cell cytotoxicity by disrupting immunological synapse formation through PCAF-mediated acetylation of lamin A-K542 [78].

## *Neuropathy*

Lamin B1 and lamin B2, encoded by *LMNB1* and *LMNB2*, are expressed ubiquitously throughout the development, whereas lamins A and C are expressed only from the midgestational period in differentiated cells [29]. Lack of farnesylation of lamins B2 and B1 is associated with severe brain abnormalities, such as abnormal layering of cerebral cortex and cerebellum neurons and faulty neuronal migration [21, 59, 63, 65]. Aberrant phosphorylation of lamins is also implicated in some neurodegenerative diseases like Alzheimer's disease. Deregulated Cdk5 triggers early and irreversible nuclear lamina dispersion through direct phosphorylation of lamins A and B1, driving neuronal death independent of apoptosis. This mechanism links elevated Cdk5 activity in Alzheimer's disease to Aβ toxicity, tau pathology, and neurodegeneration (Fig. 11.2) [16]. Lamin methylation also has an important role in neurogenic differentiation. Oxidative stress in neural stem/progenitor cells (NSPCs) activates FOXO3-GNMT signalling, leading to SAM depletion and defective nuclear lamin maturation. This triggers cGAS/STING-IFN-I activation through chromatin leakage, ultimately suppressing neurogenic differentiation and impacting NSPC regenerative potential [54].

## Conclusion and Future Perspective

Nuclear lamins are essential components of the nuclear organization and normal cellular physiology. The meshwork-like organization of lamins and their related functions are regulated by multiple factors including the post-translational modifications like phosphorylation, acetylation, farnesylation, carboxymethylation, ubiquitination, and SUMOylation. The contribution of these lamin PTMs in upholding the chromatin organization, nuclear architecture, nuclear envelope maintenance, cell cycle progression, cellular differentiation, ageing, stress response, and disease progression has been increasingly studied. In humans, although numerous residues of lamins have been identified to be post-translationally modified (92 residues in lamins A/C, 52 in lamin B1, and 51 in lamin B2) [119], the functional implications of many of these PTMs are yet to be characterized. This keeps huge avenues open to further research in laminopathies and overall nuclear architecture.

**Acknowledgement** AS and RB acknowledge the Ph.D. programme fellowship of Jawaharlal Nehru Centre for Advanced Scientific Research (JNCASR), and SVB and ND acknowledge CSIR-UGC fellowship. TKK is a recipient of the J C Bose National Fellowship. AS, RB, ND, and SVB have contributed equally to writing the manuscript. AS, RB, ND, and SVB acknowledge Transcription and Disease Laboratory, Molecular Biology and Genetics Unit, JNCASR.

## References


1. Alcorta-Sevillano N, Macías I, Rodríguez CI, Infante A (2020) Crucial role of lamin A/C in the migration and differentiation of MSCs in bone. Cells 9(6):1330. Published 2020 May 26
2. Al-Saaidi R, Bross P (2015) Do lamin A and lamin C have unique roles? Chromosoma 124:1–12
3. Amulic B, Knackstedt SL, Abu Abed U, Deigendesch N, Harbort CJ, Caffrey BE, Brinkmann V, Heppner FL, Hinds PW, Zychlinsky A (2017) Cell-cycle proteins control production of neutrophil extracellular traps. Dev Cell 43(4):449–462.e5
4. Bank EM, Ben-Harush K, Feinstein N, Medalia O, Gruenbaum Y (2012) Structural and physiological phenotypes of disease-linked lamin mutations in C. elegans. J Struct Biol 177(1):106–112
5. Bank EM, Ben-Harush K, Wiesel-Motiuk N, Barkan R, Feinstein N, Lotan O, Medalia O, Gruenbaum Y (2011) A laminopathic mutation disrupting lamin filament assembly causes disease-like phenotypes in Caenorhabditis elegans. Mol Biol Cell 22(15):2716–2728
6. Baricheva EA, Berrios M, Bogachev SS, Borisevich IV, Lapik ER, Sharakhov IV, Stuurman N, Fisher PA (1996) DNA from Drosophila melanogaster beta-heterochromatin binds specifically to nuclear lamins in vitro and the nuclear envelope in situ. Gene 171(2):171–176
7. Ben-Harush K, Wiesel N, Frenkiel-Krispin D, Moeller D, Soreq E, Aebi U, Herrmann H, Gruenbaum Y, Medalia O (2009) The supramolecular organization of the C. elegans nuclear lamin filament. J Mol Biol 386(5):1392–1402.
8. Bhattacharjee P, Dasgupta D, Sengupta K (2017) DCM associated LMNA mutations cause distortions in lamina structure and assembly. Biochimica et biophysica acta General subjects 1861(11 Pt A):2598–2608
9. Bickmore WA (2013) The spatial organization of the human genome. Annu Rev Genomics Hum Genet 14:67–84
10. Bijarnia-Mahay S, Roy G, Padiath QS, Saxena R, Verma IC (2021). LMNB1 Duplication-Mediated Autosomal Dominant Adult-Onset Leukodystrophy in an Indian Family. Ann Indian Acad Neurol 24(3):413–416
11. Bitman-Lotan E, Orian A (2021) Nuclear organization and regulation of the differentiated state. Cell Mol Life Sci 78(7):3141–3158
12. Bonne G, Barletta M, Varnous S et al (1999) Mutations in the gene encoding lamin A/C cause autosomal dominant Emery-Dreifuss muscular dystrophy. Nat Genet 21:285–288
13. Bronshtein I, Kepten E, Kanter I, Berezin S, Lindner M, Redwood AB et al (2015) Loss of lamin A function increases chromatin dynamics in the nuclear interior. Nat Commun 6:8044
14. Butin-Israeli V et al (2011) Simian virus 40 induces lamin A/C fluctuations and nuclear envelope deformation during cell entry. Nucleus 2(4):320–330
15. Buxboim A, Swift J, Irianto J et al (2014) Matrix elasticity regulates lamin-A,C phosphorylation and turnover with feedback to actomyosin. Curr Biol 24(16):1909–1917
16. Cenni V, Bertacchini J, Beretti F et al (2008) Lamin A Ser404 is a nuclear target of Akt phosphorylation in C2C12 cells. J Proteome Res 7(11):4727–4735
17. Chang K-H et al (2011) Nuclear envelope dispersion triggered by deregulated Cdk5 precedes neuronal death. Mol Biol Cell 22(9):1452–1462
18. Charar C, Gruenbaum Y (2017) Lamins and metabolism. Clin Sci (London, England: 1979) 131(2):105–111
19. Charli A, Luo J, Palanisamy B et al (2024) Mitochondrial stress disassembles nuclear architecture through proteolytic activation of PKCδ and lamin B1 phosphorylation in neuronal cells: implications for pathogenesis of age-related neurodegenerative diseases. bioRxiv [Preprint]. Published 2024 Nov 1
20. Chelsky D, Sobotka C, O'Neill CL (1989) Lamin B methylation and assembly into the nuclear envelope. J Biol Chem 264(13):7637–7643
21. Chu CT, Chen YH, Chiu WT, Chen HC (2021) Tyrosine phosphorylation of lamin A by Src promotes disassembly of nuclear lamina in interphase. Life Sci Alliance 4(10):e202101120

22. Coffinier C et al (2010) Abnormal development of the cerebral cortex and cerebellum in the setting of lamin B2 deficiency. Proc Natl Acad Sci USA 107(11):5076–5081
23. Collas P, Thompson L, Fields AP, Poccia DL, Courvalin JC (1997) Protein kinase C-mediated interphase lamin B phosphorylation and solubilization. J Biol Chem 272(34):21274–21280
24. Contu F et al (2018) Distinct 3D structural patterns of lamin A/C expression in Hodgkin and Reed-Sternberg cells. Cancer 10(9):286
25. Cortelli P, Terlizzi R, Capellari S, Benarroch E (2012) Nuclear lamins: functions and clinical implications. Neurology 79(16):1726–1731
26. Cronshaw JM, Krutchinsky AN, Zhang W, Chait BT, Matunis MJ (2002) Proteomic analysis of the mammalian nuclear pore complex. J Cell Biol 158(5):915–927
27. Dai Y, Ma Y, Li S, et al (2017). An LMNB1 Duplication Caused Adult-Onset Autosomal Dominant Leukodystrophy in Chinese Family: Clinical Manifestations, Neuroradiology and Genetic Diagnosis. Front Mol Neurosci.10–215
28. Dauer WT, Worman HJ (2009) The nuclear envelope as a signaling node in development and disease. Dev Cell 17(5):626–638
29. de Leeuw R, Gruenbaum Y, Medalia O (2018) Nuclear lamins: thin filaments with major functions. Trends Cell Biol 28(1):34–45
30. Dechat T, Korbei B, Vaughan OA, Vlcek S, Hutchison CJ, Foisner R (2000) Lamina-associated polypeptide 2alpha binds intranuclear A-type lamins. J Cell Sci 113(Pt 19):3473–3484
31. Dechat T et al (2008) Nuclear lamins: major factors in the structural organization and function of the nucleus and chromatin. Genes Dev 22(7):832–853
32. Dechat T, Adam SA, Taimen P, Shimi T, Goldman RD (2010) Nuclear lamins. Cold Spring Harb Perspect Biol 2(11):a000547
33. Denecke J, Brune T, Feldhaus T, et al (2006). A homozygous ZMPSTE24 null mutation in combination with a heterozygous mutation in the LMNA gene causes Hutchinson-Gilford progeria syndrome (HGPS): insights into the pathophysiology of HGPS. Hum Mutat 27(6):524–531
34. Dessev GN, Iovcheva-Dessev C, Goldman RD (1990) Lamin dimers. Presence in the nuclear lamina of surf clam oocytes and release during nuclear envelope breakdown. J Biol Chem 265:12636–12641
35. Desterro JM et al (1997) Ubch9 conjugates SUMO but not ubiquitin. FEBS Lett 417(3):297–300
36. Edens LJ, Dilsaver MR, Levy DL (2017) PKC-mediated phosphorylation of nuclear lamins at a single serine residue regulates interphase nuclear size in Xenopus and mammalian cells. Mol Biol Cell 28(10):1389–1399
37. Engler AJ, Sen S, Sweeney HL, Discher DE (2006) Matrix elasticity directs stem cell lineage specification. Cell 126(4):677–689
38. Farnsworth CC, Wolda SL, Gelb MH, Glomset JA (1989) Human lamin B contains a farnesylated cysteine residue. J Biol Chem 264(34):20422–20429
39. Fatkin D et al (1999) Missense mutations in the rod domain of the lamin A/C gene as causes of dilated cardiomyopathy and conduction-system disease. N Engl J Med 341(23):1715–1724
40. Favreau C, Dubosclard E, Ostlund C, (2003). Expression of lamin A mutated in the carboxyl-terminal tail generates an aberrant nuclear phenotype similar to that observed in cells from patients with Dunnigan-type partial lipodystrophy and Emery-Dreifuss muscular dystrophy. Exp Cell Res 282(1):14–23
41. Foster CR et al (2011) The role of lamin A in cytoskeleton organization in colorectal cancer cells: a proteomic investigation. Nucleus (Austin, Tex) 2(5):434–443
42. Freund A, Laberge RM, Demaria M, Campisi J (2012) Lamin B1 loss is a senescence-associated biomarker. Mol Biol Cell 23(11):2066–2075
43. Geiss-Friedlander R, Melchior F (2007) Concepts in sumoylation: a decade on. Nat Rev Mol Cell Biol 8(12):947–956
44. Ghosh S, Isma J, Ostano P et al (2024) Nuclear lamin A/C phosphorylation by loss of androgen receptor leads to cancer-associated fibroblast activation. Nat Commun 15(1):7984. Published 2024 Sep 12

45. Giorgio E, Rolyan H, Kropp L, et al (2013). Analysis of LMNB1 duplications in autosomal dominant leukodystrophy provides insights into duplication mechanisms and allele-specific expression. Hum Mutat 34(8):1160–1171
46. Glynn MW, Glover TW (2005) Incomplete processing of mutant lamin A in Hutchinson-Gilford progeria leads to nuclear abnormalities, which are reversed by farnesyltransferase inhibition. Hum Mol Genet 14(20):2959–2969
47. Gong G et al (2015) Loss of lamin A but not lamin C expression in epithelial ovarian cancer cells is associated with metastasis and poor prognosis. Pathol Res Pract 211(2):175–182
48. Goss VL, Hocevar BA, Thompson LJ, Stratton CA, Burns DJ, Fields AP (1994) Identification of nuclear beta II protein kinase C as a mitotic lamin kinase. J Biol Chem 269(29):19074–19080
49. Gruenbaum Y, Medalia O (2015) Lamins: the structure and protein complexes. Curr Opin Cell Biol 32:7–12
50. Haas M, Jost E (1993) Functional analysis of phosphorylation sites in human lamin A controlling lamin disassembly, nuclear transport and assembly. Eur J Cell Biol 62(2):237–247
51. Hamirally S et al (2009) Viral mimicry of Cdc2/cyclin-dependent kinase 1 mediates disruption of nuclear lamina during human cytomegalovirus nuclear egress. PLoS Pathog 5(1):e1000275
52. Hart GW, Slawson C, Ramirez-Correa G, Lagerlof O (2011) Cross talk between O-GlcNAcylation and phosphorylation: roles in signaling, transcription, and chronic disease. Annu Rev Biochem 80:825–858
53. Heald R, McKeon F (1990) Mutations of phosphorylation sites in lamin A that prevent nuclear lamina disassembly in mitosis. Cell 61:579–589
54. Hennekes H, Peter M, Weber K, Nigg EA (1993) Phosphorylation on protein kinase C sites inhibits nuclear import of lamin B2. J Cell Biol 120(6):1293–1304
55. Heo SJ, Driscoll TP, Thorpe SD et al (2016) Differentiation alters stem cell nuclear architecture, mechanics, and mechano-sensitivity. Elife 5:e18207. Published 2016 Nov 30
56. Hershberger RE, Siegfried JD (2011) Update 2011: clinical and genetic issues in familial dilated cardiomyopathy. J Am Coll Cardiol 57(16):1641–1649
57. Hocevar BA, Burns DJ, Fields AP (1993) Identification of protein kinase C (PKC) phosphorylation sites on human lamin B. Potential role of PKC in nuclear lamina structural dynamics. J Biol Chem 268(10):7545–7552
58. Höger TH, Krohne G, Kleinschmidt JA (1991) Interaction of Xenopus lamins A and LII with chromatin in vitro mediated by a sequence element in the carboxyterminal domain. Exp Cell Res 197(2):280–289
59. Hwang I, Uchida H, Dai Z et al (2021) Cellular stress signaling activates type-I IFN response through FOXO3-regulated lamin posttranslational modification. Nat Commun 12(1):640. Published 2021 Jan 28
60. Ikegami K et al (2020) Phosphorylated lamin A/C in the nuclear interior binds active enhancers associated with abnormal transcription in progeria. Dev Cell 52(6):699–713.e11
61. Irianto J, Pfeifer CR, Bennett RR, et al (2016). Nuclear constriction segregates mobile nuclear proteins away from chromatin. Mol Biol Cell 27(25):4011–4020
62. Isermann P, Lammerding J (2013) Nuclear mechanics and mechanotransduction in health and disease. Curr Biol 23(24):R1113–R1121
63. Johnson BR, Nitta RT, Frock RL et al (2004) A-type lamins regulate retinoblastoma protein function by promoting subnuclear localization and preventing proteasomal degradation. Proc Natl Acad Sci USA 101(26):9677–9682
64. Johnson ES, Blobel G (1997) Ubc9p is the conjugating enzyme for the ubiquitin-like protein Smt3p. J Biol Chem 272(43):26799–26802
65. Jung HJ, Nobumori C, Goulbourne CN, Tu Y, Lee JM, Tatar A, Wu D, Yoshinaga Y, de Jong PJ, Coffinier C, Fong LG, Young SG (2013) Farnesylation of lamin B1 is important for retention of nuclear chromatin during neuronal migration. Proc Natl Acad Sci USA 110(21):E1923–E1932
66. Kandert S, Wehnert M, Müller CR, Buendia B, Dabauvalle MC (2009) Impaired nuclear functions lead to increased senescence and inefficient differentiation in human myoblasts with a dominant p.R545C mutation in the LMNA gene. Eur J Cell Biol 88(10):593–608

67. Karoutas A, Szymanski W, Rausch T, Guhathakurta S, Rog-Zielinska EA, Peyronnet R, Seyfferth J, Chen HR, de Leeuw R, Herquel B, Kimura H, Mittler G, Kohl P, Medalia O, Korbel JO, Akhtar A (2019) The NSL complex maintains nuclear architecture stability via lamin A/C acetylation. Nat Cell Biol 21(10):1248–1260
68. Khanna R, Krishnamoorthy V, Parnaik VK (2018) E3 ubiquitin ligase RNF123 targets lamin B1 and lamin-binding proteins. FEBS J 285(12):2243–2262
69. Kim SJ, Yoo BC, Uhm CS, Lee SW (2011) Posttranslational arginine methylation of lamin A/C during myoblast fusion. Biochim Biophys Acta 1814(2):308–317
70. Kim Y, Zheng Y (2013) Generation and characterization of a conditional deletion allele for Lmna in mice. Biochem Biophys Res Commun 440(1):8–13
71. Kim Y et al (2011) Mouse B-type lamins are required for proper organogenesis but not by embryonic stem cells. Science (New York, N.Y.) 334(6063):1706–1710
72. Kitten GT, Nigg EA (1991) The CaaX motif is required for isoprenylation, carboxyl methylation, and nuclear membrane association of lamin B2. J Cell Biol 113(1):13–23
73. Kochin V, Shimi T, Torvaldson E et al (2014) Interphase phosphorylation of lamin A. J Cell Sci 127(Pt 12):2683–2696
74. Kowluru A (2000) Evidence for the carboxyl methylation of nuclear lamin-B in the pancreatic beta cell. Biochem Biophys Res Commun 268(2):249–254
75. Kovacs MT, Vallette M, Wiertsema P et al (2023) DNA damage induces nuclear envelope rupture through ATR-mediated phosphorylation of lamin A/C. Mol Cell 83(20):3659–3668.e10
76. Krishnamoorthy V, Khanna R, Parnaik VK (2018) E3 ubiquitin ligase HECW2 targets PCNA and lamin B1. Biochimica et biophysica acta. Mol Cell Res 1865(8):1088–1104
77. Kuchay S et al (2019) GGTase3 is a newly identified geranylgeranyltransferase targeting a ubiquitin ligase. Nat Struct Mol Biol 26(7):628–636
78. Kuny CV et al (2010) Cyclin-dependent kinase-like function is shared by the beta- and gamma- subset of the conserved herpesvirus protein kinases. PLoS Pathog 6(9):e1001092
79. Küpper K, Kölbl A, Biener D, Dittrich S, von Hase J, Thormeyer T, Fiegler H, Carter NP, Speicher MR, Cremer T, Cremer M (2007) Radial chromatin positioning is shaped by local gene density, not by gene expression. Chromosoma 116(3):285–306
80. Lammerding J (2011) Mechanics of the nucleus. Compr Physiol 1(2):783–807
81. Leukel M, Jost E (1995) Two conserved serines in the nuclear localization signal flanking region are involved in the nuclear targeting of human lamin A. Eur J Cell Biol 68(2):133–142
82. Li Y, Jiang X, Zhang Y, Gao Z, Liu Y, Hu J, Hu X, Li L, Shi J, Gao N (2019) Nuclear accumulation of UBC9 contributes to SUMOylation of lamin A/C and nucleophagy in response to DNA damage. J Exp Clin Cancer Res 38(1):67
83. Li Y, Li M, Weigel B, Mall M, Werth VP, Liu ML (2020) Nuclear envelope rupture and NET formation is driven by PKCα-mediated lamin B disassembly. EMBO Rep 21(8):e48779
84. Li Y et al (2024) Tumor cells impair immunological synapse formation via central nervous system-enriched metabolite. Cancer Cell 42(6):985–1002.e18
85. Lin F, Worman HJ (1993) Structural organization of the human gene encoding nuclear lamin A and nuclear lamin C. J Biol Chem 268(22):16321–16326
86. Liu S, Xiong F, Dou Z et al (2025) Phosphorylation of lamin A/C regulates the structural integrity of the nuclear envelope. J Biol Chem 301(1):108033
87. Luk A et al (2009) Dilated cardiomyopathy: a review. J Clin Pathol 62(3):219–225
88. Lund E et al (2013) Lamin A/C-promoter interactions specify chromatin state-dependent transcription outcomes. Genome Res 23(10):1580–1589
89. Machowska M, Piekarowicz K, Rzepecki R (2015) Regulation of lamin properties and functions: does phosphorylation do it all? Open Biol 5(11):150094
90. Maeshima K, Yahata K, Sasaki Y et al (2006) Cell-cycle-dependent dynamics of nuclear pores: pore-free islands and lamins. J Cell Sci 119(Pt 21):4442–4451
91. Maggi L et al (2016) Skeletal muscle laminopathies: a review of clinical and molecular features. Cells 5(3):33
92. Maraldi NM, Lattanzi G, Cenni V, Bavelloni A, Marmiroli S, Manzoli FA (2010) Laminopathies and A-type lamin-associated signalling pathways. Adv Enzym Regul 50(1):248–261

93. Maraldi NM, Capanni C, Cenni V, Fini M, Lattanzi G (2011) Laminopathies and lamin-associated signaling pathways. J Cell Biochem 112(4):979–992
94. Mariappan I, Gurung R, Thanumalayan S, Parnaik VK (2007) Identification of cyclin D3 as a new interaction partner of lamin A/C. Biochem Biophys Res Commun 355(4):981–985
95. Mariappan I, Parnaik VK (2005) Sequestration of pRb by cyclin D3 causes intranuclear reorganization of lamin A/C during muscle cell differentiation. Mol Biol Cell 16(4):1948–1960
96. Marmiroli S, Bertacchini J, Beretti F et al (2009) A-type lamins and signaling: the PI 3-kinase/Akt pathway moves forward. J Cell Physiol 220(3):553–561
97. McKenna T, Baek J-H, Eriksson M (2013) Laminopathies
98. Moir RD, Spann TP, Lopez-Soler RI, Yoon M, Goldman AE, Khuon S et al (2000) Review: the dynamics of the nuclear lamins during the cell cycle- relationship between structure and function. J Struct Biol 129:324–334
99. Moriuchi T, Kuroda M, Kusumoto F, Osumi T, Hirose F (2016) Lamin A reassembly at the end of mitosis is regulated by its SUMO-interacting motif. Exp Cell Res 342(1):83–94
100. Morrison LA, DeLassus GS (2011) Breach of the nuclear lamina during assembly of herpes simplex viruses. Nucleus (Austin, Tex) 2(4):271–276
101. Moss SF et al (1999) Decreased and aberrant nuclear lamin expression in gastrointestinal tract neoplasms. Gut 45(5):723–729
102. Münkel C, Eils R, Dietzel S, Zink D, Mehring C, Wedemann G, Cremer T, Langowski J (1999) Compartmentalization of interphase chromosomes observed in simulation and experiment. J Mol Biol 285(3):1053–1065
103. Muralikrishna B, Dhawan J, Rangaraj N, Parnaik VK (2001) Distinct changes in intranuclear lamin A/C organization during myoblast differentiation. J Cell Sci 114(Pt 22):4001–4011
104. Murray-Nerger LA, Justice JL, Rekapalli P, Hutton JE, Cristea IM (2021) Lamin B1 acetylation slows the G1 to S cell cycle transition through inhibition of DNA repair. Nucleic Acids Res 49(4):2044–2064
105. Murray-Nerger LA, Cristea IM (2021) Lamin post-translational modifications: emerging toggles of nuclear organization and function. Trends Biochem Sci 46(10):832–847
106. Murray LA, Sheng X, Cristea IM (2018) Orchestration of protein acetylation as a toggle for cellular defense and virus replication. Nat Commun 9(1):4967. Published 2018 Nov 23
107. Naetar N, Ferraioli S, Foisner R (2017) Lamins in the nuclear interior – life outside the lamina. J Cell Sci 130(13):2087–2096
108. Naetar N, Georgiou K, Knapp C, Bronshtein I, Zier E, Fichtinger P et al (2021) LAP2alpha maintains a mobile and low assembly state of A-type lamins in the nuclear interior. eLife 10:e63476
109. Neupane B, Pradhan K, Ortega-Ramirez AM, Aidery P, Kucikas V, Marks M, van Zandvoort MAMJ, Klingel K, Witte KK, Gründer S, Marx N, Gramlich M (2022) Personalized medicine approach in a DCM patient with *LMNA* mutation reveals dysregulation of mTOR signaling. J Pers Med 12(7):1149
110. Okada Y et al (2005) Dissociation of heterochromatin protein 1 from lamin B receptor induced by human polyomavirus agnoprotein: role in nuclear egress of viral particles. EMBO Rep 6(5):452–457
111. Okada K, Ito D, Morimoto S, et al (2024). Multiple lines of evidence for disruption of nuclear lamina and nucleoporins in FUS amyotrophic lateral sclerosis. Brain 147(11):3933–3948
112. Oldenburg A, Briand N, Sørensen AL et al (2017) A lipodystrophy-causing lamin A mutant alters conformation and epigenetic regulation of the anti-adipogenic *MIR335* locus. J Cell Biol 216(9):2731–2743
113. Padiath QS, Saigoh K, Schiffmann R, et al (2006). Lamin B1 duplications cause autosomal dominant leukodystrophy. Nat Genet 38(10):1114–1123
114. Parnaik VK, Chaturvedi P, Muralikrishna B (2011) Lamins, laminopathies and disease mechanisms: possible role for proteasomal degradation of key regulatory proteins. J Biosci 36(3):471–479

115. Perepelina K, Dmitrieva R, Ignatieva E, Borodkina A, Kostareva A, Malashicheva A (2018) Lamin A/C mutation associated with lipodystrophy influences adipogenic differentiation of stem cells through interaction with Notch signaling. Biochem Cell Biol 96(3):342–348
116. Perepelina K, Klauzen P, Kostareva A, Malashicheva A (2019) Tissue-specific influence of lamin A mutations on notch signaling and osteogenic phenotype of primary human mesenchymal cells. Cells 8(3):266. Published 2019 Mar 21
117. Pérez-Serra A et al (2015) A novel mutation in lamin a/c causing familial dilated cardiomyopathy associated with sudden cardiac death. J Card Fail 21(3):217–225
118. Peter M, Nakagawa J, Dorée M, Labbé JC, Nigg EA (1990) *In Vitro* disassembly of the nuclear lamina and M phase-specific phosphorylation of lamins by Cdc2 kinase. Cell 61:591–602
119. Prokocimer M, Davidovich M, Nissim-Rafinia M, et al (2009). Nuclear lamins: key regulators of nuclear structure and activities. J Cell Mol Med 13(6):1059–1085
120. Radsak KD et al (1991) Focal nuclear envelope lesions and specific nuclear lamin A/C dephosphorylation during infection with human cytomegalovirus. Eur J Cell Biol 54(2):299–304
121. Rao RA, Ketkar AA, Kedia N et al (2019) KMT1 family methyltransferases regulate heterochromatin-nuclear periphery tethering via histone and non-histone protein methylation. EMBO Rep 20(5):e43260
122. Rodriguez MS et al (2001) SUMO-1 conjugation in vivo requires both a consensus modification motif and nuclear targeting. J Biol Chem 276(16):12654–12659
123. Shah PP, Lv W, Rhoades JH et al (2021) Pathogenic LMNA variants disrupt cardiac lamina-chromatin interactions and de-repress alternative fate genes. Cell Stem Cell 28(5):938–954.e9
124. Shimi T et al (2015) Structural organization of nuclear lamins A, C, B1, and B2 revealed by superresolution microscopy. Mol Biol Cell 26(22):4075–4086
125. Shimi T, Pfleghaar K, Kojima S, Pack CG, Solovei I, Goldman AE, Adam SA, Shumaker DK, Kinjo M, Cremer T, Goldman RD (2008) The A- and B-type nuclear lamin networks: microdomains involved in chromatin organization and transcription. Genes Dev 22(24):3409–3421
126. Shoeman RL, Traub P (1990) The in vitro DNA-binding properties of purified nuclear lamin proteins and vimentin. J Biol Chem 265(16):9055–9061
127. Simon DN, Wilson KL (2013) Partners and post-translational modifications of nuclear lamins. Chromosoma 122(1–2):13–31
128. Snider NT, Omary MB (2014) Post-translational modifications of intermediate filament proteins: mechanisms and functions. Nat Rev Mol Cell Biol 15(3):163–177
129. Spear ED, Hsu ET, Nie L, Carpenter EP, Hrycyna CA, Michaelis S (2018) *ZMPSTE24* missense mutations that cause progeroid diseases decrease prelamin A cleavage activity and/or protein stability. Dis Model Mech 11(7):dmm033670
130. Spektor A, Umbreit NT, Pellman D (2017) Cell biology: when your own chromosomes act like foreign DNA. Curr Biol 27(22):R1228–R1231
131. Steele-Stallard HB, Pinton L, Sarcar S et al (2018) Modeling skeletal muscle laminopathies using human induced pluripotent stem cells carrying pathogenic *LMNA* mutations. Front Physiol 9:1332. Published 2018 Oct 15
132. Stierlé V, Couprie J, Ostlund C, Krimm I, Zinn-Justin S, Hossenlopp P, Worman HJ, Courvalin JC, Duband-Goulet I (2003) The carboxyl-terminal region common to lamins A and C contains a DNA binding domain. Biochemistry 42(17):4819–4828
133. Swift J, Ivanovska IL, Buxboim A et al (2013) Nuclear lamin-A scales with tissue stiffness and enhances matrix-directed differentiation. Science 341(6149):1240104
134. Thompson LJ, Bollen M, Fields AP (1997) Identification of protein phosphatase 1 as a mitotic lamin phosphatase. J Biol Chem 272:29693–29697
135. Tsai CF, Wang YT, Yen HY, Tsou CC, Ku WC, Lin PY, Chen HY, Nesvizhskii AI, Ishihama Y, Chen YJ (2015) Large-scale determination of absolute phosphorylation stoichiometries in human cells by motif-targeting quantitative proteomics. Nat Commun 6:6622
136. Turgay Y, Eibauer M, Goldman AE et al (2017) The molecular architecture of lamins in somatic cells. Nature 543(7644):261–264

137. Varga R, Eriksson M, Erdos MR, Olive M, Harten I, Kolodgie F, Capell BC, Cheng J, Faddah D, Perkins S, Avallone H, San H, Qu X, Ganesh S, Gordon LB, Virmani R, Wight TN, Nabel EG, Collins FS (2006) Progressive vascular smooth muscle cell defects in a mouse model of Hutchinson-Gilford progeria syndrome. Proc Natl Acad Sci USA 103(9):3250–3255
138. Verstraeten VL, Broers JL, Ramaekers FC, van Steensel MA (2007) The nuclear envelope, a key structure in cellular integrity and gene expression. Curr Med Chem 14(11):1231–1248
139. Wang AS et al (2015) Tissue specific loss of A-type lamins in the gastrointestinal epithelium can enhance polyp size. Differentiation 89(1–2):11–21
140. Worman HJ, Bonne G (2007) "Laminopathies": a wide spectrum of human diseases. Exp Cell Res 313(10):2121–2133
141. Yang SH, Meta M, Qiao X, Frost D, Bauch J, Coffinier C, Majumdar S, Bergo MO, Young SG, Fong LG (2006) A farnesyltransferase inhibitor improves disease phenotypes in mice with a Hutchinson-Gilford progeria syndrome mutation. J Clin Invest 116(8):2115–2121
142. Young SG, Meta M, Yang SH, Fong LG (2006) Prelamin A farnesylation and progeroid syndromes. J Biol Chem 281(52):39741–39745
143. Zhang B, Yang Y, Keyimu R, Hao J, Zhao Z, Ye R (2019) The role of lamin A/C in mesenchymal stem cell differentiation. J Physiol Biochem 75(1):11–18
144. Zhang X et al (2017) Baculovirus infection induces disruption of the nuclear lamina. Sci Rep 7(1):7823
145. Zhang YQ, Sarge KD (2008) Sumoylation regulates lamin A function and is lost in lamin A mutants associated with familial cardiomyopathies. J Cell Biol 182(1):35–39
146. Zhang D, Tang J, Xu Y et al (2022) Global crotonylome reveals hypoxia-mediated lamin A crotonylation regulated by HDAC6 in liver cancer. Cell Death Dis 13(8):717. Published 2022 Aug 17
147. Zheng M, Jin G, Zhou Z (2022) Post-translational modification of lamins: mechanisms and functions. Front Cell Dev Biol 10:864191Published 2022 May 17
148. Zhong N, Radu G, Ju W, Brown WT (2005) Novel progerin-interactive partner proteins hnRNP E1, EGF, Mel 18, and UBC9 interact with lamin A/C. Biochem Biophys Res Commun 338(2):855–861

GPSR Compliance

*The European Union's (EU) General Product Safety Regulation (GPSR) is a set of rules that requires consumer products to be safe and our obligations to ensure this.*

*If you have any concerns about our products, you can contact us on ProductSafety@springernature.com*

In case Publisher is established outside the EU, the EU authorized representative is:

Springer Nature Customer Service Center GmbH
Europaplatz 3
69115 Heidelberg, Germany

**Batch number: 09105420**

Printed by Printforce, the Netherlands